Advances in Cladistics
Proceedings of the First Meeting of the Willi Hennig Society

Edited by V. A. Funk
& D. R. Brooks

The New York Botanical Garden
Bronx, New York
1981

Published by The New York Botanical Garden, Bronx, New York 10458, U.S.A.

Issued 28 December 1981

Library of Congress Cataloging in Publication Data
Main entry under title:

Advances in cladistics.

Bibliography: p.
Includes index.
1. Biology—Classification—Congresses.
2. Cladistic analysis—Congresses. I. Funk,
V. A. (Vicki A.), 1947– . II. Brooks,
D. R. (Daniel R.), 1951– . III. Willi Hennig
Society. IV. Title: Cladistics.
QH83.A43 574′.012 81-19028
ISBN 0-89327-240-X AACR2

Printed by Allen Press, Lawrence, Kansas, U.S.A.

Distance Data in Phylogenetic Analysis

James S. Farris

Table of Contents

Introduction

Analyses based directly on a matrix of distances between pairs of taxa are rare in phylogenetic systematics—which aims primarily for the construction of synapomorphy schemes—but they are much more common in comparative biochemical studies. Some types of biochemical comparisons, such as microcomplement fixation, yield data directly in the form of distances. This made it necessary to develop methods applicable to such matrices, but once those methods were in use, it also became common practice to reduce other types of data, such as electromorph frequencies, to distance form for purposes of analysis, direct treatment of the characters then being neglected. The validity of the conclusions of both sorts of applications has almost universally been argued on grounds of the intrinsic superiority of biochemical data: their intimate connection with the genotype itself. That faith may or may not be misplaced, but without a doubt, reliance on it has served to obscure another issue. The best data are of no use unless properly analyzed.

It is my aim here to concentrate on the methodological issues posed by distance analyses. I shall begin by tracing the development of techniques in the context of immunological distance, perhaps the most common type of distance data. Later I shall show how that discussion can be extended to other kinds of molecular distances, and to distances generally.

The introduction of microcomplement fixation and immunological distances produced considerable interest in systematics. The new measure of distance, seeming, as it did, to offer a way of dating the divergence times of groups lacking an adequate fossil record, promised to shed new light on the genealogical relationships of organisms. But the near-constancy of molecular evolutionary rate on which this dating depended came to be controversial in itself. In order to be a respectable scientific theory, in any event, rate constancy could not be accepted as a matter of faith; some means would have to be found to test it. Most proposed tests relied on inspecting the apparent distribution of molecular changes on a phylogenetic tree. Some such tests were performed on trees regarded as well-established from external evidence, but most were conducted with trees derived from the molecular data themselves. There is no reason not to do so, of course, provided the tests are themselves logically sound, but some of them posed the danger of circular reasoning. The methods used to analyse immunological distances were crucial. If the method of analysis itself effectively presumed rate constancy, then apparent constancy of rate according to the resulting tree would provide no corroboration of the constant-rate theory.

Early work on deriving phylogenetic trees from immunological distance data may or may not have assumed constancy of rate. It is difficult to tell from what was published, as authors devoted little explicit discussion to methods of phylogenetic analysis. Sarich (1969a, b), for example, discusses allocation of lengths (postulated amounts of molecular evolutionary change) to the branches of a tree, but says little of how the arrangement of those branches was determined, and what he does write does little to inspire confidence that rate constancy has not been tacitly assumed.

Reconstructing Genealogies

In 1972 I proposed a way of arriving at trees from distance matrices without assuming rate constancy, a modification of the Wagner method (Farris, 1970) that was already used for phylogenetic analysis of character data. (For a description of the method, see Farris, 1972; Swofford, 1981.) On applying the new method to the data of Sarich (1969a, b), I found considerably greater variability in molecular evolutionary rate than he had concluded. In subsequent applications, distance Wagner analyses have sometimes suggested near-constancy of rate, but sometimes indicated substantial rate heterogeneity.

Workers of the immunological distance school have always strongly favored the idea that their sort of data provide a "molecular clock"—a measure of divergence directly proportional to age of common ancestry. Most available evidence, to be sure, does suggest a good correlation between molecular difference and time since divergence, although the correlation is hardly perfect. But while the imperfection of the correlation is as clear as the general correlation itself, advocates of the molecular clock in practice interpret immunological distances as if homogeneity of molecular evolutionary rate were a completely general phenomenon, as if degree of molecular divergence always indicated age of common ancestry. There is more to this insistence than purity of generalization for its own sake. The claim of universality is needed in order to justify the use of the clock concept in its most important application, settling controversial questions of phylogenetic relationship. Since distance Wagner analyses persisted in failing to support the universality of the molecular clock, it became necessary, in order to keep that clock, to find some grounds for discarding the method in favor of some other technique more inclined to favor constancy of evolutionary rate.

There is one way of analysing a distance matrix that is quite likely indeed to provide reconstructions suggesting near-constant rates of divergence: use phenetic clustering. Form the tree, that is, by repeatedly uniting mutually most similar pairs of taxa until a complete hierarchy is produced. Just this means of "phylogenetic" analysis has been used in some molecular studies (discussed below). This method has not been directly advocated by the immunological distance school, however, because, as is well known, its use quite obviously depends on the prior assumption of rate constancy. Unless rates are presumed to be constant, there is no reason to suppose that mutually most similar pairs of taxa are genealogically most closely related. Since it was desired to obtain evidence for rate constancy, a method had to be found that would yield homogeneous reconstructed rates without seeming to assume them a priori. This was the approach taken by Prager and Wilson (1978), who advocated a form of the method of Fitch and Margoliash (1967).

In the Fitch-Margoliash method, a tree is first formed by phenetic clustering, but is then modified by a series of trial-and-error rearrangements. For each trial tree, a numeric length (or perhaps "length"—it may be negative) is assigned to each branch. The lengths determine a matrix of tree-derived distances. The derived distance between a pair of taxa is the sum of the lengths of the branches that must be traversed in tracing a path on the tree from one member of the pair to the other. Preference among possible rearrangements is based on the fit of the tree-derived matrix to the data distance matrix. The fit is assessed by the *%SD* (percent standard deviation) statistic:

$$\%\mathrm{SD} = 100\sqrt{\sum_{i,j}\left[\frac{d(i,j) - p(i,j)}{(d(i,j))}\right]^2 \Big/ \left(\frac{t(t-1)}{2}\right)} \tag{1}$$

Here there are t terminal taxa, and the summation extends over all distinct pairs of taxa. The term $t(t-1)/2$ is the number of these pairs. The data distances are denoted d, the tree-derived ones p. The branch lengths for each tree are chosen with the aim of minimizing %SD for the given genealogy and data distances. The evaluation is repeated for each trial tree. Trees with lowest %SD among those produced by a large number of rearrangements are selected as most preferable estimates of the historical genealogy.

Whether the Fitch-Margoliash method makes the assumption of rate constancy depends primarly on the zeal with which possible rearrangements are investigated. It can easily be shown that when the distance matrix accurately reflects evolution without homoplasy, %SD can be made 0 for the correct genealogy regardless of the extent of variation in rates of evolution. It seems reasonable to extrapolate from this result to the conclusion that minimizing %SD over possible trees is an inference method free of the rate constancy assumption. The Fitch-Margoliash method ought to work well, then, provided the right trees are checked. The initial tree, however, is chosen in a way that presumes rate constancy. If only this tree were evaluated, or if only a few trees close to the initial one were tried, the method would in practice rely on the premise of rate constancy.

Fitch and Margoliash went to some effort to avoid the pitfall of checking too few trees. They used a computer to evaluate a large number of alternatives. But Prager and Wilson were not quite so careful. Their calculations were done by hand, and, understandably, only the initial phenetic clustering and at most a few nearby alternatives were checked. What they call the "Fitch-Margoliash" method is thus in practice quite different from the original method of that name. In order to avoid confusion it would probably be better to refer to the new procedure as the "Prager-Wilson" technique. The original terminology, however, suited Prager and Wilson's purposes far better. By arguing for their own method under the name of "Fitch-Margoliash" they were able to advocate a procedure quite likely to support the molecular clock, while preserving the appearance of freedom from the assumption of rate constancy.

Whatever Prager and Wilson's purposes may have been, the argument that they present for the superiority of their method to the distance Wagner method seems simple and direct. It is that the former gives better fit to data. The distance Wagner method, like both the Fitch-Margoliash and the Prager-Wilson procedures, assigns a length to each branch of the constructed tree, and so determines a matrix of derived distances in the same way. It does not compute branch lengths in the same way, however, so that its derived matrix will not usually be the same as that of the Prager-Wilson method, even if the genealogies themselves are alike. Prager and Wilson found empirically that the derived matrix produced by their method was usually closer to the data matrix, as measured by the %SD criterion, than was that produced by the distance Wagner method. They argued from this observation that their ("Fitch-Margoliash") method ought to be preferred for inferring phylogenies from distance data. Straightforward as this argument seems, it nonetheless contains a defect. That flaw, as one might expect, is related to the reasons for differences in branch-length calculations between the two methods.

When I first devised the distance Wagner method it was generally thought that immunological distances measured sequence differences between proteins. The distance computed from microcomplement fixation was supposed to be proportional to the number of sites having different amino acids. I incorporated some of the consequences of this theory into the distance Wagner method, my aim being to make it as realistic as possible. Two of those consequences will be relevant here. First, if a distance measure d were proportional to sequence difference, then it would necessarily obey the triangle inequality. That is, for any three taxa i, j, k, $d(i,j) \leq d(j,k) + d(i,k)$. (Distance measures satisfying this condition are called metric.) The second concerns the relationship between observed and tree-derived distances. On the sequence difference model the derived distances measure the number of amino acid substitutions occurring in the evolutionary path connecting a pair of taxa. It is possible that this number exceeds the number of sites in which the taxa presently differ; this will occur if some site changes more than once during evolution. But it is not possible that the number of presently different sites exceeds the number of substitutions in the path. A postulated phylogeny with one or more derived distances smaller than corresponding observed distances cannot then be possible. Accordingly, the distance Wagner method was designed so that observed distances would not exceed derived ones. Likewise, if branch lengths represent amounts of evolutionary change, then they cannot logically be negative. The distance Wagner method, unlike that of Fitch and Margoliash, was designed so that its branch lengths would not be negative unless the observed matrix violated the triangle inequality—a contingency that was not expected to arise.

The restriction $p \geq d$ seemed suitable in 1972, when published immunological distance matrices all seemed to be metric. But subsequent studies produced several matrices that violate the triangle inequality. Prager and Wilson argued that the restriction is unrealistic and that their method, which allows either observed or derived distances to exceed the other, is more suitable. That is correct as far as it goes, although, as we shall see, those observations raise other problems for the interpretation of immunological distance analyses.

It is not hard to see why Prager and Wilson's method of branch length calculation should give lower %SD values than does that of the distance Wagner method. The %SD criterion is a weighted squared deviation measure, the weights being $1/d^2$. As with any squared deviation measure, it is a necessary condition for optimal fit that the weighted mean of the derived distances equal that of the original distances. Inasmuch as the derived distances of the original distance Wagner method all exceed or equal the corresponding observed values, the mean of the former must exceed that of the latter unless the two matrices are identical. It is seldom possible in practice to make the matrices equal, and in all other cases it will be algebraically impossible for the branch lengths of the original distance Wagner method to optimize %SD. But of course they were never intended to do so, having been designed with another aim in mind. What Prager and Wilson believed to be an empirical result is instead a logical consequence of the properties of the %SD measure. It should hardly be surprising that a method that produces derived distances subject to a restriction does not yield such good fits as can be obtained when the restriction is dropped.

Optimal Branch Length Fitting

If Prager and Wilson's claim were merely that the original distance Wagner branch length calculations should not be used when the aim is to minimize %SD, it would be no more exceptionable than it is surprising. But the conclusion they draw goes beyond this simple tautology. They use the %SD criterion not just to select branch lengths for given genealogies, but also to defend their preference in methods for reconstructing the genealogies themselves. They argue in effect that the low-%SD trees arrived at by their method should be preferred over the higher-%SD results of the distance Wagner procedure. This conclusion of theirs has a serious defect. Suppose that for given observed distances, genealogy A, when provided with suitable branch lengths, attains lower %SD than can genealogy B with any choice of branch lengths whatever. According to the %SD criterion, at least, A is then the better-supported hypothesis of kinship. But now suppose that B is equipped with branch lengths that make %SD as small as possible for B, while A is assigned branch lengths that lead to high %SD. Interpreted naively, this would seem to imply that B is better supported than is A. But it is plain that that conclusion does not follow. The comparison of %SD values is now uninformative. Any genealogy could be made to appear poorly supported according to the %SD criterion, if it were assigned suitably ludicrous branch lengths. Accordingly, it is illogical to reject a genealogy on grounds of high %SD, unless it can be insured that branch lengths have been provided so as to keep %SD as small as possible.

Since the branch length calculations of the original distance Wagner method do not minimize %SD for its genealogies, Prager and Wilson's preference in methods is based only on uninformative comparisons. This does not mean, of course, that the distance Wagner method is superior to theirs, only that no evidence on the question has been provided. In order to evaluate validly the genealogies produced by the two methods, it is necessary to devise a way of finding %SD-minimizing branch lengths for distance Wagner trees.

It will be useful to formulate the problem of choosing branch lengths in terms of matrix expressions. Think of the lower triangular part of the symmetric matrix of observed distances as written in a single column $\mathbf{D}$ of $m = t(t - 1)/2$ rows and likewise the derived distances as an $m \times 1$ matrix $\mathbf{T}$. $\mathbf{D} - \mathbf{T}$ is then a column of deviations of derived distances from corresponding observed ones. Denote by $\mathbf{W}$ a matrix with 0 entries off the diagonal and diagonal entries $\mathbf{W}(i,i) = 1/\mathbf{D}(i)^2$. Then $(\mathbf{D} - \mathbf{T})'\mathbf{W}(\mathbf{D} - \mathbf{T})$ is proportional to the square of %SD. Each derived distance in **T** is the sum of branch lengths—which branches depends on the genealogy. To represent this structure let **B** denote a $c \times 1$ matrix whose element **B**(i) is the length of the ith branch, and write as **A** an $m \times c$ matrix in which $\mathbf{A}(i,j) = 1$ if the jth branch lies on the tree path between the pair of terminal taxa whose distance is $\mathbf{D}(i)$. Then $\mathbf{T} = \mathbf{AB}$. To minimize %SD for a given genealogy and observed distance matrix, we choose B to minimize $(\mathbf{D} - \mathbf{AB})'\mathbf{W}(\mathbf{D} - \mathbf{AB})$. As there are no restrictions on the optimization, this requires only differentiating the expression with respect to **B** (for the necessary techniques, see Anderson, 1958), then solving for a 0 derivative. The minimizing **B** is found to be

$$\mathbf{B} = (\mathbf{A}'\mathbf{WA})^{-1}\mathbf{A}'\mathbf{WD}. \tag{2}$$

This choice of **B** gives the exact optimum of %SD provided that the inverse of **A'WA** exists. It cannot exist if two columns of **A** are identical. That will occur if the tree has a basal bifurcation, as then any path between two terminals that includes one of the basal branches will include the other. In this case one proceeds by combining those two branches for purposes of calculation, in effect solving for their total length. The tree can be rerooted after calculation of **B** by the midpoint rooting method of Farris (1972), or by application of the outgroup criterion, if a suitable outgroup is available. The rooting method has no effect on %SD so long as the total of the basal branch lengths is conserved.

This means of assigning branch lengths is quite different from those of both the original distance Wagner and the Prager-Wilson methods. That it departs from the former is scarcely surprising, as the two were devised for different purposes. The Prager-Wilson (or Fitch-Margoliash) method, however, was presumably designed to give low %SD, and here the difference from the optimal calculation of (2) is more startling. It seems that Prager and Wilson's selections of genealogies within their method were based on uninformative comparisons of nonminimal %SD's, just as were their contrasts of their own trees with those from the Wagner method.

To assess the practical significance of these considerations, I have reanalysed the data sets used by Prager and Wilson (1978; cf. their Table 2). My results are summarized in Table I. The first two columns of that Table recapitulate with one exception values reported by Prager and Wilson: %SD values for their method (PW) and for their distance Wagner trees (dW). Prager and Wilson followed Wallace et al. (1973) in claiming %SD = 2.5 for the "Fitch-Margoliash" tree for the *Rana* albumin data of the latter authors, but I have been unable to verify this figure. It appears that the correct values are %SD = 3.96 for PW and 4.58 for dW, as in my Table. It is seen that the PW value is lower than that for the original distance Wagner in 8 of the 11 cases. Column PWO gives %SD for Prager-Wilson genealogies with branch lengths computed as in (2). Not all their genealogies had been published; the missing cases are marked "?." The new branch length calculations always give better %SD than the originals (PW), sometimes by an impressive margin, as in the *Rana* electromorph data. Column dWO comprises %SD values for genealogies computed by the original distance Wagner method, but with branch lengths computed by equation (2). These are all lower than the original Prager-Wilson values (PW), but that merely shows that the PW branch lengths were poorly chosen. The informative comparisons are those that contrast Prager-Wilson and distance Wagner genealogies through their optimal %SD values. Two of these comparisons cannot be made (?), and three of them give the same %SD because the two methods arrived at the same tree. Of the six remaining, the distance Wagner genealogy gives the better fit in five cases, the Prager-Wilson tree in one instance. Whereas Prager and Wilson's uninformative com-

Table I

%SD for eleven data sets used by Prager and Wilson (1978). Explanation in text

	PW	dW	PWO	dWO	SSO
		Immunological distances			
Five carnivores	1.5	1.4	1.26	0.89	0.89
Xenopus	19.4	21.8	14.39	10.64	13.93
Seven carnivores	6.8	8.7	?	4.01	4.04
Rana	3.96	4.58	3.29	3.29	3.29
Conifers	11.1	41.8	10.23	10.64	10.23
Hylids	9.9	21.1	7.02	6.46	6.63
Birds	12.1	29.6	10.06	9.93	10.96
		DNA annealing			
Primates	1.9	1.7	1.08	1.08	1.08
		Nei's distances			
Rana	21.6	26.2	12.03	10.89	11.8
Drosophila	8.5	23.5	?	7.84	7.95
		Sequence differences			
Vertebrate lysozyme c	9.8	9.7	4.5	4.5	4.11

parisons gave the impression that their method gave a better fit to data in 8 of 11 cases, informative comparisons with optimal branch lengths reverse this conclusion, showing a stronger ratio in favor of the distance Wagner method.

Improving Fit

While the distance Wagner method thus seems to be considerably more effective than Prager and Wilson believed, it is clear that it does not necessarily produce the best fitting genealogy. It is natural to inquire what sort of method might do better still. The answer to this question depends primarily on how much computational effort can reasonably be expended. A best fitting genealogy can always be identified by checking each possibility, but this is not feasible for more than eight taxa or so, as the number of possible trees increases rapidly with number of terminal taxa. At the other extreme, simple procedures such as distance Wagner are quite practical for a hundred taxa, their computational effort with a properly designed algorithm being proportional only to the square of the number of taxa. [Prager and Wilson complain that the distance Wagner method is more laborious than the phenetic (UPGMA) clustering used in their method. It would appear that they were expending unnecessary effort on the former, as the computational work of the latter is proportional to at least the 2.5 power of number of taxa.] No such simple method is known, however, that can generally assure finding the best fitting tree.

One way of improving chances of finding a best fitting tree is to modify a simple tree building procedure so that it provides several trees, instead of just one. Mickevich (1978b) used a distance Wagner program that automatically checks each of the possibilities when a tie is encountered in the criterion for placing a taxon. The method of Swofford (1981) relies on the same idea, although it discards some trees if they become inconveniently numerous. It is likewise easy to check the results of different tree building procedures, although for this to do any good, the methods must be both economical and reasonably effective—and there are not many such procedures.

One technique that does seem useful for this application is the special similarity clustering of Farris (1977, 1979a). In this method the distance matrix is transformed into a similarity matrix with elements $a(i,j) = (d(i,r) + d(j,r) - d(i,j))/2$. Here r denotes the reference point. For distance data this is chosen as one of the terminal taxa. For character data, r can be taken as a point in the character space. (An equivalent method has been "independently" discovered by Klotz et al., 1979, who use a "converted difference" equal to $-2a$.) The transformation discards points of similarity common to i, j, and r from the similarity between i and j. If r has all plesiomorph states and the right distance (Manhattan; see below) is used, the special similarity a counts putative synapomorphies, and so, homoplasy aside, indicates genealogical relationships regardless of variation in evolutionary rates (Farris et al., 1970). If r is a terminal taxon, the same applies, except that the root of the tree may be misplaced.

A tree can be derived from a special similarity matrix simply by applying a clustering method such as UPGMA. The procedure is somewhat more laborious than a properly designed distance Wagner, but not prohibitively so, and it has the advantage that it requires less computer memory. For data showing homoplasy the results of clustering may be sensitive to choice of the reference point. Farris (1980) suggests using each of the terminal taxa as a reference point, retaining the trees of best fit. I have included the results of UPGMA clustering by special similarity in Table I. Column SSO gives %SD for trees produced by this method and assigned branch lengths by equation (2). It is seen that this procedure improves on the dWO values in two cases, but generally does not perform as well as distance Wagner, although it betters the Prager-Wilson method just as often. The strategy of picking the better of SSO and dWO, however, surpasses the Prager-Wilson value in every case. It seems that the two phylogenetic methods complement one another nicely.

It is worth noting in passing that special similarity clustering outperforms distance Wagner on the lysozyme c sequence differences. As I have pointed out before (Farris, 1980), the former method effectively assumes that homoplasious changes into the state of the reference point do not occur. Generally this would be an unrealistic premise, and the Wagner method, which does not assume it, should give better performance. With amino acid sequences, however, such changes might be relatively rare. On a random substitution model this would occur because of the number of possible states at each site. The more restrictive method might then be expected to give better results under those circumstances.

Methods that improve the fit of a tree by a series of rearrangements of its branching sequence offer a way of achieving better fits than those of the simplest methods, but with relatively moderate increase in computational effort. In order to obtain this benefit, however, the method must be carefully designed. The original method of Fitch and Margoliash investigates possible rearrangements almost exhaustively (if allowed to run long enough), but it checks rearrangements differing slightly from the initial phenogram before evaluating any tree that differs more strongly from the phenetic clustering. Recalling that Prager-Wilson trees are generally quite close to phenograms, it is then seen from the results of Table I that the Fitch-Margoliash method is not particularly well designed. As the optimal tree is likely to be rather different from the phenogram, a better choice of starting point would probably both increase chances of success and decrease the number of alternatives that need to be tried. A considerable reduction in the number of alternatives tested could also be achieved by adopting a more flexible means of choosing them, not just checking all the small rearrangements first. Rearrangement methods of this general type have been used in phylogenetic analysis of character data (see Mickevich, 1978b), but there has been little work on developing efficient rearrangement methods for distance data. I would suppose that a carefully designed method would be able to obtain better fitting trees than can present simple methods, and with quite reasonable computational effort. I would expect that the distance Wagner method would continue to play a valuable role, both as an efficient means of approximate analysis and as a source of promising starting points for rearrangement techniques. In the two cases in Table I in which the distance Wagner method does not do best, its tree differs only slightly from the better fitting tree found by special similarity clustering.

Measures of Fit

For the purpose of identifying best fitting genealogies, branch lengths are chosen to optimize the fit measure primarily in order that the fits of alternative trees may meaningfully be compared. But for other purposes branch lengths are important in their own right. The relative lengths of sister branches are used to assess apparent homogeneity of evolutionary rates, and this logically requires that the branch lengths be interpretable as amounts of evolutionary change. Inasmuch as conclusions on rate constancy tend to be controversial, it is useful to investigate just how well this interpretation of branch lengths is justified.

The results already discussed bear on this question in a troublesome way. It has already been seen that different methods of branch length calculation can lead to quite different %SD, which is to say that they give different suites of branch lengths. If branch lengths are selected to optimize a fit measure, as is necessary in order to compare fits of genealogies, then they may well depend on the measure of fit employed. Only one fit measure has been discussed so far, but %SD is not the only measure available, and its use may seem arbitrary to an extent. While the general objective of close conformity between observed and reconstructed distances seems plausible, the details of how closeness is measured seem less obvious. One might, it would seem, just as well use $(d - p)^2$ as a measure of departure, rather than $(1 - p/d)^2$, the quantity used in %SD. Or, for that matter, $|d - p|$ or $|1 - p/d|$. Further, one might conceivably choose branch lengths either with or without the restriction $p \geqslant d$. There are still other possibilities, but these few are enough to show the extent of the problem.

Although it seems obvious simply from an algebraic standpoint that different criteria may lead to different preferences in branch lengths—or even genealogies—the problem posed by this observation has received little attention. Prager and Wilson, for example, use just one set of branch lengths in their method, though they use more than one fit measure, with the result, as already seen, that their comparisons of degrees of fit are meaningless. They do offer an argument in favor of a fit measure ("F") based on $|d - p|$, but their reasoning comes only to the observation that F sometimes gives a substantially lower percentage value than %SD. As they put it (p. 137), "the F value suggests a satisfactory fit to the input data, while the SD suggests a much poorer fit." The motivation for their preference, then, seems to be a desire to publish "good" fits between trees and observed distances, as revealed by a "small" percentage departure, this being taken, I suppose, as an indication of the value of the data as evidence on phylogeny. But it is nonsense to prefer F to %SD simply because the former yields smaller values. Values of the two statistics cannot be directly compared, inasmuch as they are not in the same scale. That F is smaller than %SD, moreover, can hardly imply much about the value of the data, as that observation has little empirical content. It is little more informative than is observing—tautological—that for several distinct $x > 1$, $\sqrt{\sum x^2}$ exceeds Σ x.

The choice of whether to use the restriction $p \geqslant d$ is based on a property of the distance, nonmetricity, and it seems plain that other methodological choices should take account of the nature of the data whenever possible. The choice between $p - d$ and $1 - p/d$ also corresponds to a property of the data. In selecting branch lengths or genealogies each deviation $p - d$ represents in effect the degree to which a part of the data provides evidence against the phylogenetic reconstruction. Just about any measure would incorporate the idea that larger deviations provide stronger contrary evidence. For any one $p - d$ this seem unexceptionable. But summing these deviations effectively averages contributions from different observed distances. If $p - d$ is used, then it is effectively supposed that all deviations of the same magnitude provide equally strong evidence, regardless of the value of d. If $1 - p/d$ is used, it is supposed instead that the strength of evidence depends on d. If one thinks that a deviation $p - d = 1$ is equally disturbing whether d is 1 or 100, then one ought to use $p - d$, whereas if large deviations are even more disturbing when d is small than when it is large, then $1 - p/d$ is appropriate. Naturally it seems unlikely that large distances are as accurate as small ones in an absolute—not relative—scale. Further, it is usually found in practice that the larger absolute deviations are associated mostly with the larger observed distances. On the interpretation of tree-derived distances as amounts of evolution, this would be expected. Large observed distances correspond to long evolutionary paths in which there has been a greater chance for homoplasious changes to accumulate than in the shorter paths showing small observed distances. For these reasons, it seems unlikely that Prager and Wilson's preference for $p - d$ over $1 - p/d$ is well justified.

It is less clear that an argument based on properties of distances can provide a choice between $(1 - p/d)^2$ and $|1 - p/d|$, but a different line of reasoning does seem relevant to this issue. It was seen before that meaningful comparison of fit values requires assurance that optimal branch lengths have been chosen for the genealogies considered. One should use a fit measure, then, only if a means of optimal branch length allocation is available for that measure. It develops that there is an interaction between this pragmatic requirement and the restriction $p \geqslant d$.

Equation (2) shows how to compute optimal branch lengths for $(1 - p/d)^2$. A similar result can be derived for $(p - d)^2$:

$$\mathbf{B} = (\mathbf{A}'\mathbf{A})^{-1}\mathbf{A}'\mathbf{D}, \tag{3}$$

the notation being as for (2). This is the same as (2) with $\boldsymbol{W}$ taken as the identity matrix, so that all deviations are weighted equally. [The method of (3)—but not that of (2)—had been reported by Cavalli-Sforza and Edwards, 1967. The method of (3) also provides optimal branch lengths for cophenetic correlation, a criterion used by Farris, 1979b, who tabulated results for the data sets of Table I.] Either of these methods allows some $p < d$, so that they may be used only when the restriction $p \geqslant d$ is not enforced. If the restriction were used, no available method, so far as I am aware, can guarantee optimal branch lengths for $(p - d)^2$ or for $(1 - p/d)^2$.

Just the opposite obtains for absolute deviations. If $p \geqslant d$ is required, optimization for $|p - d|$ or $|1 - p/d|$ can be formulated as a linear programming problem (see Hillier and Lieberman, 1967, for techniques). Without the restriction, however, that method is not applicable, nor is any other presently known method. It would seem that in the absence of any obvious theoretical reason for preferring either the squared or the absolute deviation measure, one should rely on the latter only when it is realistic to require $p \geqslant d$, while the former should be used only when that restriction is not enforced. At least until a wider variety of optimization methods become available, then, the squared measure should be used for nonmetric distances, as violation of the triangle inequality makes the restriction unrealistic.

It will be worthwhile to comment on one further fit measure, the total length of the tree—i.e., the sum of its branch lengths. This criterion is suggested by analogy with the parsimony criterion used in phylogenetic analysis of character data, which also requires minimization of tree length. There, the length of a tree corresponds to the number of ad hoc hypotheses of homoplasy needed to defend a genealogical hypothesis. The analogy might seem to be strengthened by the application of the distance Wagner method to distance analyses, the Wagner method having originated as a procedure for parsimony analysis. I had pointed out before (Farris, 1972), however, that the parsimony criterion cannot be carried over into distance problems; the interpretation of length as counting ad hoc hypotheses certainly does not apply in this case. Indeed, some of the computational steps in the original distance Wagner method have the deliberate effect of making the length of the tree nonminimal. The length criterion has nonetheless been advocated by authors of modifications of the distance Wagner method (for example, Waterman et al., 1977). If the restriction $p \geqslant d$ is enforced, length minimization is a linear programming problem, solvable by the same means as that of minimizing total $|p - d|$. It is plain that the length criterion can only be sensibly used when $p \geqslant d$ is required, as without the restriction, any tree would admit of a trivial optimum. Like the absolute deviation measure, then, the length criterion is not applicable to nonmetric distances.

Clocks and Ultrametrics

While the discussion of the previous section suggests some firm conclusions on what branch lengths—or at least means of choosing them—ought to be preferred, it might well be observed that any consideration of the relative merits of alternative branch lengths tacitly presumes that those quantities can be interpreted meaningfully. Certainly that implicit presumption has been made in the literature to date—in my own papers, as well as those of others. It turns out, however, that that premise is open to serious question, and the question is raised by the same property that was prominent in selecting a fit measure: the triangle inequality.

That immunological distances are nonmetric—do not conform to the triangle inequality—seems well established. It would hardly be questioned by Prager and Wilson, who relied on that observation in their attempt to do away with the distance Wagner method. There is irony here, for not only did their attempt fail, but the premise of nonmetricity undermines as well their underlying aim of bolstering the molecular clock hypothesis. A distance measure that violates the triangle inequality cannot logically show clocklike behavior. Indeed, as we shall see later on, metricity is required in order that distances be interpretable as amounts of evolutionary change.

To first see why a nonmetric distance cannot be fully clocklike, suppose that a distance measure d shows a constant rate of divergence. Given three taxa i, j, k, two, say i and j, will be genealogically most closely related, and so will have an ancestor common to them but not to k. Since i and j diverged more recently from each other than either did from k, $d(i,j) \leqslant d(i,k)$, $d(j,k)$. Further, i and j diverged from k at the same time, inasmuch as at that time they were the same line. Then if d accurately reflects time, $d(i,k) = d(j,k)$. It is readily seen from these relations that if d is clocklike, then for any three taxa $d(i,j) \leq \max(d(i,k),d(j,k))$. (This condition is called the ultrametric inequality.) If no two of i, j, k were most closely related—a trifurcation—a clocklike d should give all three pairwise distances equal, and the inequality would still be satisfied. This has been pointed out before (for example, Jardine and Sibson, 1971; Farris et al., 1979). Since for positive d, the maximum of any two d values is no greater than their sum, it is seen that any ultrametric distance is also metric. Conversely, any distance that violates the triangle inequality must also be nonultrametric, and so cannot show a constant rate of divergence. Indeed, since ultrametrics are contained within the set of metrics, a distance showing strong departure from metricity cannot even be close to ultrametric.

A reasonable reply would be that immunological distance, while not entirely clocklike, is still pretty well correlated with antiquity of common ancestry. And of course that is true, as far as it goes, but the distinction between clocklikeness and reasonably strong correlation is crucial for phylogenetic analysis. If immunological distance is well correlated with time since divergence, but not perfectly so, then there will be some cases in which taxa mutually most similar immunologically are nonetheless not genealogically most closely related. Prager and Wilson no doubt expect such cases to be rare, and perhaps that is so, but even so the clock idea cannot be used in the recognition of kinship. One would never know which cases were the exceptions without analysis of each case by a method that does not rely on the clock assumption. (A striking example of how distance data that seem generally well correlated with time may nonetheless seriously misplace some taxa is given by Farris et al., 1979.) While the clock model may well serve as a rough descriptive generalization on molecular evolution, it is thus too rough to be useful as a premise in phylogenetic analysis.

Immunological distances are not the only measures of molecular distance to which clocklike behavior has been attributed. The genetic distance of Nei (1972) has been widely used in analysis of frequency data obtained from electrophoretic studies. Its popularity is apparently based on Nei's argument that his measure shows a linear relationship to antiquity of common ancestry. Encouraged by this premise, most analyses of Nei's distance have arrived at "genealogies" simply by phenetic clustering. To be sure, the measure does have clocklike behavior—in the mean, and under Nei's model, which supposes that differentiation of populations is entirely a homogeneous genetic drift process. But that sort of assurance of clocklikeness is no more than enough to raise suspicions. Even granting homogeneity of drift, the stochastic variation in the process can still be large enough so that mutually most similar taxa may not be genealogically most closely related. (This was shown by Cavalli-Sforza and Edwards, 1967.) Further, since the assurance of clocklikeness is based on a theoretical model, the question remains of whether the postulated process is approached closely enough in nature to be practically useful. Empirical studies have turned out much the same as for immunological distance. There does seem to be a reasonably good correlation between Nei's distance and time since divergence, but exceptions to the general relationship certainly occur. Baverstock et al. (1979), for example, show that an analysis free of the clock assumption leads to modification of a scheme constructed by phenetic clustering according to Nei's distance. Proponents of Nei's version of the clock nonetheless persist in interpreting this distance as if it were purely clocklike—and for a familiar reason: if exceptions are admitted, the method loses its claim to the role of direct indicator of phylogenetic kinship.

Nei's distance is like immunological distance in another respect as well: it is nonmetric. Why this is so can be seen from a simple example. Suppose that for a single locus with two alleles, population A has gene frequency 0.9; B, 0.1; and C, 0.5. Then the Nei's distances are AB = 1.52, AC = BC = 0.25. AB exceeds AC + BC by a factor of 3. Nonmetric Nei's distances have often been obtained empirically as well, as in the passerine data of Avise et al. (1980). Like immunological distance, then, Nei's distance cannot be strictly clocklike; its correlation with antiquity of common ancestry must be subject to exceptions.

If nonmetricity implied only a departure from strict rate constancy, it would be of relatively little interest. The property would show only the falsehood of a hypothesis that was not supported by available data anyway. The methodological consequences would be equally mild, simply that phenetic clustering of such distances should not be regarded as phylogenetic reconstruction. The analysis should be performed instead with methods that do not presume rate constancy—but again, that was largely acknowledged already. But nonmetricity turns out to be somewhat more important than that. The branch length fitting methods that are free of the constant rate assumption depend on interpreting branch lengths or their sums as evolutionary changes. Nonmetricity presents a logical problem for this interpretation.

Path Length Interpretations

The triangle inequality originated as a way of characterizing a distinction between physically realizable and nonrealizable abstract spaces. It is named after a property of ordinary, Euclidean space. On a Euclidean triangle with vertices A, B, C no two sides may in combination be exceeded in length by the third. If they were, this would mean that the line that passes directly from, say, A to B would be longer than the two-sided path from A to B by way of C. By length of a path in a physical space is meant the amount that must be traveled to traverse it, and by distance between two points the length of the shortest path between them. If by some measure AC + BC were less than AB, then that measure could not represent amount of travel or length of path.

The interpretation of evolutionary distances as amounts of change seems readily applicable to some kinds of molecular distance. The length of a path connecting two proteins in evolution, for example, is just the number of substitutions along it, and sequence difference is a metric, as can readily be shown. The observed difference between two sequences, then, satisfies one requirement for being interpretable as a path length: it can logically be regarded as the total of a series of changes. The nonmetricity of immunological and Nei's "distances," however, implies that they cannot represent any sort of path-length measure; it is impossible for them to represent the total of a series of changes. This poses a fundamental problem for analysis of such distances. In fitting branch lengths to a distance matrix, it is necessarily assumed that the lengths can be taken as amounts of evolutionary change—else, what are they?—and the distances as path lengths.

The nature of this difficulty can be readily exemplified in the case of Nei's distance. Suppose that A and B are sister taxa and that their subtended branches have the same fitted length (as they should on a clock model). Their common ancestor, then, should be equally different from each. If A and B have gene frequencies as in the example above, the Nei's distance between them is 1.52. The lengths of branches leading to A and B will sum to 1.52 if chosen for conformity of derived to observed distances, and if equal, will each be 0.76. Population C, as above, seems a good common ancestor for A and B. In gene frequency it is intermediate between them, conforming to the equal divergence premise of the present example. But the Nei's distance between C and A or B is only 0.25.

The fitted branch lengths of 0.76 are then not the Nei's distances between the terminals A and B and their common ancestor. Conversely, if the branch lengths were picked so that they were Nei's distances, they would not add up to the observed distance between A and B. There is nothing distinctive about C in this regard. No choice of a gene frequency for the common ancestor will allow Nei's distances as branch lengths to add up to the observed AB distance, unless the ancestor is taken to be identical to A or to B. In that last case one of the subtended branches would have length 0—and fitted branch lengths of 0 for terminal branches rarely occur in analyses of real data. It seems that if Nei's distances are analysed by fitting branch lengths, the resulting branch lengths cannot be interpreted as Nei's distances, or, so far as I know, anything at all. If the branch lengths are chosen to add up for a good fit, they cannot be Nei's distances. If they were somehow chosen as Nei's distances between ancestor and descendant, they would not add up to observed distances among the terminals. While branch length fitting has been often used for Nei's distances—I did it myself (Farris, 1974), before these properties occurred to me—it would seem that this measure is not suitable for such analysis.

No such direct example can be produced for immunological distance, as it is not computed from observed sequence information. A similar conclusion, however, can be obtained from consideration of any three-taxon case in which the distance violates the triangle inequality. Farris et al. (1979) discuss an instance drawn from the immunological distances of Case (1978), the distances between X (*Rana aurora*), Y (*R. muscosa*), and Z (*R. pipiens*) being XY = 12, XZ = 89, and YZ = 49. Selecting branch lengths to add up to any such observed distances necessarily produces a negative value; in this case that subtending Y is −14. It is plain that the negative value cannot be either an immunological distance or the amount of change between an ancestor and a descendant, and what that value does represent is thus a complete mystery. If the branch lengths, however, were somehow made into immunological distances, then they could not add up to the observed distances among these three taxa. Like Nei's distance, then, immunological distance does not seem amenable to analysis by branch length fitting.

It is worth noting that this form of argument can be applied to Nei's distance as well. If A, B, and C from the previous example are taken as terminal taxa, choosing branch lengths to add up to the observed distances between them requires assigning a value of −0.51 to the branch subtending C. Negative fitted lengths also arise in real analyses, as in the passerine distance Wagner tree of Avise et al. (1980).

Fitch and Margoliash (1967) had regarded negative branch lengths with apparent equanimity, interpreting them as reversals in sequence evolution. I objected to them in my 1972 paper on the grounds that they had no possible interpretation in terms of physically realizable—that is metric—distances. Prager and Wilson (1978) seem to follow my previous argument in part. They recommend preference for putative genealogies requiring fewer negative branch lengths, a suggestion that had also been made by Cavalli-Sforza and Edwards (1967). Scanlan et al. (1980) use a modified form of analysis in which negative branch lengths are set to 0—with the result (which they do not acknowledge or defend) that fit to the data is worsened. My earlier position was based on the premise that immunological distances could be interpreted as path lengths. It is now clear that this supposition is not justified. The suggestion of Prager and Wilson has a technical defect to which I shall turn below. Both that suggestion and the procedure of Scanlan et al. seem intended primarily to preserve the appearance of suitability of immunological distances for branch length fitting, by avoiding the obvious impossibility of negative distances. But both merely disguise symptoms. The real difficulty is that immunological distances, being nonmetric, cannot be interpreted as path lengths at all. The negative branch lengths differ from the others only in the obviousness of their uninterpretability. If immunological distances are not path lengths, the positive branch lengths can no more be immunological distances than are the negative ones. And if branch lengths fitted to an immunological distance matrix cannot be immunological distances, then, it would seem, they cannot be anything whatever.

These observations make it clear why using unrestricted optimization of branch lengths is desirable with nonmetric distances. All the measures of fit rely on conformity of the derived to the observed matrix. To the extent that any of these methods is justified at all, it is by the principle of fit. Certainly, then, fit should not be sacrificed without compelling reason. The reasons for eschewing negative lengths and $p \leq d$ were originally that these restrictions were necessary for physical interpretation of branch lengths and derived distances. With nonmetric distance, however, physical interpretation is in any case impossible, and any restriction on optimization merely trades off fit for no return.

Euclidean Distances

One might think that these conclusions would apply only to those individual matrices of immunological or Nei's distance that observedly violate the triangle inequality, that apparently metric matrices could still meaningfully be analysed. There are two defects to that view, however. First, any general consideration of immunological distance necessarily supposes that "immunological distance" is one particular kind of measure, that it always assesses difference in the same way. Otherwise more than one quantity is combined under a single name. That immunological distance is sometimes observed to be nonmetric implies that that way of assessment, whatever it may be, is not a path length measure. Once that is established, then its further interpretation as a path length measure is gratuitous. As an analogy, consider the orbit of Mercury. On finding that it does not obey Newton's gravitational mechanics, one hardly concludes that Einsteinian relativity applies to Mercury, but not, say to Pluto, for which no non-Newtonian effects are yet known. The same applies to Nei's distance, for which, in any case, the general reasons for its nonmetricity are plain from algebraic analysis. Second, metricity of a distance, even if general, is no guarantee of its suitability for branch length fitting.

As an illustration of this second point, I shall use Rogers' (1972) distance, another measure of genetic distance. For any one locus, this measure is proportional to the Euclidean distance between populations in allele frequency space, and so it can be established algebraically that this distance is always metric. To see why branch length fitting cannot be applied, consider a one-locus, three-allele case in which each of three taxa E, F, G is fixed for a different allele. The Rogers' distance between any pair is 1.0. Fitting branch lengths to an unresolved tree assigns length 0.5 to each of the three branches. The derived distance between any two taxa is 1.0; the derived distances fit the observed ones perfectly. This seems unproblematical until one considers the internal node of this tree. This hypothetical ancestor, X, say, is linked by a branch of length 0.5 to each of the three terminals. Suppose that those branch lengths are Rogers' distances. As this is a Euclidean metric, ordinary analytic geometry applies. All pairwise observed distances are the same, so the three terminals form the vertices of an equilateral triangle. Since EF = 1.0 and EX = FX = 0.5, X must lie at the midpoint of the side EF. But the same reasoning shows that X also must lie at the midpoints of sides EG and FG. Any postulated gene frequency whatever for X then implies a contradiction. The common ancestor X cannot exist anywhere in the frequency space. There is nothing unique about this example. The same result obtains for any data unless the three points are colinear.

Of course, this contradiction results from the premise that the branch lengths are Rogers' distances. If that premise—evidently impossible—is dropped, the contradiction vanishes. But then it is replaced by the familiar dilemma. If the branch lengths are chosen to fit the observed distances, they cannot be Rogers' distances. And if they are not Rogers' distances, what are they? But if it were arranged in some way that the branch lengths were Rogers' distances, then they could not add up to the observed distances. In this respect Rogers' distance is quite like Nei's. Neither is suitable for analysis by branch length fitting.

While nonmetricity leads to unrestricted optimization of branch lengths, it is now seen that metricity need not justify restrictions. Metrics such as Euclidean distances no more yield physically interpretable branch lengths than do nonmetrics, and so neither can they provide any reason for imposing restrictions on optimization, sacrificing fit in the hope of interpretability.

The problems of Euclidean distances are not limited to gene frequency data. A further example will provide some instructive observations. Suppose that for four taxa J, K, L, M there are 56 morphological characters, of which 20 are autapomorphies of J, 30 of M, 2 of K, 2 of L, and 1 each synapomorphies of J + K and of L + M. By assigning numerical values to the features, 0 for plesiomorphies, 1 for apomorphies, Euclidean distances can be calculated between the taxa.

	J	K	L	M
J	0			
K	4.7	0		
L	4.89	2.45	0	
M	7.21	5.83	5.66	0

These data are largely uninformative, but what synapomorphies there are define the tree ((J,K),(L,M)). Fitting branch lengths to this tree to minimize %SD yields lengths of 3.34, 1.26, 1.32, 4.34, and −0.08, respectively, for the branches subtending J, K, L, M, and the combined interior branches. The value of %SD is 4.41.

While the branch length fitting procedure for this tree does not seem to fit the distance matrix particularly well, the tree requires no homoplasy for the original characters, and so, on grounds extrinsic to %SD, is the right tree. The behavior of the fitted branch lengths in this case has implications for several issues already discussed.

As mentioned before, Prager and Wilson suggest avoiding genealogies that require negative branch lengths. This example indicates that that may be unwise. In this case it would lead to discarding the correct genealogy. The suggestion of Scanlan et al. that negative branch lengths ought to be set to 0 would leave the tree unresolved, although in fact there are uncontradicted synapomorphies supporting the resolution. Both ideas, it might be observed, amount to a methodological confusion. If negative branch lengths are, reasonably, considered nonsense, then one ought to use a fitting method which precludes them, not simply patch the restriction onto an unrestricted optimization after the fitting. This is required in order that the derived matrix be optimal for the conditions imposed, and that in turn is required, as already has been seen, in order that fit values may be compared. Fitting with the branch lengths required to be nonnegative can be accomplished by linear programming to minimize, say, total $|p - d|$. This fitting method gives branch lengths, in the previous order, of 3.04, 1.65, 1.67, 3.99, and 0.18. The total of $|p - d|$ is 1.5. The problem of negative lengths then disappears, although this is not to say that this procedure is well justified, as we shall see. The second fitting of necessity uses the restriction $p > d$, and so Prager and Wilson would, if these were immunological distances, presumably object to it. But that objection would be just another sort of confusion. Their basis for it is that immunological distance is nonmetric, but nonmetricity, as already seen, produces negative branch lengths with an unrestricted fitting procedure. The requirement that branch lengths be nonnegative seems to have no justification beyond the obvious, that negative values are nonsense as amounts of change. But if the lengths are fitted to a nonmetric matrix, they are not interpretable as amounts of change in any event.

That last confusion might be encapsulated as the erroneous premise that the matrix analysed is amenable to branch length fitting. Euclidean distances, though metric, are not well suited to this type of analysis, and the example illustrates several new sorts of difficulties that can arise from misapplication of the technique. The character data underlying these distances show no homoplasy for the tree, yet neither fitting procedure achieves an exact fit of p to d. Imperfect fit is normally taken to imply the existence of homoplasy; in this case that conclusion would be an artifact of technique. The fitted lengths also give a most misleading impression of relative rates of evolution. M has 15 times as many autapomorphies as does its sister taxon L, and so has evolved 15 times as rapidly since their separation, but the %SD fitted branch lengths show a ratio of only 3.4. The second fitting does even worse, the ratio of lengths of the same two branches being but 1.9. The Euclidean distances appear much more clocklike than the underlying characters. On the other hand, the total number of synapomorphies is equal to the number of autapomorphies of K, but the ratio between the corresponding branch lengths is 9 in the second fitting. In conjunction with the Euclidean distance, the effect of the branch length fitting seems to be to exaggerate the lengths of terminal branches relative to internal ones.

Manhattan Distances

All these misleading values are in this case a direct result of using the Euclidean distance. If the Manhattan distance (see Farris, 1970) is used instead, the distance matrix becomes

	J	K	L	M
J	0			
K	22	0		
L	24	6	0	
M	52	34	32	0

Branch length fitting by either method now gives branch lengths, in the same order, of 20, 2, 2, 30, and 2, and there is a perfect fit of p to d. The implications of the fitted branch lengths are now entirely consistent with the character data. The ratios of branch lengths are precisely the relative rates apparent in the characters. This corresponds to my earlier demonstration (Farris, 1967a) that the Manhattan distance is, in present terms, amenable to a path length interpretation.

Sequence differences are analogous to Manhattan distances. Data in which no site shows more than two amino acids (or base pairs) among taxa is directly comparable to the suites of two-state characters just discussed, and all the same observations apply. If there are more than two conditions at any one site, however, a new difficulty arises. Suppose taxa T, U, V each have a different state at one site. For that site, the pairwise differences are all 1, and fitting an unresolved tree gives a length of 0.5 for each branch. This implies that the common ancestor has a state differing by 0.5 from each of the three states in the data; this is plainly impossible. In my paper (Farris, 1972) I suggested ways of minimizing the effects of this phenomenon, but no procedure can eliminate the paradox. In real data there will almost always be sites showing several conditions, and so in practice sequence differences are not really suitable for branch length fitting, although, to be sure, they come far closer than the Euclidean distance.

With sequence differences computed from observed sequences, the logical problem posed by branch length fitting of the differences can be avoided simply by analysing the sequences directly as character data. Sometimes, however, sequences are compared indirectly, as in DNA annealing studies (for example, Benveniste and Todaro, 1976). In this case the distances must be analysed somehow, and branch length fitting seems to be the only method available. The markedly different behavior of the Manhattan and Euclidean metrics, however, suggests some questions on the interpretation of DNA distances. This measure seems to be metric or very close to it, and indeed in branch length fitting analyses it often seems close to ultrametric—clocklike. The clock idea is, if anything, even more appealing here than with immunological distances. Just how safe is this interpretation? The metricity, first of all, might be spurious. Some data sets of immunological and Nei's distances appear to conform to the triangle inequality, but it is now known that those measures are nonmetric. Nonmetric DNA distances may be awaiting discovery. But supposing that these distances are indeed metric, what kind of metric are they? Is the relationship of the experimentally determined distance to the DNA itself analogous to that between characters and a Manhattan distance, or is it perhaps like a Euclidean distance? It is not possible to resolve this question by inspection of the distances themselves; either interpretation is consistent with the triangle inequality. The issue is crucial for deciding how DNA distances should be analysed. If they were like Euclidean distances, then it would not be proper to use branch length fitting on them. If that method were used anyway, for lack of any alternative, %SD would probably be the most reasonable fit measure to apply, but then it might be necessary to accept negative branch lengths in order to find the right genealogy. If DNA distances were like the Euclidean metric, moreover, the fitted branch lengths might give a quite misleading picture of relative rates of evolution. Underlying DNA sequence evolution with greatly varying rates might appear through a Euclidean-like distance as nearly clocklike. If DNA distances were like sequence or Manhattan distances, however, then few of these problems would arise. Some experimental means will have to be devised for determining the properties of DNA distances before analyses of such data can be safely interpreted.

Similar considerations apply to the prospects of immunological distance. It might be maintained that the nonmetricity of some existing immunological distances is due to experimental error. Means might be found to improve technique, or perhaps the distance might be modified in some way so as to become metric. If this could be accomplished, however, it would still have to be assured that the improved distance resembled sequence difference rather than Euclidean or some other unsuitable measure in its properties. The claim of correspondence to sequence difference has been made for present immunological distances, but their nonmetricity shows that this position is unjustified.

Although Manhattan distances are always logically capable of being interpreted as path lengths, the interpretations can sometimes be misleading. For underlying data such as

	Characters		
Taxa	1	2	3
N	1	1	1
O	1	0	0
P	0	1	0
Q	0	0	1

the Manhattan distance is 2 between each pair of taxa. Fitting an unresolved tree gives a length of 1 for each of the four branches, and $p = d$. Exact fits are usually taken to imply the absence of homoplasy, and by extension small %SD is often used as evidence of the high reliability of the data as an indicator of phylogeny. It is seen that that conclusion does not generally follow. The unresolved tree requires two steps for each character: all the apomorphies are parallelisms. If fit measures anything about homoplasy, it is the variability of it among pairs of taxa, not the total amount of it. The total of the fitted branch lengths, likewise, is 4, whereas the characters could not have evolved on the unresolved tree with fewer than six changes. Finally, if different data were considered, say, four characters, each taxon having a single autapomorphy, the observed distances would be the same as before, and so the fitted branch lengths would be the same as well. Fitting branch lengths from distances—like the distances themselves—cannot distinguish between the two underlying character sets. This means that the process of computing the distances must discard information present in the character matrix. That is of course an obvious and often cited conclusion, but its implications for phylogenetic analysis seem to be often overlooked. If character data are available, they should be analysed directly rather than by way of a distance matrix. There is nothing to be gained by deliberately discarding evidence on kinship. The same examples apply to sequence differences, and the conclusion applies even more forcefully, as that measure has the additional drawbacks already discussed.

The Manhattan distance can also be used on frequency data, but in that application it has defects analogous to those of Rogers' distance. Returning to the example of E, F, G used for Rogers' distance above, the pairwise Manhattan distances are all 2. Branch length fitting to the unresolved tree yields all three lengths 1, and the derived matrix matches the observed matrix exactly—just as for Rogers' distance. Euclidean geometry cannot be applied with this metric, but it is still possible to discover a contradiction in finding possible gene frequencies for the ancestor X. Since X is linked by a branch of length 1 to each of the terminals, and if those branch lengths are Manhattan distances, then the gene frequency of X must be 0 for all three of the alleles present in the various terminals. It is possible, of course, that an ancestor lacks all the alleles of its descendants, but one hardly wants to use a method that forces this conclusion automatically. In any event, on the reasonable assumption that X had some allele, its Manhattan distance from each of its descendants must be 2 if it lacks all their alleles. This leads to the same difficulty as for all the other genetic distance measures. If the branch lengths add up to the observed Manhattan distances, then they cannot themselves be Manhattan distances, and if they are Manhattan distances they cannot add up. It seems that no measure of gene frequency distance can be logically analysed by branch length fitting.

Conclusion

It seems that the only general conclusion one can draw is that nothing about present techniques for analysing molecular distance data is satisfactory. The distance Wagner method seems to be the best available method for arriving at genealogies efficiently—certainly it is far more effective than Prager and Wilson imagined—but, at least for minimizing %SD, further improvement seems possible. But using any method that fits branch lengths to a distance matrix presupposes that the distances are suitable for this sort of analysis. That supposition seems unjustified for any distance measure now in use.

None of the known measures of genetic distance seems able to provide a logically defensible method, and it appears that some altogether different approach will have to be adopted for analysing electrophoretic data. Several authors (Baverstock et al., 1979; Mickevich and Johnson, 1976; Mickevich and Mitter, 1981) have already discarded the genetic distance concept, using methods of phylogenetic character analysis directly on electromorphs. This will no doubt prove controversial, both because it obviates the clock and because it uses relatively coarse frequency information, often just the presence of alleles. I doubt that those sorts of objections need be taken seriously. The nonmetricity of Nei's distance shows that there never was an electrophoretic clock, at least one precise enough to show kinship directly. The value of detailed frequency information in genetic distance measures is likewise questionable. Supposing that it were useful, one would require a logically defensible means of analysis to make use of it, and none seems possible. There is, moreover, good reason to doubt that details on frequencies are useful at all. At any systematically worthwhile level, say subspecies or higher, it is nearly always true that a locus that differs appreciably between two taxa differs just about completely; entirely separate sets of alleles will be present (Avise and Ayala, 1976; Ayala et al., 1974; Zimmerman et al., 1978). This being the case, there is not much comparative information in the frequencies beyond simple presences, and there is correspondingly little reason to mourn the loss of details on frequencies. Direct phylogenetic analysis of alleles as characters, moreover, avoids the information loss that attends reducing character data to distances.

Sequence data have sometimes been analysed by way of distances, although perhaps more often they have been treated by character analysis methods (for example Goodman et al., 1979). While sequence differences do not seem to offer the severe problems of interpretability of branch lengths that plague genetic distances, neither is there any good reason to rely on distance techniques. Reducing character data to distances, once again, simply wastes evidence on kinship.

Immunological distances, like genetic distances, cannot be truly clocklike, nor can they be analysed by branch length fitting, but there is no recourse to underlying character data in this case. A possible solution is suggested by an analogy with regression analysis, which branch length fitting resembles algebraically. If the original form of a variable does not meet the assumptions of a regression analysis, very likely some transformation of it will do so. Perhaps a way of transforming immunological distances can be found that will make them more amenable to analysis by branch length fitting. As the present distance is nonmetric and suitable distances would need to be metric, the transformation would have to nonlinear, and this would require recalibrating the immunological clock. That prospect may seem alarming or amusing, depending on one's point of view. But it is certainly preferable to retaining the present distance, analysis of which yields branch lengths with no possible physical interpretation.

Acknowledgements

I would like to thank C. Mitter for his help with the proof reading of this paper. This research was supported by grant number DEB-78–24647 from the National Science Foundation.

On the Utility of the Distance Wagner Procedure

David L. Swofford

Table of Contents

Introduction

The problem of choosing among methods for estimating phylogenetic trees from information contained in distance matrices is familiar to most biochemical systematists. This is due both to a plethora of available "matrix methods" (Cavalli-Sforza and Edwards, 1967; Farris, 1972; Felsenstein, 1973; Fitch and Margoliash, 1967; Li, 1981; Moore et al., 1973; Sarich, 1969a, 1969b; Sneath and Sokal, 1973) and to conflicting opinions in the literature regarding the relative merits of different approaches (e.g., Farris, 1972; Nei, 1978a; Prager and Wilson, 1978; Sneath and Sokal, 1973).

Unfortunately, the dilemma cannot be easily resolved through direct comparison of various methods, since many criteria are available for choosing among trees. Thus, a method that consistently provides "better" trees by one criterion may perform poorly when evaluated by another. To date, goodness of fit has been the most widely employed means of evaluating trees. This is reflected in the heavy use of the cophenetic correlation in the phenetic literature and on a number of more recent measures by phylogenetic systematists: the "percent standard deviation" (Fitch and Margoliash, 1967), *f* value of Farris (1972), *F* of Prager and Wilson (1976) and the "Moore Residual Coefficient" (Moore et al., 1973). These statistics measure the degree to which "output" ("tree" or "path-length" distances) between taxa reflect the corresponding "input" distances. Goodness of fit measures are often used to choose between alternative trees generated by a single method (e.g., Fitch and Margoliash, 1967; Moore et al., 1973) or to compare different methods (Farris, 1979b; Prager and Wilson, 1978).

Recently, some authors have suggested that the Fitch-Margoliash (1967) procedure (subsequently referred to as the F-M method) is the best available for tree estimation from distance data (Nei, 1978a; Prager and Wilson, 1978). In particular, Prager and Wilson claim to demonstrate, on the basis of goodness of fit measures, that the F-M method is superior to the most widely-used alternatives, the unweighted pair-group method of cluster analysis (UPGMA; Sokal and Michener, 1958) and distance Wagner procedure of Farris (1972). I certainly have no quarrel with the inferiority of the UPGMA method (and most other agglomerative or divisive clustering schemes). It is well established that dendrograms constructed by these phenetic clustering methods can be interpreted reasonably in a phylogenetic sense only when rates of evolutionary divergence are relatively homogeneous across phyletic lines (Ashlock, 1972; Farris, 1972), a condition all too often unsatisfied in real data sets. However, I find little empirical justification for the notion that the F-M method is preferable to the distance Wagner procedure. Indeed, I will show here that the same data sets and trees used by Prager and Wilson (1978) to demonstrate the supposed superiority of the F-M method can in fact be used to support exactly the opposite conclusion. That is, when the two methods are compared fairly, the distance Wagner procedure outperforms the F-M method.

In the following discussion I do not question the suitability of the various types of biochemical data used in the estimation of phylogenies; nor do I attempt to justify the use of distance coefficients when other methods of analysis may be more appropriate. These problems are addressed by J. S. Farris in another paper in this volume (see also Farris et al., 1979). My motivation in comparing tree building procedures follows simply from the realization that biochemical systematists will undoubtedly continue to rely on distance measures, whether by necessity (immunological and DNA hybridization data) or by choice (electrophoretic and nucleotide or protein sequence data).

Sources of Data

Most of the data sets employed below are taken from references cited in Table 1 of Prager and Wilson (1978). I have focused on these data sets for two reasons. First, they represent a cross-section of the types of biochemical information currently being used in systematics, including immunological, electrophoretic, and DNA hybridization data. Second, since they were used by Prager and Wilson to argue that the F-M method is superior to the distance Wagner procedure, any claim of bias in my results due to the selection of favorable data sets is not valid.

I have not used the lysozyme c sequence data referred to by Prager and Wilson because the original reference (Jolles et al., 1976) does not provide a complete distance matrix for the taxa included in Prager and Wilson's tree. Data for *Rana* proteins were taken from Case (1978) rather than Case (1976). Otherwise, the data sets used here are identical with those analyzed by Prager and Wilson (1978).

Evaluation of Trees

I will make use of three goodness of fit statistics in my comparison of tree-building procedures:

1. The "percent standard deviation" of Fitch and Margoliash (1967), a weighted least-squares criterion:

$$SD = \left\{ \sum_{i=1}^{n} \sum_{j=1}^{i} \left[\frac{(I_{ij} - O_{ij})100}{I_{ij}} \right]^2 \Big/ \left[\frac{n(n-1)}{2} - 1 \right] \right\}^{\frac{1}{2}}$$

2. The f statistic of Farris (1972):

$$f = \sum_{i=1}^{n} \sum_{j=1}^{i} |O_{ij} - I_{ij}|$$

3. The F value of Prager and Wilson (1976), which is a simple function of Farris' f (above):

$$F = 100f \Big/ \sum_{i=1}^{n} \sum_{j=1}^{i} I_{ij}$$

In each of these measures, n is the number of OTUs and I_{ij} and O_{ij} are, respectively, the "input" and "output" distances between taxa indexed by i and j. The input and output distances refer to the values in the original phenetic distance matrix and the path-length distances calculated from the completed tree, respectively. Output distances are thus closely related to the *patristic* distances of Farris (1967). The difference between the output distance and the input distance, $O_{ij} - I_{ij}$, reflects the amount of *homoplasy* between a pair of taxa (Farris, 1972).

Some workers have suggested that the ability to recover the "true" tree in computer simulation studies is a better criterion for evaluating tree-building procedures than goodness of fit (Kidd and Sgaramella-Zonta, 1971; Kidd et al., 1974; Li, 1981; Tateno, 1978). While I agree that the ability to recover a true tree (or to find a tree with minimal distortion of the true tree) is a desirable property, I hesitate to employ this as a primary criterion. If the simulation generates a tree exhibiting high degrees of convergence, parallelism, and/or reversal, the "most likely" tree for the resulting phenetic distance matrix will certainly not be the true tree. In this case, the inability to recover the true tree should not reflect adversely on the overall performance of a method. Furthermore, it is not impossible that a deficiency in a particular method might be compensated for by some property of the distribution on which the simulation is based and thus lead to erroneous conclusions about the desirability of a method. Because of these considerations, I have chosen goodness of fit as the primary criterion for evaluating trees. In particular, the three measures described above were employed because of their common usage in the literature and their consistency with the statistics used by Farris (1972) and Prager and Wilson (1978).

Distance Wagner Procedure and Modifications

The distance Wagner algorithm has been described in detail by Farris (1972) and will be reviewed here only briefly. Wherever possible Farris' original terminology and notation are used. I will be concerned only with the construction of unrooted bifurcating trees. Although Farris (1972) described a number of rooting methods permitting phylogenetic interpretation of an undirected tree, these do not, in general, affect goodness of fit measures and are not relevant here. I follow Lundberg (1972) and others in using the term *network* synonymously with *unrooted tree,* although in the terminology formalized in graph theory, "network" is actually more inclusive. Readers wishing a more complete description

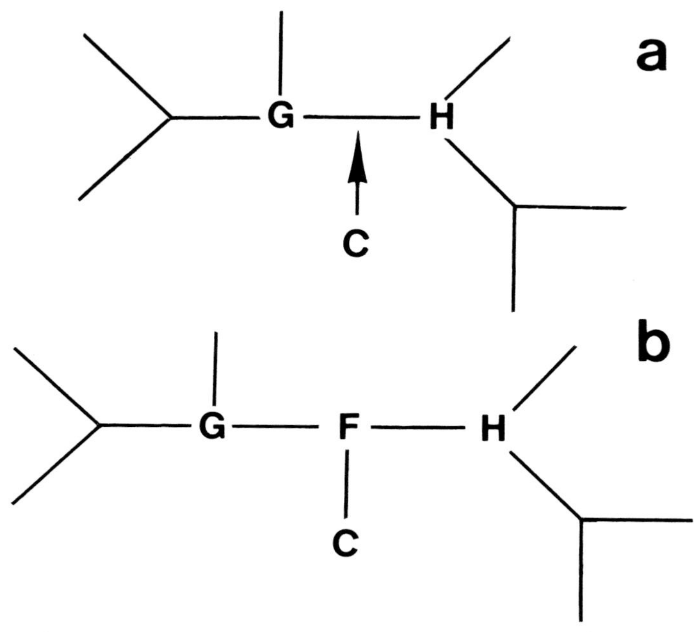

Fig. 1. Addition of a previously unplaced OTU to a Wagner network. a. OTU "C" selected to add next to branch bounded by nodes G and H. b. HTU "F" created and new branches constructed.

of this terminology as applied to phylogenetic trees should consult Kidd and Sgaramella-Zonta (1971) or Foulds et al. (1979).

Basically, the algorithm is as follows. At any stage in the construction of a tree, the original taxa (Operational Taxonomic Units or OTUs) will have been partitioned into two groups: those already connected to the network and those remaining to be attached. One OTU, denoted C, from the set of unattached OTUs is selected to join the network next (Fig. 1a). A branch of the network, bounded by nodes denoted G and H, is selected to minimize the distance

$$D[C,(G,H)] = \tfrac{1}{2}[D(C,G) + D(C,H) - D(G,H)], \quad (1)$$

where, in general, $D(I,J)$ refers to the "phenetic distance" between the real or hypothetical taxa I and J and is obtained either from the original "input" distance matrix or from values calculated during a previous cycle. [Equation (1) is an application of what is generally referred to as the "triangle distance formula."] The G-H branch is then split by the creation of a new node (Hypothetical Taxonomic Unit or HTU), designated F, and new branches connecting F to C, G, and H are constructed (Fig. 1b). The entire process continues until all OTUs have been added to the network.

Of course, the above paragraph is an oversimplified description, and the reader may have noticed at least two ambiguities: (1) How do we select an OTU to add next? (2) How do we calculate the phenetic distances from unplaced OTUs to newly created HTUs? The latter problem arises since the algorithm must reference only a distance matrix, and only the OTU × OTU portion of this matrix is known initially. Thus, as new HTUs are created, the distance between these nodes and the remaining unattached OTUs must be approximated before subse-

Table I

Matrix of albumin immunological distances for seven mammalian carnivores

Genus	Ca	Ur	Pr	Mu	Ph	Za	Fe
Canis (Ca)	0						
Ursus (Ur)	32	0					
Procyon (Pr)	48	26	0				
Mustela (Mu)	51	34	42	0			
Phoca (Ph)	50	29	44	44	0		
Zalophus (Za)	48	33	44	38	24	0	
Felis (Fe)	98	84	92	86	89	90	0

quent steps can be performed. To do this, I use the method of greatest lower bound approximation of these distances proposed by Farris (1972).

The selection of OTUs for next addition to the network poses a somewhat more complicated problem. Farris (1972) actually describes three criteria for determining which OTU should be connected next, which I have denoted *I*, *II*, and *III*:

Criterion *I*. Select for next addition the unplaced OTU C to minimize $m(C,G,H)$ for adjacent tree nodes G and H, where

$$m(C,G,H) = 2 \min\{D[C,(G,H)],D[G,(C,H)],D[H,(G,C)]\}, \quad (2)$$

where the distances on the right hand side of the equation are calculated using the triangle distance formula (1).

Criterion *II*. Select for next addition the unplaced OTU C to minimize an (upper bound) approximation of $m(C,G,H)$, namely

$$\hat{m}(C,G,H) = \min[D(C,G),D(C,H),D(G,H)]. \quad (3)$$

Criterion *III*. Select for next addition the unplaced OTU C which minimizes $D[C,(G,H)]$ for all possible values of C, G, and H, where G and H are adjacent tree nodes and $D[C,(G,H)]$ is calculated using (1).

Criterion *III* was originally proposed only as a means of choosing between OTUs when a tie occurs under criteria *I* or *II*. However, Farris (1974) and other workers (e.g., Prager and Wilson, 1978) have apparently used this as the primary addition criterion. The theory behind these criteria is not important to the present discussion, and readers are referred to Farris (1972) for details.

During the course of writing a computer program to perform the distance Wagner procedure, I noticed a property of Farris' method which at first seemed quite disconcerting: application of different addition criteria resulted in trees with not only different branch lengths, but often with different topologies as well. The effect of using different addition criteria can be shown through the example presented by Farris (1972). The data matrix for this example consists of albumin immunological distances for seven mammalian carnivores (Sarich 1969a, 1969b). The first step of the distance Wagner algorithm results in the joining of *Phoca* and *Zalophus* (Fig. 2a), since the distance between these taxa is the smallest entry in the original phenetic distance matrix (Table I). We must now select an OTU to add next from the set of unplaced OTUs {*Canis, Ursus, Procyon, Mustela, Felis*} (Fig. 2b). Application of Criterion *I* [exact $m(C,G,H)$ calculation (2)] results in the addition of *Mustela* to the tree next, since

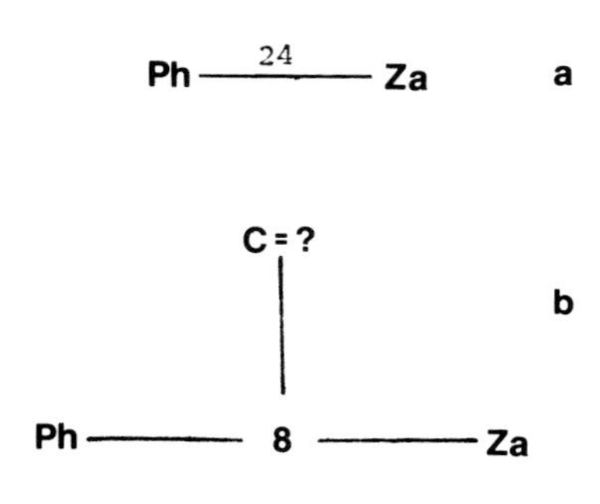

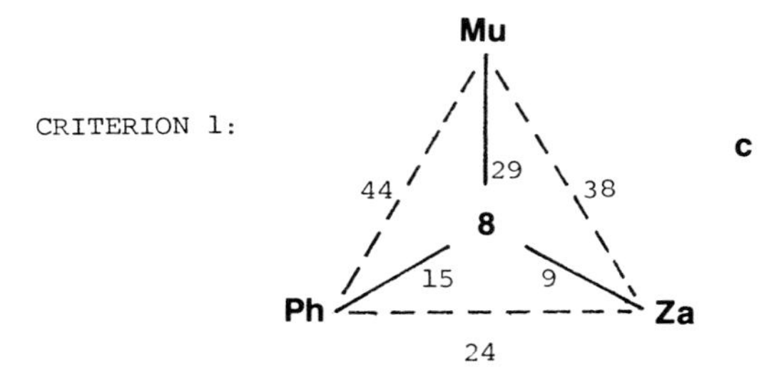

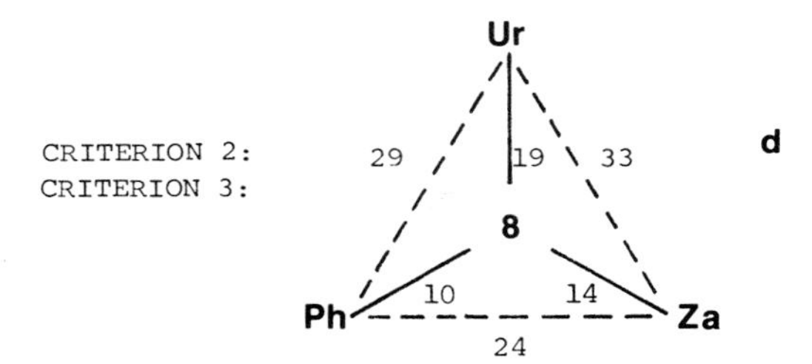

Fig. 2. Initial stages in distance Wagner tree construction for carnivore albumin data (abbreviations as in Table I). Dashed lines indicate phenetic distances from Table I; solid lines indicate distances calculated using expression (1). See text for details.

$$m(Mustela, Phoca, Zalophus) = 2 \min\{29, 15, 9\} = 18$$

minimizes $m(C,Phoca,Zalophus)$ for all possible values of C (Fig. 2c). Thus, a new HTU, denoted *8*, is created, and the lengths of branches (*8,Phoca*), (*8,Zalophus*), and (*8,Mustela*) are calculated as 15, 9, and 29, respectively (Fig. 2c). Farris (1972), however, used Criterion *II* to approximate the "more exact" $m(C,G,H)$ calculation. In this case, ties occur under Criterion *II*, and *Ursus* is selected to add next since, from equation (4),

$$D[Ursus, (Phoca, Zalophus)] = \tfrac{1}{2}(29 + 33 - 24) = 19$$

minimizes $D[C,(Phoca,Zalophus)]$ (Fig. 2d). The lengths of branches (*8,Phoca*), (*8,Zalophus*), and (*8,Mustela*) then equal 10, 14, and 29, respectively.

Table II

Branch lengths obtained using different addition criteria[a]

Branch	Addition criterion		
	I	*II, III*	Multiple
a	9.0	14.0	9.0
b	15.0	10.0	15.0
c	17.0	18.0	17.0
d	69.0	68.0	69.0
e	7.5	5.5	7.5
f	18.5	20.5	18.5
g	8.5	9.5	9.5
h	3.5	6.5	2.5
i	3.5	0.5	2.5
j	26.0	27.0	27.0
k	4.5	3.5	4.5

[a] Distance Wagner procedure for carnivore albumin data.

Continued application of Criterion *I* results in the sequential addition of *Canis, Felis, Procyon,* and *Ursus* to the 3-taxon network of Figure 2c. This addition sequence is quite different from the sequence specified by Criteria *II* and *III*: *Procyon, Mustela, Canis,* and *Felis*. The final trees in this case are identical in topology (Fig. 3) but differ considerably in branch length (Table II). The *f* value for the tree calculated by Farris (1972), using Criterion *II*, is 59. If, however, Criterion *I* had been used, an *f* of only 55 could have been obtained. This does not imply that Criterion *I* is generally superior. In fact, as shown in Table III, there is no indication that any one of the three criteria is optimal. The number of data sets in which a given criterion results in an *F* value equal to the minimum *F* value for criteria *I, II,* and *III* is 2, 5, and 7, respectively. Furthermore, each criterion produces a unique minimum *F* value at least once (e.g., data sets 1, 8, and 9). The effect of using different addition criteria is even more dramatic for *SD* values. For example, *SD* varies from 21.5 (Criteria *II, III*) to 46.1 (Criterion

Table III

Goodness of fit of distance Wagner trees using different addition criteria

Data set	Addition criterion							
	I		*II*		*III*		Multiple[a]	
	F	*SD*	*F*	*SD*	*F*	*SD*	*F*	*SD*
1. Carnivore albumins	4.9	10.1	5.2	8.7	5.2	8.7	4.5	9.9
2. *Xenopus* albumins	16.6	46.1	11.8	21.8	11.8	21.8	11.8	21.8
3. Ranid albumins	2.2	3.8	1.8	3.5	1.8	3.5	1.8	3.5
4. Conifer seed proteins	14.5	29.5	29.6	41.8	29.6	41.8	10.9	19.4
5. Hylid albumins	22.2	29.7	16.5	21.1	16.5	21.1	11.4	16.9
6. Bird transferrins	25.0	35.5	38.1	48.0	23.3	29.7	23.3	29.7
7. Primate DNAs	1.0	3.3	0.9	1.4	0.9	1.4	0.9	1.4
8. *Rana* proteins	15.9	31.4	15.1	29.9	17.1	24.3	15.1	24.3
9. *Drosophila* enzymes	16.4	24.1	16.2	22.2	14.6	21.3	12.7	18.8

[a] When more than a single tree is found using multiple addition criteria, the one with lowest *F* or *SD* value is listed.

I) for data set 2, and from 29.7 (Criterion *III*) to 48.0 (Criterion *II*) for data set 6.

These results indicate that the ability of the Farris method to find a tree with optimal goodness of fit is heavily dependent on the sequence of addition of OTUs to the network. However, we have just seen that none of the three addition criteria is, in itself, sufficient to determine the best addition sequence. One way to circumvent this problem is to use all three addition criteria simultaneously, rather than restricting the addition sequence to that specified by any one criterion. In elaborating on this idea, it is useful to define a *partial network* as a network in which some of the OTUs in the data set have yet to be connected.

This "multiple addition criterion" procedure diverges from that of Farris (1972) in that more than one partial network may be generated during each cycle of the construction of the final tree. Thus, at the beginning of any cycle j, k partial networks will exist, each with x OTUs attached and $n - x$ OTUs still unplaced, where n is the total number of OTUs represented in the original matrix and $k \geq 1$. Now, for any of these partial networks, some OTU C must be selected from the $n - x$ remaining OTUs to add next. However, since we are using all three addition criteria simultaneously, Criterion *II* may select a different "C" than Criterion *I*, provided that $n - x \geq 2$. Likewise, if $n - x \geq 3$, Criterion *III* may select a different OTU for next addition than does either of criteria *I* or *II*. Thus, for every partial network existing at the beginning of the jth cycle, up to three partial networks will have been generated by the beginning of the $(j + 1)$th cycle (subject, of course, to the availability of unplaced OTUs). Each of the $3k$ possible networks will then be treated independently in subsequent steps; in all other respects the algorithm is identical to the original Farris (1972) procedure.

It may seem at first that the number of partial networks might become prohibitively large, since this number can theoretically be increased by a factor of three during each cycle. In practice, this does not occur, for two reasons. First, criteria *I*, *II*, and *III* rarely select three *different* OTUs to be added next, and often the same OTU is specified by all three criteria. Second, partial networks often converge to the same topology. The branch lengths for these partial networks usually differ, however, since the common topology was arrived at through different addition sequences. In this case, the number of partial networks held for subsequent steps can be reduced by saving, from all partial networks with the same topology, only the one with the closest fit to the input data (as evaluated by, e.g., the F value). The number of partial networks can be reduced even further by ranking them in order of decreasing goodness of fit and saving only some arbitrary number, say three or five, with closest fit to the input data.

When this technique is applied to the carnivore albumin data (Table I), the addition sequence of the best-fitting network (*Mustela, Ursus, Felis, Procyon, Canis*) is different from that given by any of the three single criteria. Furthermore, the f value is reduced even further, to 51. In other words, the estimate of the total amount of homoplasy has been reduced by 8 immunological distance units from that of the tree originally constructed by Farris (1972).

Table III also shows the results of application of this new method to the Prager and Wilson data sets. Note that in four cases (including the carnivore albumin data), the best tree obtained using multiple addition criteria has a closer fit to the input data than all three single-criterion trees. However, for all but two of these data sets, the topology obtained using multiple criteria is identical to at least one of the single-criterion trees. This is not true in general, however, for I have found that as the number of taxa increases, the chance decreases that any single-addition-criterion approach will find a tree with a topology identical to the best fitting multiple-addition-criterion tree.

Comparison of Distance Wagner, F-M, and UPGMA Methods

The question addressed by Prager and Wilson (1978) is simply this: Of the three most commonly used methods for constructing trees from distance matrices, which is most appropriate for various types of biochemical data? One way to answer this question is to compute trees using all three methods for a variety of data sets and to determine if any one method consistently performs better in terms of some goodness of fit statistic. Prager and Wilson, of course, did exactly this and concluded that the F-M method is the most appropriate for most types of biochemical data. However, their logic in reaching this conclusion is flawed, as I shall show below.

The original distance Wagner procedure and the modified method described above are constrained in that all output (patristic) distances are required to be equal to or greater than the corresponding input (phenetic) distances. In other words, "negative homoplasies" are not allowed. The rationale for this restriction is aptly presented by Farris (1972) (see also Waterman et al., 1977):

> If two OTUs differ by an observed amount, d, it is not physically possible for the evolutionary pathway connecting them to contain fewer than d changes.

Since the validity of this statement cannot be questioned, an important problem arises. How are we to compare goodness of fit measures for trees calculated by different procedures when the distance Wagner procedure requires only non-negative homoplasies and other methods (including both F-M and UPGMA) do not? Clearly, the distance Wagner procedure cannot be expected to perform as well under the constraints just mentioned; yet the branch lengths it estimates are almost certainly biologically more meaningful. Other workers have recognized the need to distinguish between the estimation of topologies and of branch lengths (Li, 1981; Nei, 1978a). Prager and Wilson apparently did not realize the importance of this concept.

At any rate, in order to make a *meaningful* comparison between the distance Wagner procedure and other methods on the basis of goodness of fit criteria, either we must subject these other methods to the constraint of non-negative homoplasies or we must relax this restriction for the distance Wagner trees. Theoretically, either approach could be taken, simply by altering branch lengths after the topology of a tree is determined. Practically, it is simpler to recalculate branch lengths of distance Wagner trees so as to permit negative homoplasies, and I describe a method for doing so below.

Optimization of Distance Wagner Trees

The method of branch length optimization described here is designed to minimize Prager and Wilson's F value. Methods for least-squares calculation of branch lengths have been described (Cavalli-Sforza and Edwards, 1967; Kidd and Sgaramella-Zonta, 1971), but these do not seem appropriate in terms of F since this statistic measures absolute rather than squared deviations. Although I am unable to provide statistical justification for my approach, it appears to be quite effective. It is also relatively simple and, unlike least-squares optimization, can be performed without the aid of a computer.

Before I begin the description, a few definitions are necessary. A *terminal node* is one that is connected to the network by a single branch. A node is said to be *internal* if it lies in the interior of the network and is connected by branches to exactly three other nodes, which may themselves be terminal or internal. Likewise, a *terminal branch* connects a terminal node to an internal node and an *internal branch* connects two internal nodes. (My usage of "branch" is synonymous with the terms "interval," "segment," and "link" of some authors.) Initially, the terminal nodes consist of all OTUs in the data set, and the internal nodes represent the "HTUs" that were created during the tree-building process. *Cladistic difference* (Farris, 1967a) refers here to the number of internal nodes lying on the path between a pair of terminal nodes.

The optimization procedure can be described by the following sequence of steps. Note that all that is required initially is a tree topology and an associated OTU × OTU distance matrix.

1. Select any pair of terminal nodes, denoted X_i and Y_i, having a cladistic difference of unity. Let Z_i denote the internal node lying on the path from X_i to Y_i and let x_i and y_i denote the optimal lengths of the terminal branches leading to X_i and Y_i.
2. For all terminal nodes T_{ij} other than X_i and Y_i, estimate $D(T_{ij}, Z_i)$ using the triangle distance formula:

$$D(T_{ij}, Z_i) = \tfrac{1}{2}[D(T_{ij}, X_i) + D(T_{ij}, Y_i) - D(X_i, Y_i)]. \tag{4}$$

$D(T_{ij}, Z_i)$ will be used in subsequent steps and can be recorded by augmenting the original distance matrix.
3. For each of the T_{ij} in step 2, estimate x_i and y_i as

$$x_{ij} = D(T_{ij}, X_i) - D(T_{ij}, Z_i) \tag{5}$$

and

$$y_{ij} = D(T_{ij}, Y_i) - D(T_{ij}, Z_i). \tag{6}$$

4. Calculate x_i and y_i as weighted averages of the x_{ij}'s and y_{ij}'s, respectively:

$$x_i = \sum_j w_j x_{ij} \Big/ \sum_j w_j, \tag{7}$$

and

$$y_i = \sum_j w_j y_{ij} \Big/ \sum_j w_j, \tag{8}$$

where w_j is the "weight" of the jth 3-way comparison, calculated as

$$w_j = v_{X_i} \cdot v_{Y_i} \cdot v_{Z_i}. \tag{9}$$

Note that $v = 1$ for all X, Y, and T representing original OTUs. Otherwise, v has a value calculated in a previous performance of step 7 (below).
5. Repeat steps 1–4 for all other pairs of nodes having a cladistic difference of unity.
6. Distances between the Z_i's selected during steps 1–5 will be needed in subsequent cycles. Therefore, complete the augmentation of the distance matrix begun during previous performances of step 2 by evaluating

$$D(Z_k, Z_l) = \tfrac{1}{2}[D(Z_k, X_l) + D(Z_k, Y_l) - D(X_l, Y_l)], \quad k \neq l. \tag{10}$$

7. For all X-Y-Z triads just considered, delete the terminal nodes X_i and

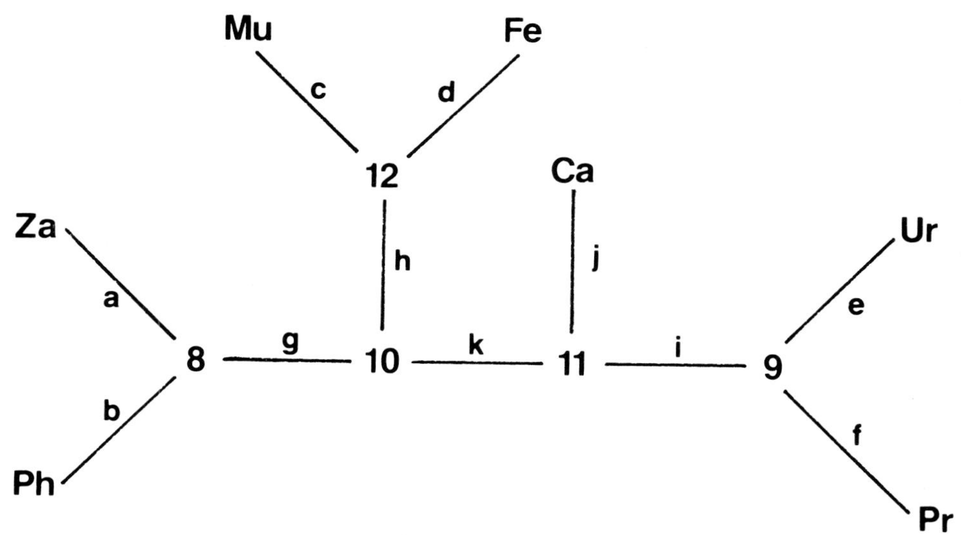

Fig. 3. Distance Wagner network for carnivore albumin data (abbreviations as in Table I).

Y_i and destroy the terminal branches leading to these nodes ("pruning," in the sense of Felsenstein, 1973). Z_i is now a terminal node. Calculate the weight for Z_i as

$$v_{Z_i} = v_{X_i} + v_{Y_i}. \quad (11)$$

Collapse the distance matrix by removing the rows and columns corresponding to the deleted nodes.

8. If more than a single branch remains, return to step 1. Note that if only three branches are left, only one repetition of steps 1–4 will be required to calculate the remaining branch lengths.

To illustrate this method, I will use Sarich's (1969a, 1969b) carnivore albumin data (Table I) and the tree topology given in Figure 3. The objective is to estimate the optimal branch lengths $a, b, \ldots, k$.

The first application of step 1 results in the selection of, say, *Zalophus* and *Phoca* as two terminal nodes X_1 and Y_1 with cladistic difference unity. Z_1 is then HTU *8*. We must now calculate the distances from all other terminal nodes (in this case, all other OTUs) to *8* and the corresponding estimates of branch lengths a and b. For example, letting T_{11} = *Mustela* and using (4),

$$\begin{aligned} D(T_{11},Z_1) = D(\textit{Mustela},8) &= \tfrac{1}{2}[D(\textit{Mustela},\textit{Zalophus}) \\ &\quad + D(\textit{Mustela},\textit{Phoca}) \\ &\quad - D(\textit{Zalophus},\textit{Phoca})] \\ &= \tfrac{1}{2}(38 + 44 - 24) \\ &= 29. \end{aligned}$$

The estimated branch lengths are then (5,6)

$$\begin{aligned} x_{11} &= D(\textit{Mustela},\textit{Zalophus}) - D(\textit{Mustela},8) \\ &= 38 - 29 \\ &= 9 \end{aligned}$$

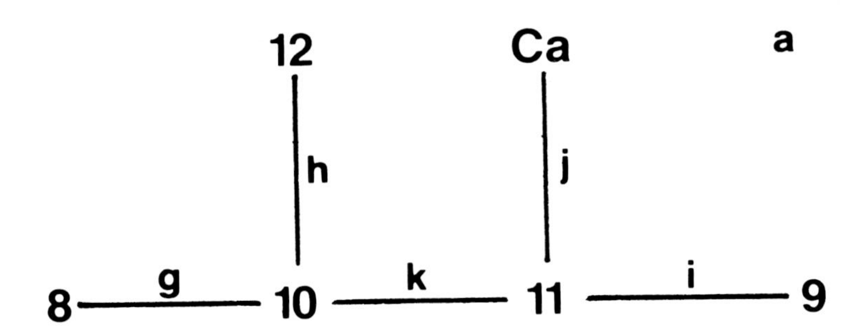

Fig. 4. Collapsed trees obtained during optimization process (abbreviations as in Table I). See text for details.

and

$$\begin{aligned} y_{11} &= D(\textit{Mustela}, \textit{Phoca}) - D(\textit{Mustela}, 8) \\ &= 44 - 29 \\ &= 15. \end{aligned}$$

Similar calculations are performed letting T_{1j} represent, in turn, *Felis*, *Canis*, *Ursus*, and *Procyon*. The branch length a equals the weighted average of the x_{1j}'s (7):

$$\begin{aligned} a = x_1 &= \frac{1^3(9) + 1^3(12.5) + 1^3(11) + 1^3(14) + 1^3(12)}{1^3 + 1^3 + 1^3 + 1^3 + 1^3} \\ &= 11.7, \end{aligned}$$

where the weights 1^3 are calculated using (9). Of course, in this case the weighting is trivial, since all terminal nodes are OTUs and thus have equal weight. Similarly, $b = x_2$ is calculated, using (8), as $(15 + 11.5 + 13 + 10 + 12)/5 = 12.3$.

At this stage, there are two other pairs of OTUs having cladistic differences of unity: {*Mustela*,*Felis*} and {*Ursus*,*Procyon*}. By calculations analogous to those

Table IV

Distance matrix after first optimization cycle[a]

	Canis	*8*	*9*	*12*
Canis	0			
8	37	0		
9	27	12.5	0	
12	31.5	10.25	7	0

[a] Distances from *Canis* to other nodes calculated using expression (4); other entries calculated using expression (10).

Table V

Optimized branch lengths for tree of Figure 3

Branch										
a	b	c	d	e	f	g	h	i	j	k
11.7	12.3	18.6	67.4	7.2	18.5	7.875	2.375	1.25	25.75	3.375

above, the branch lengths *c, d, e,* and *f* are determined as 18, 68, 7.5, and 18.5, respectively.

We must now collapse the tree of Figure 3 to the form given in Figure 4a by destroying the appropriate terminal branches. Before this is done, however, $D(8,9)$, $D(8,12)$, and $D(9,12)$ are calculated, since they will be needed later. For example, $D(8,12)$ can be calculated (10) either as:

$$\begin{aligned} D(8,12) &= \tfrac{1}{2}[D(8, \mathit{Mustela}) + D(8, \mathit{Felis}) - D(\mathit{Mustela}, \mathit{Felis})] \\ &= \tfrac{1}{2}(29 + 77.5 - 86) \\ &= 10.25 \end{aligned}$$

or, equivalently, as

$$\begin{aligned} D(8,12) &= \tfrac{1}{2}[D(12, \mathit{Zalophus}) + D(12, \mathit{Phoca}) - D(\mathit{Zalophus}, \mathit{Phoca})] \\ &= \tfrac{1}{2}(21 + 23.5 - 24) \\ &= 10.25. \end{aligned}$$

The distance matrix resulting from the first performance of steps 1–7 is shown in Table IV. We now calculate branch lengths *g, h, i,* and *j* (Fig. 4a). Note that, from (11), $v_8 = v_9 = v_{12} = 2$. First, we may let $X_1 = 8$, $Y_1 = 12$, and $Z_1 = 10$. Letting $T_{11} = 9$, we calculate [using (4), (5), and (6)]:

$$\begin{aligned} D(9,10) &= \tfrac{1}{2}[D(9,8) + D(9,12) - D(8,12)] \\ &= \tfrac{1}{2}(12.5 + 7 - 10.25) \\ &= 4.625, \end{aligned}$$

$$\begin{aligned} x_{11} &= D(9,8) - D(9,10) \\ &= 12.5 - 4.625 \\ &= 7.875, \end{aligned}$$

and

$$\begin{aligned} y_{11} &= D(9,12) - D(9,10) \\ &= 7 - 4.625 \\ &= 2.375. \end{aligned}$$

We now let $T_{12} = \mathit{Canis}$. By similar calculations, $x_{12} = 7.875$ and $y_{12} = 2.375$. The weighted average of x_{1j}'s is then

$$\begin{aligned} g = x_1 &= \frac{(2)(2)(2)(7.875) + (2)(2)(1)(7.875)}{(2)(2)(2) + (2)(2)(1)} \\ &= 7.875. \end{aligned}$$

The weighting in this case is unnecessary since $x_{11} = x_{12}$, but this is not generally true. Branch length *h* is calculated in a similar manner as 2.375. We now let $X_2 = 9$, $Y_2 = \mathit{Canis}$, and $Z_2 = 11$. By the same process as above, we compute the branch lengths $i = 1.25$ and $j = 25.75$. The distance $D(10,11)$ is needed:

Table VI

F values of Fitch-Margoliash, UPGMA, and optimized distance Wagner trees

			Distance Wagner (criterion)			
Data set	F-M[a]	UPGMA[a]	(*I*)	(*II*)	(*III*)	(Multiple[b])
1. Carnivore albumins	3.4	7.8	2.5	2.5	2.5	2.5
2. *Xenopus* albumins	6.5	8.4	5.7	5.7	5.7	5.7
3. Ranid albumins	2.1	4.3	1.7	1.7	1.7	1.7
4. Conifer seed proteins	8.2	9.4	8.1	8.8	8.8	8.1
5. Hylid albumins	7.0	11.1	6.7	5.4	5.4	5.4
6. Bird transferrins	8.6	12.8	8.9	10.4	8.5	8.4
7. Primate DNAs	1.3	1.4	0.7	0.7	0.7	0.7
8. *Rana* proteins	9.8[c]	18.3[d]	8.1	8.1	8.4	8.1
9. *Drosophila* enzymes	6.7	8.2	6.6	7.4	6.4	6.3

[a] From Prager and Wilson (1978) unless otherwise indicated.
[b] The best-fitting multiple addition criterion tree is listed.
[c] From Case (1978).
[d] Recalculated using Nei distance matrix of Case (1978).

$$\begin{aligned} D(10,11) &= \tfrac{1}{2}D(10,9) + D(10,Canis) - D(9,Canis) \\ &= \tfrac{1}{2}(4.625 + 29.125 - 27) \\ &= 3.375. \end{aligned}$$

Collapsing as before, we delete nodes *8, 9, 12,* and *Canis,* leaving a single branch (Fig. 4b). We then let branch length k equal $D(10,11) = 3.375$.

The branch lengths resulting from the optimization procedure are shown in Table V. The f value for the optimized tree is only 27.8, corresponding to an F of 2.5%. This represents a 52% improvement over the F value that Prager and Wilson used for comparison with the F-M and UPGMA methods.

Extension to the Prager and Wilson Data Sets

F values for the optimized distance Wagner trees computed from the nine Prager and Wilson (1978) data sets (see Table III) are presented in Table VI, along with F values for the F-M and UPGMA trees calculated by Prager and Wilson. In every case, the F value for the best-fitting multiple-addition-criterion distance Wagner tree is less than those for the corresponding F-M and UPGMA trees.

The objection could be raised that my results are not comparable to those of Prager and Wilson on the grounds that I used a distance Wagner procedure that is highly modified from Farris' (1972) original method. Two arguments might be advanced to support this claim. The first is that the "multiple-addition-criterion" approach, not available to Prager and Wilson, results in better trees than does the Farris (1972) method. Of course, I believe this is true; otherwise I would not have described the modified procedure in full detail. However, this argument is irrelevant in the present context. It is clear from comparison of Prager and Wilson's (1978) Table 2 with my Table III that "Criterion III" was used in their distance Wagner calculations. For five of the nine data sets in Tables III and VI, the topology given by Criterion III is identical to that of the best fitting multiple-addition-criterion tree. Moreover, even in the exceptions, data sets 4, 6, 8, and 9, optimized trees can be obtained using single addition criteria that have closer

Table VII

F values for optimized Fitch-Margoliash trees (see text)[a]

Data set	F-M before opt.	F-M after opt.	Distance Wagner
5. Hylid albumins	7.0	6.3	5.4
6. Bird transferrins	8.6	8.0	8.4
8. *Rana* proteins	9.8	9.5	8.2

[a] Optimized multiple criterion distance Wagner trees provided for comparison.

fits to their respective input data than do the F-M and UPGMA trees. For these nine data sets, then, the multiple-addition-criterion procedure is most useful in minimizing *F* under the constraint of non-negative homoplasies and less important in determining tree topologies. However, as I pointed out above, the modified procedure becomes more important in the estimation of tree topology as the number of taxa increases. The mean number of taxa included in the Prager and Wilson (1978) data sets analyzed here is only 8.9.

The second possible argument is that the optimization procedure applied above gives the distance Wagner procedure an unfair advantage over the other methods. This does not seem likely, since this method of optimizing branch lengths is similar in many respects to the method of branch length calculation used in the F-M procedure. Nevertheless, to test the validity of the argument, we must optimize in the same manner trees generated by the F-M method. Unfortunately, in many cases this is impossible, since the topologies for the F-M trees calculated by Prager and Wilson (1978) are not available. The F-M and distance Wagner procedures produced identical tree topologies for the conifer seed protein data (data set 4; Prager et al., 1976a). In two additional cases, data sets 3 and 7, we can infer that the tree topologies estimated by the two methods are identical, since the cluster distortion coefficient (Farris, 1973) between them is zero (cf. Prager and Wilson, 1978, p. 136). Three other F-M trees are available: those of Maxson and Wilson (1975) for hylid albumins, Ho et al. (1976) and Prager et al. (1976b) for bird transferrins, and Case (1978) for *Rana* proteins. The results of optimization of these trees are shown in Table VII. Although the optimized F-M tree for the bird transferrins has a slightly lower *F* value, the *F* values for the distance Wagner trees are considerably better for the other two data sets. (The reason for the poorer performance of the distance Wagner trees for the bird transferrins is largely due to the fact that this data set violates the triangle inequality.)

Discussion

The results presented above clearly indicate that, contrary to the conclusions of Prager and Wilson (1978), the F-M method is not superior to the distance Wagner procedure. In light of this, several of Prager and Wilson's specific statements require further critical consideration.

Assumptions of Distance Wagner vs. F-M Methods

Prager and Wilson claim that the distance Wagner procedure is "inferior" in terms of goodness of fit because indirect biochemical techniques (i.e., immunology, electrophoresis) "do not give minimum estimates of distance, but the Farris method assumes that they do." Farris' method, of course, makes no such assumption. Prager and Wilson apparently do not understand the distinction between phenetic and patristic distance. For instance, they state that ". . . as a consequence of their not giving minimum estimates of distance, data obtained by these indirect techniques at times violate the triangle inequality, while the Farris (1972) method requires satisfaction of this inequality." (Indeed, as pointed out earlier, biological reality requires satisfaction of this inequality.) This statement suggests that the non-minimal distances to which Prager and Wilson refer are the observed *phenetic* distances. In actuality, Farris' (1972) method simply assumes that input distances are *accurate* estimates of contemporary phenetic distance and are minimum estimates only of the *actual* amount of change which has occurred during evolution (i.e., of patristic distance). Yet, even if this were not true, there is no basis for Prager and Wilson's argument that, since electrophoresis does not provide minimum estimates of distance, the distance Wagner procedure is theoretically inappropriate for electrophoretic data. Electrophoresis does, in fact, provide minimum estimates of distance, at least for the particular set of loci analyzed, provided that the distance measure used is a metric (i.e., obeys the triangle inequality). Nei's (1972, 1978b) distances do not satisfy this criterion. There are, however, several distance measures that are metrics (e.g., Cavalli-Sforza and Edwards, 1967; Prevosti *in* Wright, 1978; Rogers, 1972), and their use with the distance Wagner procedure is quite appropriate.

Failure of a data set to satisfy the triangle inequality is therefore nothing more than a shortcoming of the data set and is not relevant to the selection of tree-building methods (see also Farris et al., 1979). It should also be pointed out that the averaging involved in the F-M and UPGMA methods (and the optimization procedure described herein) is actually the factor responsible for the generation of "negative homoplasies," rather than overestimates of distance in the original matrix. Consequently, the interpretation of "input" distances exceeding "output" distances (i.e., negative homoplasy) as indicating the overestimation of "true" distances is simply an abandonment of the concept of independent evolution along different phyletic lines and is thus tantamount to incorporating an assumption of homogeneous evolutionary rates.

Practical Aspects

The F-M method involves calculating an initial tree by a form of cluster analysis and subsequently altering its topology in an attempt to improve the value of some goodness of fit measure. The best F-M tree is defined as the one tree providing the closest fit to the original distance matrix according to the criterion chosen (e.g., F or SD value). Thus, Prager and Wilson's comparison of tree-building procedures obviously involves circular logic. They suggest that the F-M method is superior because it provides trees with lower F values, yet they *define* the optimal F-M tree as the tree having minimum F. Of course, then, the F-M method will always find the best tree for a given goodness of fit criterion if enough tree topologies are tried and if branch lengths are selected to optimize this criterion. However, this concept is not a useful one, since for more than six or seven taxa, the number of possible trees becomes so large that evaluation of all reasonable trees becomes impossible, even by computer (Cavalli-Sforza and Edwards, 1967; Felsenstein, 1978; Kidd et al., 1974; Kidd and Sgaramella-Zonta, 1971). Thus, the "trial-and-error" approach of Fitch and Margoliash (1967) and even more systematic rearrangement ("branch-switching") algorithms such as those of Kidd and Sgaramella-Zonta (1971) and Moore et al. (1973) actually consider only an exceedingly small proportion of the possible topologies when large numbers of taxa are involved. As stated by Sgaramella-Zonta and Kidd (1973),

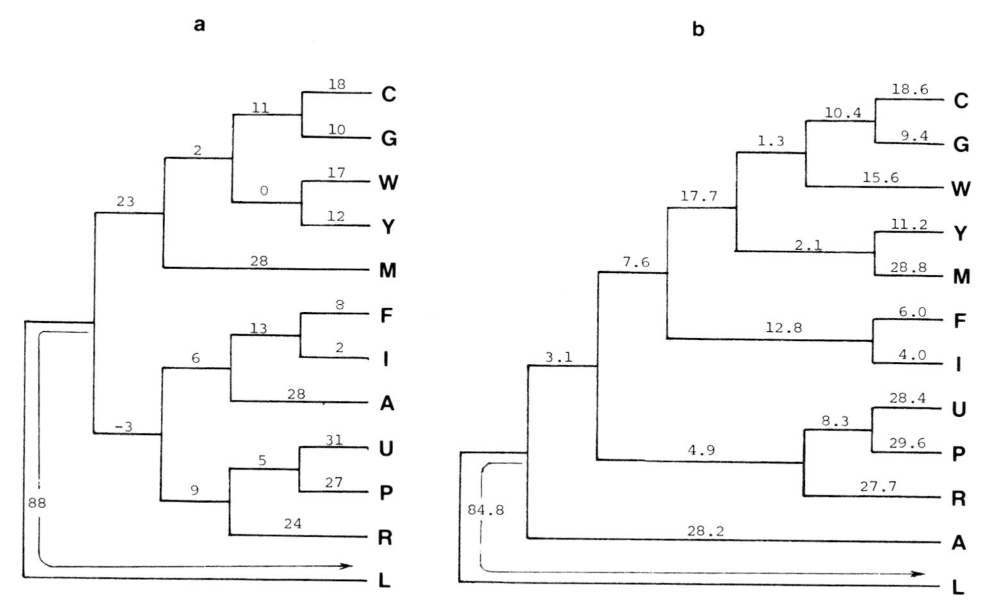

Fig. 5. Trees constructed from hylid albumin data of Maxson and Wilson (1975). a. Fitch-Margoliash tree constructed by Maxson and Wilson. b. Distance Wagner tree constructed using multiple addition criteria and branch length optimization. Tree is rooted arbitrarily to facilitate comparison.

> this [rearrangement procedure of Kidd and Sgaramella-Zonta (1971)] generally produces trees that are better by our statistics than the input tree. However, there is no guarantee that this improvement is anything but a local optimization that is highly dependent on the input tree.

I suspect that this fact alone accounts for the similarity of the F-M and UPGMA trees noted by Prager and Wilson (1978). Since the initial F-M tree topology is generally close to that given by the UPGMA procedure, there is a tendency to test alternative topologies without deviating greatly from the original tree. Thus, if the initial topology is far from optimal, it is quite unlikely that the researcher will be diligent enough to eventually arrive at an optimal tree. I illustrate this point with two simple examples.

The first example involves the F-M tree calculated by Maxson and Wilson (1975) and subsequently used by Prager and Wilson (1978). This tree is presented alongside the corresponding optimized distance Wagner tree in Figure 5. The considerable difference in the topologies of the two trees is evident in the mean distortion coefficient (calculated by Prager and Wilson) of 0.23. Note especially the differences in the placement of the species represented by *F*, *I*, and *A*. The F-M and UPGMA trees are quite similar (cluster distortion coefficient = 0.06). Yet, the optimized distance Wagner tree actually has a much better fit to the input data: $F = 5.4$ for the distance Wagner tree vs. 7.0 and 6.3 for the unoptimized and optimized F-M trees, respectively. Furthermore, the negative branch length in the F-M tree is absent in the distance Wagner tree. Thus, the distance Wagner tree is almost certainly more appropriate. On the basis of their F-M tree, Maxson and Wilson (1975) concluded (tentatively) that a group consisting of *F*, *I*, *A*, *U*, *P*, and *R* is distinct evolutionarily from the other species considered. However, as the tree of Figure 5b indicates, this assemblage may not be monophyletic.

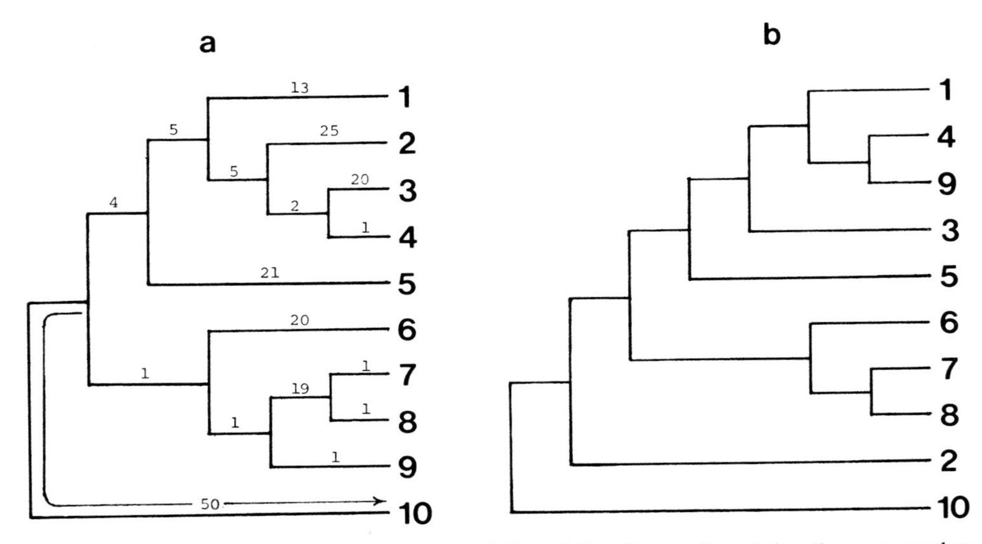

Fig. 6. Hypothetical trees of Moore et al. (1973). a. "True" tree. Branch lengths were used to construct distance matrix in Table VIII. b. Topology of UPGMA tree calculated using distance matrix in Table VIII.

The conclusions reached by Maxson and Wilson (1975) regarding the placement of *Acris* ("*A*") are even more disturbing. They claim, presumably on the basis of either the tree of Figure 5a or their "collapsed" tree (their p. 6), that *Acris* is cladistically a member of a holarctic assemblage of *Hyla* lineages [including all species of Fig. 5 except the Australian *Litoria* ("*L*")]. Since *Acris* is morphologically, ecologically, and behaviorally distinct from typical *Hyla*, Maxson and Wilson attempt to explain its placement in terms of accelerated rates of organismal evolution and "major adaptive shifts," when in reality their data do not even support the premise that *Acris* belongs in the *Hyla* assemblage (see placement of *A* in Fig. 5b).

This is not intended as a specific criticism of Maxson and Wilson; it merely indicates what can go wrong using the F-M procedure. Of course, the inability of Maxson and Wilson to find a better F-M tree than the one they published is

Table VIII

Hypothetical distance matrix from tree of Figure 6

	1	*2*	*3*	*4*	*5*	*6*	*7*	*8*	*9*	*10*
1	0									
2	43	0								
3	40	47	0							
4	21	28	21	0						
5	39	56	53	34	0					
6	43	60	57	38	46	0				
7	44	61	58	39	47	41	0			
8	44	61	58	39	47	41	2	0		
9	25	42	39	20	28	22	21	21	0	
10	72	89	86	67	75	71	72	72	53	0

understandable when one considers that there are almost 655 million unrooted bifurcating trees for 12 taxa (Felsenstein, 1978).

The second example is taken from Moore et al. (1973). These authors describe an iterative procedure for rearranging tree topologies in an attempt to minimize their "Moore Residual Coefficient." A hypothetical distance matrix (Table VIII) was constructed from path-length distances taken from a known tree (Fig. 6a). Starting from an initially quite incorrect topology obtained by UPGMA clustering (Fig. 6b), the correct tree was recovered after evaluating 160 different topologies. It seems unlikely that a researcher applying the F-M method to the initial tree of Figure 6b could recover the correct tree by trial-and-error. Interestingly, the original distance Wagner procedure of Farris (1972) immediately finds the correct tree.

Conclusions

Since there are over 2.2×10^{20} possible unrooted trees for only 20 taxa (Felsenstein, 1978) a Fitch-Margoliash approach in which even 100 different trees are tested involves the evaluation of only a tiny fraction of the possible trees. Consequently, it seems obvious that some defined algorithm, independent of the assumption of homogeneous evolutionary rates, would be a more efficient and effective means of estimating phylogenies when more than just a few taxa are involved. The distance Wagner procedure provides such a method, and the results presented here indicate that it performs effectively.

This does not imply that the distance Wagner algorithm will always find the "best" tree. In fact, the only way to verify that a tree is optimal for a given criterion is to evaluate all possible trees. Unfortunately, as I have indicated, this is not generally feasible and is frequently virtually impossible. Thus, rearrangement algorithms (Kidd and Sgaramella-Zonta, 1971; Moore et al., 1973) may be useful in conjunction with the distance Wagner procedure. In this case, the utility of the distance Wagner method is in selecting an initial tree that, hopefully, is not too far removed from the optimal topology. But whether used alone or in conjunction with rearrangement algorithms, the distance Wagner procedure is likely to be much more effective than trial-and-error alterations of what is essentially a similarity-clustered phenogram.

Acknowledgements

I would like to express my sincere appreciation to R. B. Selander for thoughtful consideration and constructive criticism of the ideas I have developed in this paper. Mary F. Mickevich first gave me the idea of using different addition criteria simultaneously. I also thank S. H. Berlocher and G. S. Whitt for helpful advice and encouragement.

Treating Polymorphic Characters in Systematics: A Phylogenetic Treatment of Electrophoretic Data

M. F. Mickevich and C. Mitter

Table of Contents

Introduction

In most systematic studies, the recognition of character states is relatively straightforward. Characters are chosen so that character states are invariant for all individuals within a taxon, and represent one ontogenetic stage. For characters which may vary among individuals of one ontogenetic stage within an indivisible taxon, however, the choice of coding is less obvious. Although such data are widely collected and used in systematic studies, there has been almost no discussion of how to code these characters for analysis. In this paper, the problem of how to best define character states for characters which have more than one condition within a terminal taxon will be addressed.

A common type of such data are those obtained by protein electrophoresis. Allozyme data are good material for investigating the general question raised in this paper for several reasons. There are many sets of such data, making the possibility of discovering general conclusions about the properties of different coding methods more likely. Moreover, electrophoretic data are readily amenable to systematic analysis because they consist of easily homologized loci observed over many taxa. A large number of hypotheses have been advanced regarding the biochemical and genetic bases of electrophoretic variation and its evolutionary significance. These hypotheses have been used to justify many procedures for estimating phylogenies or cladograms. Cladists, on the other hand, recognise that it is desirable to base systematic procedures on general scientific principles rather than idiosyncratic assumptions concerning particular types of data, and further, that systematic methods and their resulting classifications must be scientific, in that these reflect or summarize nature in the best or truest fashion. In this paper, coding procedures developed from extrinsic considerations, such as statements concerning the evolutionary genetics of allozymes, will be compared with a procedure developed from a purely systematic viewpoint. All procedures will then be judged according to optimality criteria.

Optimality Criteria

The choice of one systematic method over another should be made on scientific grounds. In systematics, our observations consist of sets of characteristics of taxa. On the basis of scientific procedure, the method which makes the simplest summary statement consistent with the data at hand and which gives the most repeatable results would be the best sytematic method. Criteria chosen to assess the different methods must reflect the degree of simplicity and repeatability attained by each method.

One criterion, introduced by Mickevich (1978b) and discussed by Nelson (1979) is taxonomic congruence. Taxonomic congruence is the degree of agreement between classifications estimated from different sources of information on the same organisms. Although an artifical division of one allozyme data set would make a congruence test physically possible, the essential ingredient of a congruence study, logically different sources of information, would in this case be missing. The results obtained would therefore be meaningless. When data from different sources become available, congruence tests will be appropriate.

However, consideration of simplicity results in the development of criteria applicable to single data sets. Coding procedures can be judged on their ability to attain these criteria. The criteria recognized and used in this paper are consistency, "boldness," and "Occam's Probative."

Consistency is directly related to simplicity: it is defined to be the degree to which the character distributions implied by a classification correspond to the data. A measurement referred to as the consistency index (Kluge and Farris, 1969) assesses this for a cladogram and a data set. For character i

$$C.I.i = Si/Pi.$$

$C.I.i$ is the consistency index of character i, Si is the number of steps for character i required by the observations and Pi is the number of steps required by the classification. Each state of a character defines a possible group, comprised of the taxa that show it. If each state need evolve but once according to the tree, then each group suggested by a state is also contiguous in the classification, and the classification is entirely consistent with the character. In this case $Pi = Si$ and C.I.i is 1.0. If the states each are required by the tree to have evolved twice, then each group suggested by a state is broken into two parts in the classification, and the classification disagrees to that extent with the character. For example, if a set of taxa have observed states 0, 1, and 2, the number of groups defined by that character is 3. This means that the minimum number of transformations required to explain the data, and hence the most efficient or best-fitting explanation, has just two transformations. Pi, the number of steps, is established by the optimization procedure of Farris (1970). This method, henceforth referred to as Farris optimization, is the exact solution to the problem of determining the best fit or most parsimonious set of implied changes for characters on any classification, cladogram or phylogeny.

Complete consistency of a character, however, does not guarantee that the character offers a unique statement corroborating the relationships between taxa. For example, the cladogram in Figure 1a has a character with states x, y, and z. Character code z is plesiomorphic because it is found both in the outgroup and within the group. However, all transformations of codes x and y from z are possible on the cladogram (Fig. 1). These are drawn in Figure 1b. Note also that the consistency index for each of the possible transformations is 1.0. While the character is thus consistent with the tree, it is also consistent with any tree, and so is vague or uninformative as evidence on kinship. It is desirable for characters to offer unambiguous evidence on the relationships to be inferred, even when it turns out that the evidence is misleading. When characters do so we shall call them "bold." Coding procedures which result in transformation series with high boldness should be judged superior to those which yield codings admitting of many seemingly equally valid possibilities. A method offering multiple explanations for one observation can hardly be judged as offering any explanation at all.

A third criterion which is a measure of simplicity but is not assessed by either consistency or boldness will be referred to herein as "Occam's Probative." This is the ability of a procedure to avoid postulating an unobserved and therefore untestable condition, in order to explain the data at hand. For example, Occam's Probative is failed if a procedure requires the simultaneous presence of a set of mutually exclusive states, or the converse, the absence of any state for a character (Fig. 2). The latter assertion is not only untestable but highly illogical, for the recognition of a character depends on observing one of its states.

Coding Methods

Electrophoretic data are usually collected from local populations within previously-defined taxa. Loci are generally homologized on the basis of chemical criteria such as reaction specificity and general similarity of electrophoretic mobility. At times the pattern of intra-population variability is also considered. The data reported in most studies are frequencies of electrophoretic mobility classes (electromorphs or "alleles") under a single set of running conditions, at presumed homologous loci, within a number of population samples. Three coding procedures will be discussed which can be advanced on the grounds of a priori plausibility. Two of these, but not the third, are based on the initial assumption, general in the literature, that a particular electromorph represents the same allele (amino acid sequence) wherever it occurs in the data.

The first approach, referred to as the "independent allele" model, was introduced by Mickevich and Johnson (1976). Its rationale is as follows. There is no a priori reason to rule out the occurrence of any combination of the observed alleles. Since the number of alleles at a locus can and does vary widely among taxa, one may regard the appearance or disappearance of individual alleles as independent events. Thus, each allele becomes a separate character whose state

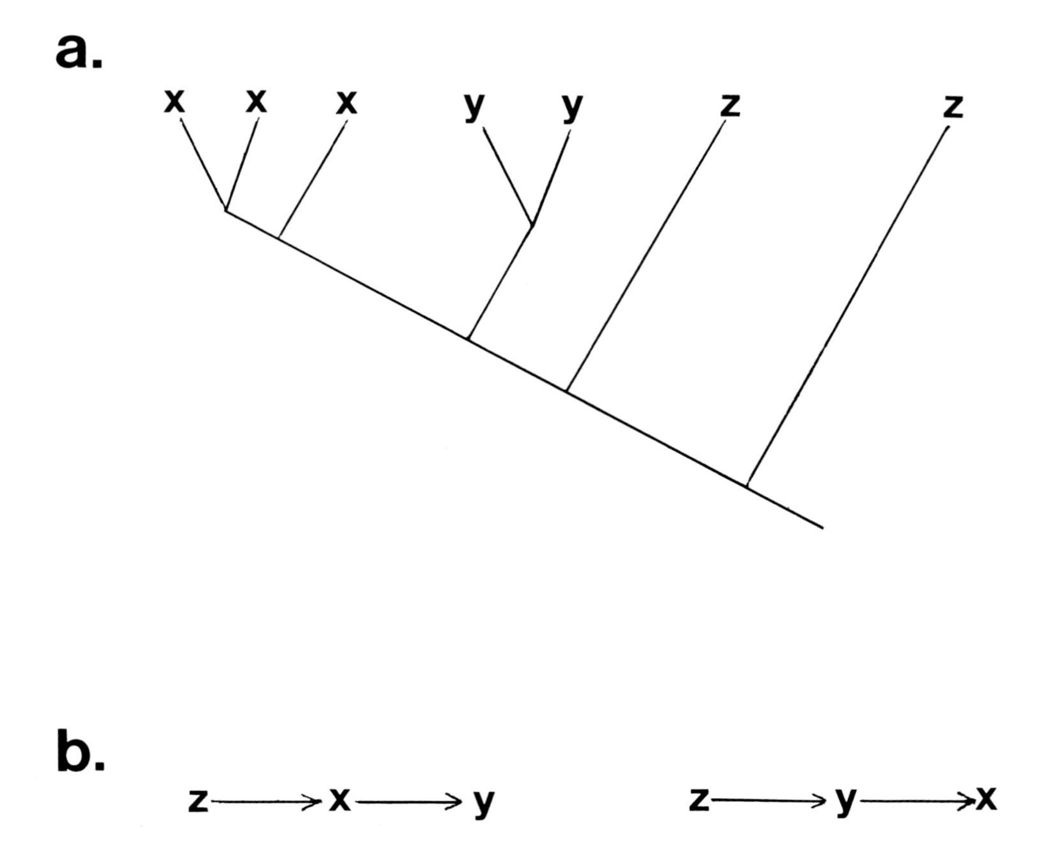

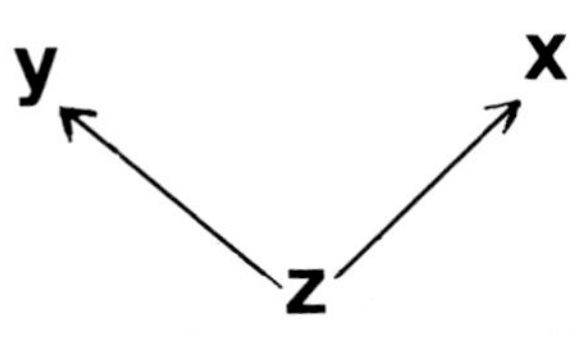

Fig. 1. Hypothetical tree with superimposed character states (1a) which supports equally well three different character state orderings (1b).

in each taxon is to be scored in some way. We have investigated two kinds of scoring. If frequencies are regarded as cladistically informative, they can be treated as quantitative characters (electromorphoclines?). If, on the other hand, the cladistic informativeness of frequency changes is suspect or demonstrably small (see Mickevich and Johnson, 1976; Farris, 1981), each allele can be treated as a binary character to be scored merely as present or absent.

The second coding method, newly described herein, is the shared allele model. This method derives from the recognition that the locus is the character undergoing change during evolution. Thus each locus becomes a single character whose

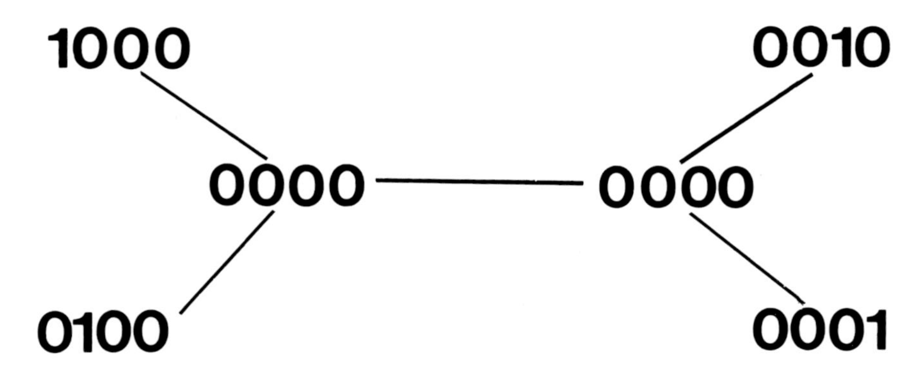

Fig. 2. Example of a character failing "Occam's Probative" under the "independent alleles" procedure.

states are combinations of alleles. When more than two such states occur in the data, the problem arises as to how their ordering is to be inferred. This can be done by invoking the assumption of genetic identity of electromorphs. Under this approach, the evidence for relationship between allele combinations (states) is taken to be the sharing of seemingly identical alleles. A character state order is inferred by applying a method analogous to the parsimony criterion. Thus, related states (those with shared alleles) are connected in such a way as to minimize the total amount of allelic evolution. This procedure joins each state to the state(s) with which it shares the most alleles. The transformation series so produced can then be coded for use in cladogram construction.

The next approach is purely systematic. The characters are recognizably different enzymes commonly assumed to be loci. The states are defined to be the observed allelic combinations. The ordering of these states is derived from a cladistic procedure defined below and called Transformation Series Analysis (Mickevich, 1981).

The above three coding procedures can result in different data sets for the same set of observations. Figure 3 illustrates these differences. Figure 3a depicts observations of populations X, Y, and Z for electromorphs a, b, c, and d, and their observed frequencies. Figure 3b is the data set generated by application of the independent allele method. Z and Y differ in four one-step characters, and therefore are said to have an observed difference of four. Under the shared allele method, taxon Y is intermediate between X and Z because it shares an allele with each of these populations. The most parsimonious transformation series is described by 1–2–3, where code 1 represents allele combination AB, code 2 combination BC, and code 3 combination CD. The purely systematic method would simply recognize the different allele combinations as different states. These are represented in Figure 3c, by the symbols 9, &, and $. Note that the differences between these symbols are impossible to estimate because the systematic method is capable of recognising different character conditions only.

Types of Characters

Boldness is recognized by the allocation of a locus character to one of the types of cladistic characters which can result from either establishing the transformation under certain a priori restraints such as in the shared allele method or by deriving

a.

alleles	taxa: X	Y	Z
a	.5	.0	0
b	.5	.5	0
c	0	.5	.5
d	0	0	.5

b.

X	Y	Z
1	0	0
1	1	0
0	1	1
0	0	1

c.

locus codes under:	X	Y	Z	
shared alleles	1	2	3	1—2—3
transformation series ("systematic" method)	9	&	$	

Fig. 3. 3a. Hypothetical allele frequencies for three taxa. 3b. Frequencies of 3a after "presence-absence" coding. 3c. Codings of allele combinations from 3b under the "systematic" and "shared alleles" procedures. Character state orderings under latter method are shown at right. Symbols for transformation series codes were chosen arbitrarily.

the transformations directly from the cladogram with Transformation Series Analysis. Several distinct types have been recognized by Mickevich (1981).

The first recognizable type is the additive character. Figure 4a exemplifies an additive character for four states such as can be derived under Transformation Series Analysis. Figure 4b is an example of an additive character derived by the shared allele method. Additive characters offer a unique hierarchical arrangement

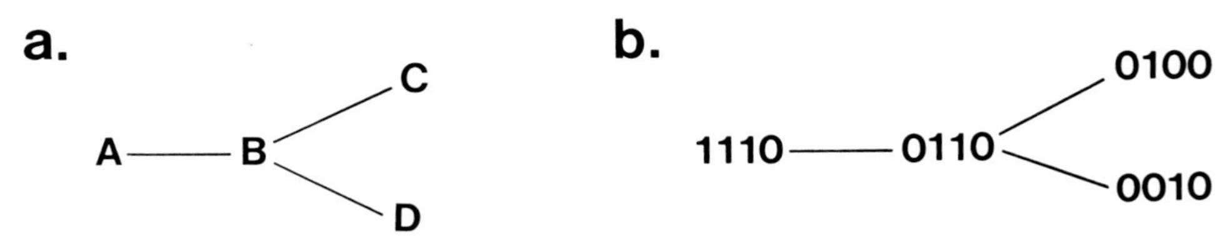

Fig. 4. Examples of additive characters under transformation series (4a) and shared alleles (4b) procedures.

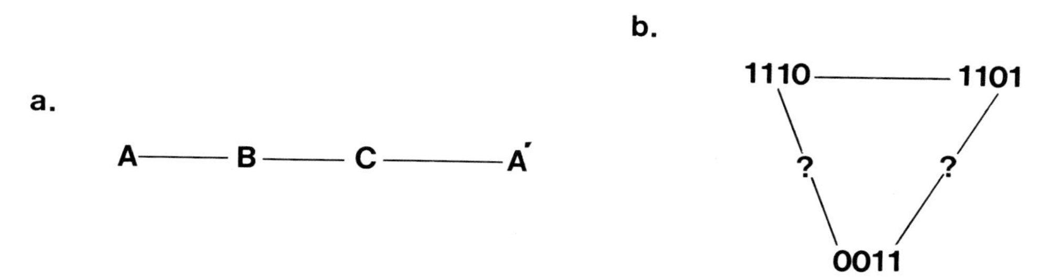

Fig. 5. Examples showing treatment of convergent characters under transformation series (5a) and shared alleles (5b) procedures.

of taxa. Therefore, additive characters are optimally bold. The observed frequency of additive loci under the shared allele method and Transformation Series Analysis will measure the degree to which these methods meet this optimality criterion.

There is another type of character, a convergent character (Fig. 5), which may indicate boldness. Transformation Series Analysis derives character state trees from a cladogram, and character states which were thought to be homologous will be divided into non-homologous components. Transformation Series Analysis therefore recognises a convergent character as an additive character which has the quality of boldness. The shared allele method, on the other hand, derives transformations in accordance with its internal criterion. This may result in conditions drawn in Figure 5b. In this example, it is impossible to resolve the relationships implied by this character, thus a convergent character under the shared allele model cannot be judged as a bold character.

Two other recognizably different types of character, non-additive (Fig. 6) and disjoint (Fig. 7) characters, do not uniquely define relationships among taxa. Non-additive characters are those in which the evidence maintains that any state can transform into any other state. Such characters are certainly not bold. Figure 6b, a non-additive shared allele character, illustrates how this can happen. Any two allele combinations are separated by just two changes. Therefore all transformations are equally likely. Disjoint characters (Fig. 7) are characters with subsets of states where the relationships of states within the subsets are uniquely defined but the relationships between the subsets are unresolvable. This can be better explained by observing the condition of a shared allele character illustrated in Figure 7c. Note that no pair of allelic combinations shares alleles with another. Therefore, there is no evidence for a unique statement of relationships, and this type of character cannot be considered bold.

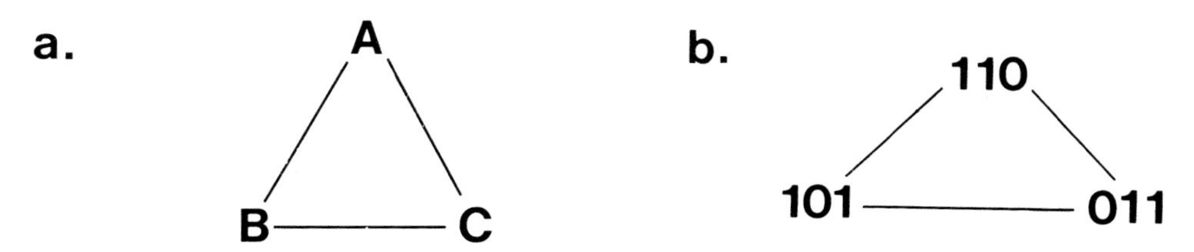

Fig. 6. Depiction of non-additive characters under transformation series (6a) and shared alleles (6b) procedures.

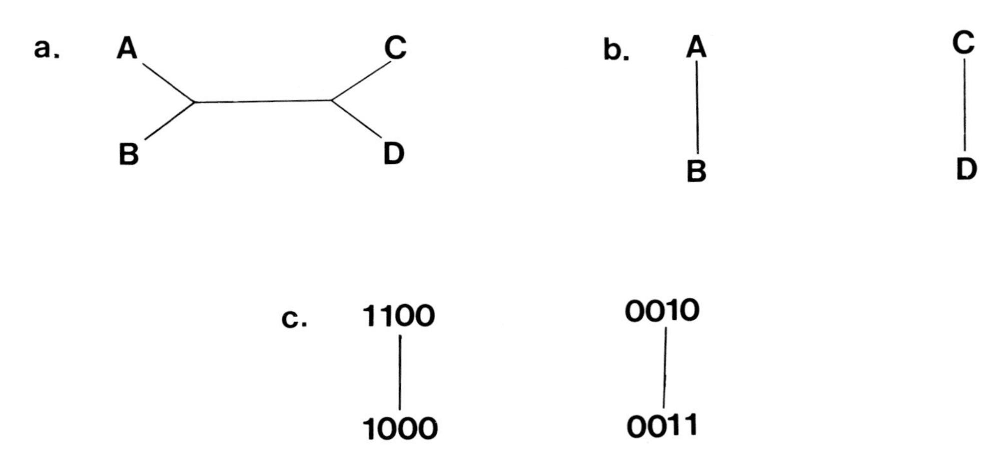

Fig. 7. Hypothetical tree and character states (7a) leading to a disjoint transformation series (7b). 7c. A disjoint character under the shared alleles procedure.

Cladogram Construction

The independent and the shared allele methods implicitly define character transformations in the process of recognising character codes. These characters will define a cladogram, at times called a Wagner tree (Farris, 1970). The programs used in this study were briefly discussed elsewhere (Mickevich, 1978b) and were written by Mickevich. Basically, one first Farris-optimizes the possible cladograms for a data set and then chooses the most parsimonious one(s).

The systematic method, however, does not delineate cladistic characters (transformations) when it recognises different character conditions. These are instead obtained with a new cladistic method referred to as Transformation Series Analysis. An outline of Transformation Series Analysis follows.

I. From a beginning set of character hypotheses, obtain a cladogram. Cladograms are minimum-length Wagner trees (Farris, 1970).

II. From this cladogram calculate the transformation series for each character (Mickevich, In prep.).

III. Calculate a cladogram from the new set of character transformations. If the new cladogram is different from the previous one, go to step II. Stop if the cladograms are identical.

Mickevich (1981) demonstrated that the solution cladogram resulting from Transformation Series Analysis does not depend on the initial starting set of character hypotheses. Identical or near-identical solutions always result despite opposite or near-opposite starting hypotheses. Therefore, the first set of character hypotheses need not be a concern.

The results of transformation series analysis consist of a set of character state transformations and a cladogram. Sometimes a series of cladograms and their corresponding sets of transformation series form a stability cycle. At other times, there may be more than one solution cladogram. However, for the data sets used in this study, there was only one solution tree.

The Data

The data listed in Table I, were obtained from a search of the literature through 1976. An effort was made to assemble reports containing allelic frequencies for

Table I

Data sets used

No.	
1	*Anolis*, Bimini (Webster et al., 1972). Loci deleted: EST-1, MDH-1, LDH-1, LDH-2, DH-X, PT-D, EPT-1.
2	*Drosophila, mulleri* group (Zouros, 1974). Taxa deleted: *mojavensis* BII, San Lorenzo, Tortuga and San Ildefonso islands; locus tested: EST-5.
3	*Drosophila*, Maui (Ayala, 1976).
4	*Anolis, roquet* group (Yang et al., 1974).
5	*Anolis, bimaculatus* group (Gorman and Kim, 1976).
6	*Peromyscus polionotus* (Selander et al., 1971). Taxa deleted: 2–4, 6–13, 15, 17–24, 26, 27.
7	*Astyanax* (Avise and Selander, 1972). Taxon deleted: 5.
8	*Peromyscus, boylii* group (Avise et al., 1974). Taxa deleted: 1, 5–11.
9	*Plethodon cinereus* (Highton and Webster, 1976). Taxa deleted: 1, 2, 4, 5, 7, 18–22, 24.
10	*Ptomophagus hirtus* (Laing et al., 1976).
11	*Lepomis* spp. (Avise et al., 1977) (see text).
12	Centrarchid genera (Avise et al., 1977) (see text). Taxa deleted: *Centrarchus, Acantharcus*.
13	*Stephanomeria exigua* (Gottlieb, 1975). Taxa deleted: 7129, 7136, 7066, 7125, 7132.
14	*Menidia* (Johnson, 1975).
15	*Thomomys talpoides* (Nevo et al., 1974). Taxa deleted: WC, DH, MT.

at least six loci, within at least six local populations, in as wide a variety of organisms as possible. Initially twenty-four such data sets were considered. Some of the locus by population data matrices were incomplete. In such cases, loci and taxa were excluded in order to obtain the largest complete subset of the original data. If it was apparent on inspection that too few loci were used to classify too many taxa, a subset of taxa which included those populations with the largest sample sizes was chosen so as to represent the entire geographical range of the group.

Cladograms (Wagner trees) were constructed for both the frequency and presence/absence data sets obtained under the independent allele model. In most cases the same cladogram was found under both codings. Sometimes, however, these disagreed. In such cases the branch lengths less than 1.0 on the frequency based trees were set to zero. These branches were then collapsed, making the trees less resolved. In most cases, this procedure removed all the differences in branching sequence between the frequency-based and the presence/absence-based trees. Thus, the differences between the frequency data sets and the presence/absence data sets were the result of a lack of cladistic information rather than the presence of incongruent information. For those sets of taxa which still had two different trees, inspection revealed highly disparate sample sizes. In

Table II

Data sets dropped from consideration due to the existence of large numbers of equally good trees

a	*Hyla* spp. (Case et al., 1975).
b	*Lacerta* spp. (Gorman et al., 1975).
c	*Phlox* spp. (Levin, 1975).
d	*Bufo arenarum* (Matthews, 1975).
e	*Drosophila* spp. (O'Brien and MacIntyre, 1969).
f	*Bufo* spp. (Rogers, 1973).
g	*Drosophila bipunctata* complex (Yang et al., 1972).
h	*Drosophila ananassae* (Johnson, 1971).

Table III

Percentage of loci meeting optimality criteria ("boldness" or "Occam's Probative"; see text)

Data set	Transformation series analysis	Independent alleles	Shared alleles
Anolis, Bimini	57	36	10
Drosophila mulleri group	77	77	30
Drosophila, Maui	45	29	13
Anolis roquet group	75	84	33
Anolis bimaculatus	74	95	54
Peromyscus polionotus	100	100	100
Astyanax	73	100	40
Peromyscus boylii group	92	69	14
Plethodon cinereus	69	87	61
Ptomophagus hirtus	100	100	100
Lepomis	83	46	10
Centrarchids	70	20	0
Stephanomeria	71	100	86
Menidia	71	86	40
Thomomys	100	100	54

these cases the disagreement between the trees was thought to be caused by sampling error. It seems that, as suggested by Farris (1981), the frequency data provide no comparative information beyond the presences of alleles. Accordingly, in what follows only the presence/absence data sets will be considered.

In addition to the data sets listed in Table I, other sets of taxa were considered. However, for these (listed in Table II), a large number of shortest trees were found. Generally, these sets of taxa were distinguished on the basis of few allelic differences or were composed of populations with very small sample sizes. These sets of taxa were eliminated from the study.

Results

The cladograms and their character distributions (obtained by Farris optimization) for each data set generated by the three different coding methods form the evidence on the merits of methods. However, the evaluation must also take into account the properties of each method. For example, the independent allele method could never result in a non-bold character interpretation, producing as it does a set of characters with only two states. Under such restraints only one transformation is possible: 0–1. Therefore the boldness criterion could never fail. On the other hand, the remaining two coding methods by their very nature preclude the possibility of failing Occam's Probative.

Although the independent allele method produces a set of two-state characters, the locus is the basic homologous unit. It is, after all, by a common reaction that the electromorphs for an allozyme locus are recognised as such. Therefore, it is possible that in some situations the independent allele method will be unrealistic, because it does not prohibit a hypothetical intermediate (electromorphotype?) without alleles at a locus, should the optimization procedure require it. An example of this kind is presented in Figure 2. Furthermore, if we take a more philosophical viewpoint it can be seen that in many cases the independent allele model forces an interpretation which is not supported by the observations. For example, if for a locus, the data contain populations which have either alleles A and B, or C and D, but no other combinations of these four alleles, there is no

Table IV
Consistency indices

Data set number	Transformation series	Independent alleles
1	96	90
2	95	90
3	95	74
4	89	68
5	73	46
6	88	79
7	87	76
8	90	76
9	86	71
10	90	88
11	93	71
12	93	83
13	89	73
14	61	50
15	85	70

observation which supports the possibility that individual alleles can arise or disappear independently. Therefore this method will be judged by counting the number of loci within each data set that satisfies Occam's Probative.

Table III presents the percentage of loci which are bold or fulfill Occam's Probative for each method. Clearly the shared allele method is markedly inferior to the other two methods. However, Transformation Series Analysis and the independent allele model are equally valid under these criteria. Table IV contains consistency indices for the data sets and cladograms for these latter coding methods. Transformation Series Analysis resulted in consistently higher indices and therefore had a better fit to the data, or a greater degree of simplicity. It is important to point out that all the characters within a data set were used to estimate these consistency indices. Under the independent allele model there are some two-state characters in which one of the states is an autapomorphy. These singleton characters have been included in the indices for this method even though the consistency of such characters is necessarily 1.0. The Transformation Series Analysis data sets do not contain any singleton characters. Thus the indices in Table IV contain a bias in a direction opposite to the observed trend.

Discussion and Conclusions

The systematic method, which recognises any distinct combination of alleles as a state and which defines the order of these states with a cladistic procedure, satisfies our optimality criteria better than do the alternative coding methods. These conclusions have implications for three aspects of systematic research. First, this study puts forth a procedure for coding allozyme data for use in phylogenetic systematics. Second, this method can be generalized for use where indivisible taxa have multiple conditions for the same character. Third, the paper is itself an illustration of an experimental approach to the problems of classification and systematics.

Allozyme Data

It is appropriate to consider some of the properties of these methods which might cause them to give different results according to our criteria. A fundamental

a.

allele combination	code
1 1 0	A
0 1 0	B
0 1 1	C

b. A A B B C C

c.

shared allele model: A—B—C

transformation series analysis:

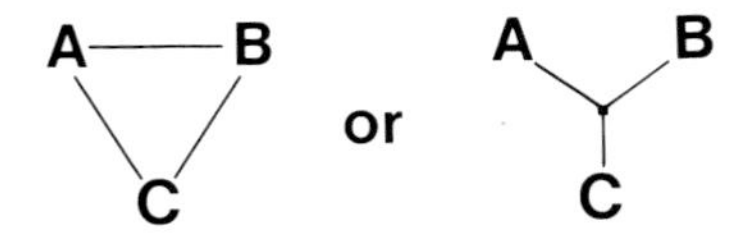

Fig. 8. Example in which shared alleles procedure makes a "bolder" statement about character state orderings than does transformation series analysis. Allele combinations (8a) are superimposed on tree (8b), yielding orderings shown in 8c. The character is non-additive under Transformation Series Analysis.

difference between the shared allele method and the Transformation Series Analysis is that the latter infers transformations from the distributions of states on a cladogram. By contrast, the shared allele method infers transformations from the allelic combinations. Thus, it might appear that the shared allele method is at a disadvantage with respect to boldness. However, cases can be found in which the allele combinations make a unique statement about taxon relationships whereas Transformation Series Analysis does not. The cladogram structure in Figure 8 does not resolve the relationships between any of the sets of taxa with codes A, B, and C (for an explanation see Mickevich, 1981). The allele composition of these character states defines a nested relationship illustrated in Figure 8. Thus, in theory, it could have turned out that abandoning the assumption of allelic homology, as does the systematic method, would lead to a loss of cladistic information and hence a less powerful method. The conclusion reached here is then empirical rather than tautological.

It might also be suggested that the transformation series by its very nature would yield higher consistency indices than the independent alleles method. Here again it is possible to devise cases in which the opposite is true. For example, in Figure 9, the transformation series gives a lower consistency index, in essence because it treats all transformations with equal weight despite the fact that some involve more alleles than others. In the example the numerous allele changes between states A and D are equivalent to the single change between A and C. It is therefore theoretically possible for the systematic method of recognising character states followed by Transformation Series Analysis to result in lower consistencies for characters (loci) than the independent alleles method. Our results show that this possibility does not prevail in nature.

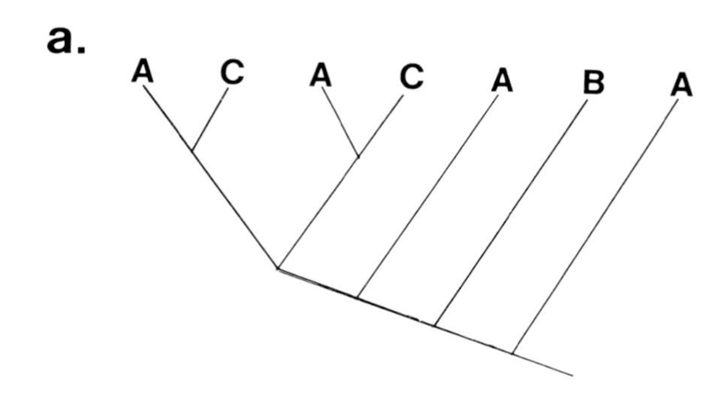

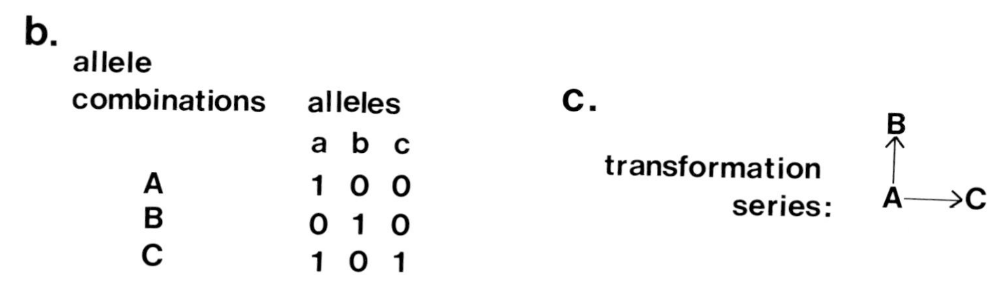

Fig. 9. Example in which consistency index is higher under "independent alleles" procedure than under Transformation Series Analysis. 9a. Hypothetical tree with superimposed allele combination codes. 9b. Allele composition of codes in 9a. 9c. Character interpretation under transformation series, for which consistency index is ⅔. Consistency index is ¾ under independent allele method.

The fact that Transformation Series Analysis on sets of alleles results in higher consistencies than a method based on the assumption of homology of individual electromorphs means that sets of alleles are better indicators of relationships than are individual alleles. This in turn implies that electromorphs are not genetic homologs. This finding is consonant with recent experimental results which demonstrate that electromorphs defined under a single set of conditions often prove to be heterogeneous collections of molecules when analyzed in other ways (Coyne et al., 1979). If all of the genetic variants within the same mobility class were phylogenetically homologous (sensu Hennig, 1966) consistencies would not be affected. However, the results of Coyne et al. (1979) suggest that electrophoretic mobility as measured under any set of conditions may be multiply convergent. Further independent corroboration of the desirability of using entire sets of alleles as codes could come from biochemical studies showing loci more efficiently accounted for under Transformation Series Analysis than under the independent allele model to be those with the largest degree of electrophoretic convergence.

Coding Polymorphic Features

The "systematic" method, which we prefer on the basis of our optimality criteria, recognizes character states solely as combinations of attributes—"morphs"—that characterize taxa. When such combinations are found within indivisible taxa, the character is a classical polymorphic feature. Characters called "variable" but which in fact are monotypic within subgroups of the taxa being classified are specifically excluded from our definition of a polymorphic character.

Because of the very general nature of the systematic coding method it can be used to analyze many kinds of polymorphic features. For example, a set of inversions all affecting the same region of a homologous chromosome can be coded in the same way as a set of alleles at a locus. Color polymorphisms, phytochemical data, or developmental sequence variation could all readily be converted to a set of data for phylogenetic reconstruction. Of course, it would remain interesting to develop coding procedures for such characters based on plausible rules which are thought to govern their evolution. These could then be compared with the systematic method. It is always possible that these special coding methods will give better results than the systematic method. This would imply that the underlying hypotheses accepted for the data are reasonable and provide useful cladistic information. Comparison of results of coding methods thus provides tests of these hypotheses.

Scientific Procedure in Systematics

Much has been made in this paper of scientific procedure. It is therefore appropriate to contrast our means of developing a cladistic method with the approach generally favored by workers concerned specifically with "population genetic" data. A commonly-used measure for estimating relationships based on gene frequency data is the genetic distance of Nei (1972). Nei's major assumption in deriving this measure is that population differentiation is a homogeneous genetic drift process. Having discovered that populations undergoing such a process would diverge on the average at a constant rate, Nei proposed to reconstruct genealogy by phenetic clustering on his genetic distance measure. The effect of Nei's proposal is to produce an arrangement in conformity with his assumptions, rather than in conformity with the data. Whether the resulting tree effectively or efficiently explains the distribution of particular alleles over populations is quite irrelevant to Nei's procedure.

A scientific procedure seeks to provide classifications (phylogenies) which give the best fit to the data rather than conforming to particular theories of evolution. We have attempted to develop optimality criteria which estimate the fit of a set of characters to a phylogenetic hypothesis. These by no means will turn out to be the total set of such criteria, but we hope that this paper will serve as a step toward the further development of a science of classification.

Acknowledgements

We wish to thank J. S. Farris and K. Timmers for their help with the manuscript and Il Fondatore for his aid in typing. This research was supported by grant number DEB-78–24647 from the National Science Foundation.

Theoretical Cladistics

Classifications as Languages of Empirical Comparative Biology

Daniel R. Brooks

Classification systems have been devised according to a wide variety of criteria in order to represent various theoretical or operational constructs. As Hennig (1966) noted, every special classification system is valid for its particular purpose if it satisfies the motivation(s) of its originator(s). That does not mean that all classification systems comprise valid scientific or empirical tools or that any of them can function properly as a general reference system of comparative biology. The current revolution in systematics concerns formulation of an empirical approach or methodology for establishing a general reference system. In order to discuss such a topic logically, it is imperative that we find criteria not affected by differences in personal motivations or in theoretical states of the art. Hull (1979) pointed out,

> Scientists have limited patience when it comes to discussing arguments . . . . Invariably the advocates of a particular methodology can know what they need to know while their opponents can never hope to know what they need to know.

Clearly, such arguments are self-defeating and pointless. I am, therefore, going to single out one trait characteristic of all classifications regardless of personal bias.

A classification, regardless of the manner in which it was contrived or of anything else it may be perceived to represent, is a linear representation of information about characteristics of organisms arranged in a particular hierarchical sequence or configuration. Now, when a sequence of symbols is ordered according to a set of constraints, the set of constraints comprises a language and the sequence of symbols is a message in that language. Thus, because classifications represent information in the form of a linear sequence arranged according to a set of rules (no matter what the rules may be), they can be viewed as symbolic languages. But why should classifications be viewed as languages rather than anything else? First of all, the linguistic criterion is one which may be applied equally to all classifications and thus provides the *possibility* for an objective resolution of the problem of selecting from among many one technique for the general reference system. Given that we are all human beings, I will not say that this provides the *certainty* of an objective resolution. Secondly, much debate over differing approaches in systematics has centered around the idea of maximum information-content classifications (Duncan and Estabrook, 1976; Estabrook, 1971; Farris, 1979b; Hennig, 1966; Mayr, 1969; Sneath and Sokal, 1973; Sokal and Sneath, 1963). In my opinion, there is no context or rationale for discussing information-content outside of a linguistic, information-theoretic construct. However, I am not going to advocate an approach which arranges some data at hand and then minimizes residual variance (Duncan and Estabrook, 1976; Gower, 1974). And lastly, as Gatlin (1972) pointed out,

> A linear sequence of symbols is the indispensible vehicle of communication between higher organisms. Of course, two- or three-dimensional objects can convey information and in some instances are superior to language; but they are limited in a way that the one-dimensional vehicle is not. Any two- or three-dimensional object and, in fact, any general situation can be described by language and thus reduced to the one-dimensional case.

Now the question is, what is the best kind of classification from the standpoint of language? This is a classic question in information theory. I will suggest some criteria for building such a language. The optimum classification would exhibit the following traits: (*1*) stored information would be represented in the most informative manner, meaning the state of lowest entropy for the data set; (*2*) ambiguous information, that which increases the entropy from an ideal lowest level to that minimum possible when there are ambiguities, must be included in the classification; (*3*) there must be assurances that the message will be transmitted accurately even in the presence of ambiguous information; and (*4*) there must be a mechanism for discerning when potential information accrued corroborates the original message and when it contradicts the original message to such a degree that a new minimal entropy message exists. The four criteria are primarily technical, but have important philosophical ramifications. What will we have when we formulate such a system?

If a language is to be useful as a tool in empirical studies it must function as a cybernetic language. Cybernetic languages differ from digital languages, such as speech, by transmitting messages derived by precluding all alternative messages rather than by describing the occurrence of an event or phenomenon without regard for alternatives. This attribute of our optimal language is required because a cybernetic explanation is the least probable one (sensu Popper, 1968) and therefore the most vulnerable to testing and refutation (hypothetico-deductive approaches). The first reference to hypothetico-deductive approaches in systematics which I encountered was in a point of view by Farris (1967b), although Hennig (1966) seemed to have been aware of Popperian approaches. No doubt others will claim prior recognition, either for themselves or for a favorite historical figure. Nonetheless, it was Wiley's (1975) discourse on Popper and systematics which aroused general interest in treating classifications as testable hypotheses in an empirical sense. There is clearly a growing demand among systematists that systematics be studied as a science and not as an intuitive art form (Mayr, 1969) or as sets of classes of observations (Sneath and Sokal, 1973; Sokal and Sneath, 1963). Indeed, classification techniques are now being judged according to whether or not they conform to the criteria of hypothetico-deductive approaches, namely, that they stay the same when corroborating data are added and that they change according to some objective pre-determined criteria when accrued data refutes them. Language can inform or deceive. We need falsification as a means of telling us if we have formulated a deceiving message.

ELEMENTS OF THE ENTROPY CONCEPT

HIGHER ENTROPY	*LOWER ENTROPY*
Random	Nonrandom
Disorganized	Organized
Disordered	Ordered
Mixed	Separated
Equiprobable events	D_1 (divergence from equiprobability)
Independent events	D_2 (divergence from independence)
Configurational variety	Restricted arrangements
Freedom of choice	Constraint
Uncertainty	Reliability
Higher error probability	Fidelity
Potential information	Stored information

Fig. 1. Elements of the entropy concept in information theory (from Gatlin, 1972).

So, how do we go about formulating an optimal language of classification and comparative biology? We begin with basic information capacities—*stored information* and *potential information*. As is shown in Figure 1, potential information is associated with a state of higher entropy whereas stored information is associated with a state of lowered entropy. In any information storage or retrieval process, such as language, potential and stored information are generally viewed as antagonistic forms combined in some sort of optimal arrangement. The truth of this statement is immediately seen when one considers language in general. It is well known that in any digital language, such as speech, single letter frequencies diverge markedly from equiprobability. A certain amount of ordering is necessary for the formulation of words and sentences. If there were no constraints and every possible letter combination occurred with random frequency, potential information would be maximal but there would be no way to detect error because error detection and correction are based on recognition of restricted and forbidden combinations. On the other hand, we can reduce entropy to the point that stored information becomes maximal and transmission is highly reliable, but the message variety is so low that we cannot say anything. In the extreme case we would have a monotone, a sequence of only one letter. There is an apparently insurmountable problem at this point because we want maximal information content as well as maximal message variety, not an optimal blend of some of each. It turns out that it is possible to increase the reliability of a message or reduce its entropy without the loss of message variety if the entropy is reduced in a particular manner, namely, by increasing divergence from independence (D_2 in Fig. 1) of the data while keeping a priori divergence from equiprobability (D_1 in Fig. 1) relatively constant.

How does the above translate into formulating a classification? First, the greatest variety a classification message may exhibit is depiction of terminal taxa in

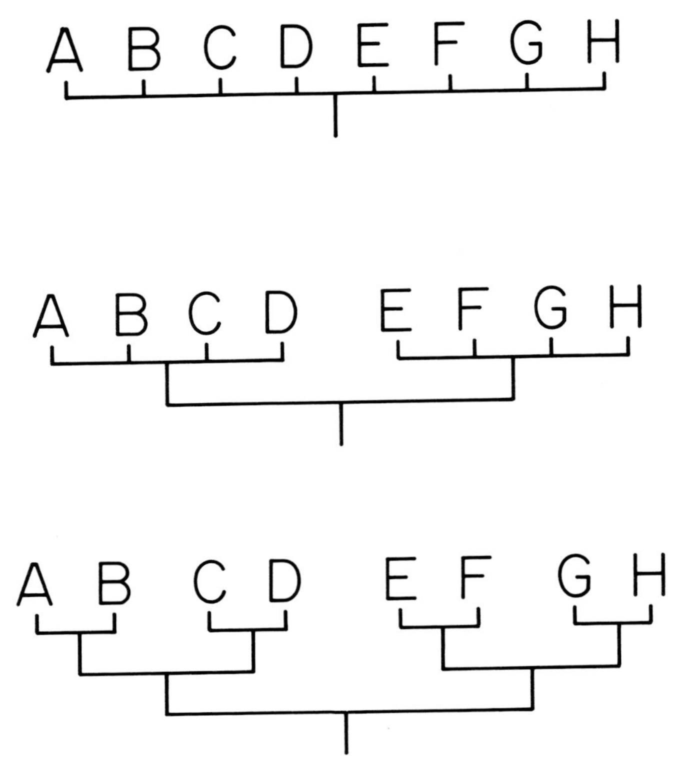

Fig. 2. Diagrammatic representation showing proposition that differing branching diagram topologies can exhibit different degrees of informativeness regardless of the data upon which the diagrams are based. Top depicts information for group *ABCDEFGH* and descriptors for each terminal taxon; middle depicts information for groups *ABCDEFGH*, *ABCD* and *EFGH* as well as descriptors for each terminal taxon; bottom depicts information for groups *ABCDEFGH*, *ABCD*, *EFGH*, *AB*, *CD*, *EF*, and *GH* as well as descriptors for each terminal taxon.

a fully dichotomous branching sequence (Fig. 2). Secondly, the lowest entropy message possible is one in which the entire data set is described using the smallest number of symbols possible. That is called increasing the information density of each symbol; Gatlin (1972) termed increased information density S-redundancy or Shannon's redundancy. Gatlin stated,

> . . . information density is a capacity to combat error. It measures how much the entropy has been lowered from its maximum value and is thus a measure of the ordering [of] . . . the system. This kind of redundancy we will call S-redundancy. S-redundancy . . . is a direct measure of all the rules which define error in a language.

Figure 3 depicts the manner in which a data set can be described in the most informative/lowest entropy configuration. In this case a data set containing 120 information symbols can be described fully by 15 symbols. However, there is only one classification in which that 15-symbol message can be displayed and that happens to be a fully dichotomous branching sequence. Thus, a classification message which comprises the lowest entropy message displayed in a fully di-

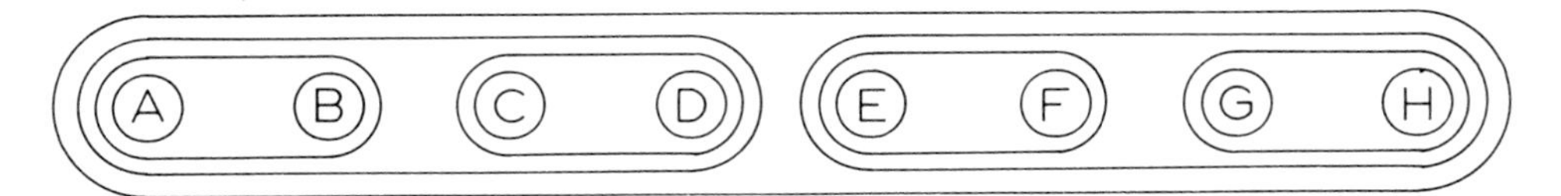

Fig. 3. Diagrammatic representation showing minimum-entropy classification for data set. The smallest number of symbols necessary to represent data set is 15 (one for each subset shown); resulting classification (Fig. 4) is unambiguous and fully dichotomous, and maximally information-dense.

chotomous branching sequence represents maximal stored information without sacrificing any potential information.

For such a language to be useful in empirical studies, there must be some way of detecting error and computing a new message with reduced error content. Thus, we must ask another question, "What is the fidelity of the message?" We have already seen that redundancy is a measure of the constraints on a language which make error detection feasible. There is another aspect of the redundancy concept which aids in error correction. That is Gatlin's R-redundancy or repetitive redundancy. The simplest way to get a message across in the presence of interfering noise is to repeat it many times. Thus, if we have a data set with some ambiguities, we want to make sure that the ambiguities do not cloud the predominant, unambiguous message. If we cloud the message and it is correct, we have lost understanding; if we cloud the message and it is incorrect, we make it more difficult to correct the errors.

There are two ways to accomplish an increase in R-redundancy in systematics. The first is weighting certain characters, in effect making them redundant to the magnitude of the weight. This is a convention for preserving the integrity of one's own personal favorite classification even when refutations pile up. It cannot be used in empirical studies because it will not allow detection of error; it explains error away. Additionally, weighting characters will change a classification only if there are ambiguities present, and one therefore risks clouding the message further rather than making things clearer. The second manner by which one can increase R-redundancy in a classification is to sample increasingly larger data sets and to constantly compute the lowest entropy message. In this manner error can be detected and corrected according to some objective criterion. Error can be reduced, conflicts between two points of view may be resolved at any given time according to the criterion of which one best represents known information, and a reproducible classification can be maintained in the presence of corroborating data. Error is thus defined not as some degree of departure from a theoretical construct, but as a departure from most informative representation of observations.

Two works have examined the information content of cladograms from viewpoints other than the one presented in this study. Nelson (1979) discussed cladistic analysis from the standpoint of comparing and choosing among competing classifications. He viewed information content expressed in two forms: *component* (*C*) and *term* (*T*) information. According to Nelson, *components* of a cladogram are branch points (nodes) defined by the terminal taxa to which they lead. Terminal taxa from each node comprise Nelson's *terms,* whose information is the individual diagnoses or set of characters for each taxon. The information content of all components is the number of components for a given cladogram expressed as a fraction of the total number possible, the total being two less than the number of terminal taxa. The information content of all terms is the sum of all information for all components. While this brief outline scarcely does justice to Nelson's analysis it is sufficient to demonstrate an empirical connection with the ideas presented herein. Nelson's *C* refers to the degree of departure from a fully dichotomous branching sequence and is thus the same quantity as S-redundancy as applied to classification languages. Nelson's term information (*T*) can be viewed as a measure of ambiguity in diagnoses for taxa depicted by a classification. Maximal *T* is obtained under Nelson's model when the maximal number of character distributions possible for all terminal taxa reflect back to the same sequence of components. Thus, Nelson's *T* is the same quantity as R-redundancy. Optimization of term information is accomplished by increasing repetitive nature of character-state distributions.

Farris (1979b) discussed the information content of phylogenetic classifications and concentrated primarily on the information density aspect of representing a data set. However, he earlier (Farris, 1969) presented a study outlining a successive approximations approach to character weighting which bears directly on this study. Farris (1969) made a number of observations of import. First, differential weighting of characters can affect the outcome of a systematic analysis only if there are ambiguous character-state distributions in the data set; second, the effect of any characters on the shaping of a tree may be calculated as can the effects of weighting any of those characters; and third, the number of changes in a tree per character times the weight of the character summed over all characters will equal unity if there are no ambiguities in the data set. Therefore, the optimal tree will be one which minimizes the summation of character effects and weightings because it will be the one most closely approximating an unambiguous message. In symbolic representation,

let t = tree
c = character
n_{tc} = number of changes in a tree for a character
w_c = weight of the character
W = ambiguity of classification message

thus,

$$W = \Sigma\, n_{tc} w_c$$

Because n_{tc} refers to changes in tree shape or message variety, it is equivalent to S-redundancy—it is a measure of the ordering of the data set; w_c is clearly equivalent to R-redundancy as discussed previously—it is a measure of information repetition. Therefore, it is apparent that three different approaches to examining the information content of classifications have produced three independent proofs of the same equation such that we now observe that

$$W = \Sigma\, n_{tc} w_c = \Sigma\, CT = \Sigma\, SR$$

Hennigian systematics achieves minimum entropy for a given data set by maximizing both S- and R-redundancy. In simpler terms, cladistic analyses portray data sets in terms of dichotomous branching sequences described in the fewest terms possible. It is impossible for any classification which differs from the most parsimonious representation of the data set (or the set of equally-parsimonious alternatives for highly ambiguous data) to be as informative as a cladogram (see

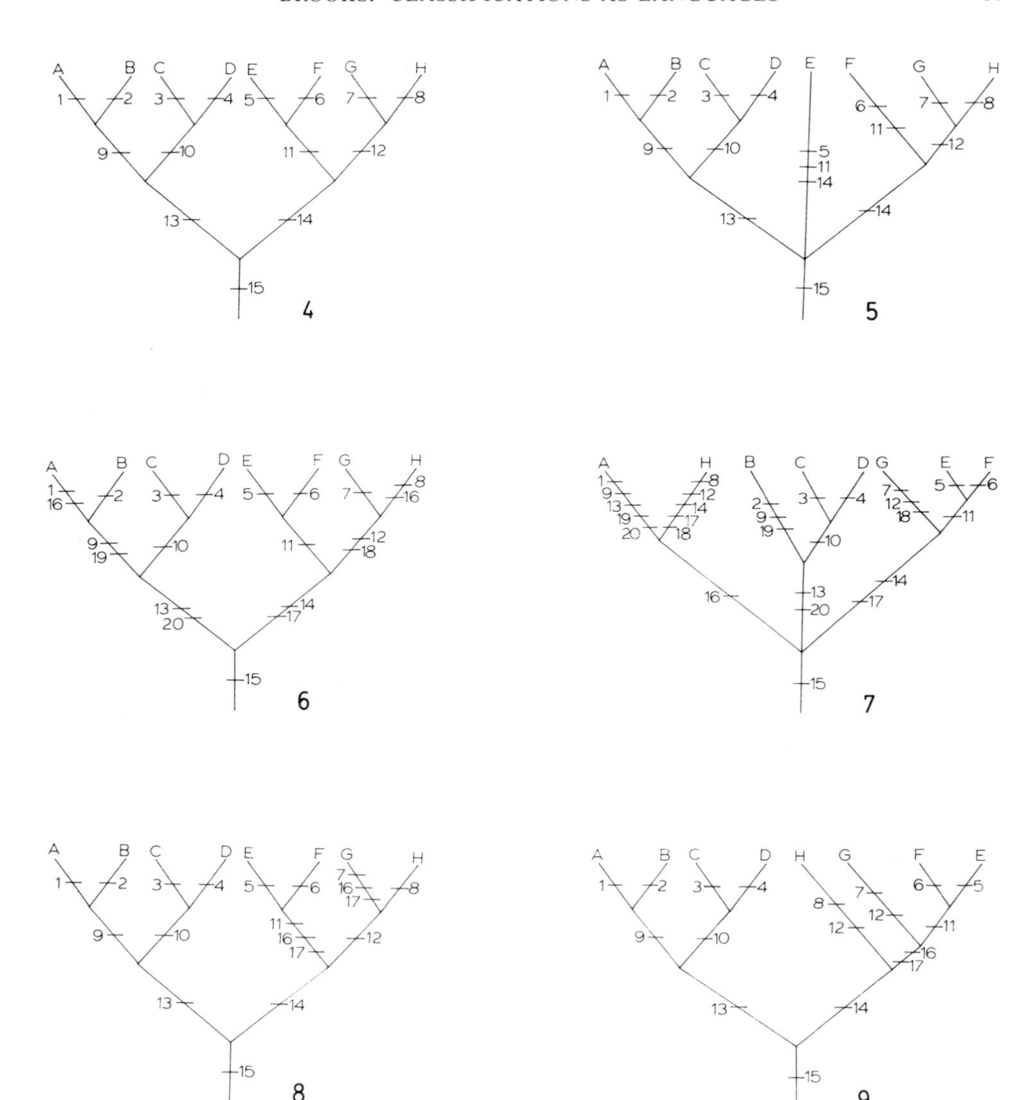

Figs. 4–9. Comparison of cladograms depicting changes in S- and R-redundancy arising from differences in choice of branching sequence. In Figures 4 and 6, S- and R-redundancy are maximized for the data set; Figure 4 has no ambiguities whereas Figure 6 has one (character 16). Figures 5 and 7 depict changes in redundancy resulting from changes in branching patterns. In Figure 5, removal of Taxon E from the group *EF* results in lowered S-redundancy (an unresolved trichotomy appears) and in lowered R-redundancy (the data set requires 17 symbols, including the ambiguous 11 and 14, for complete description rather than 15 with no ambiguities). In Figure 7, an attempt to resolve the ambiguity of character 16, by forming the group *AH,* also results in lowered S- and R-redundancy. Figure 8 depicts a cladogram with maximal S-redundancy and a data-set description comprising two ambiguities. Figure 9 depicts an alternative to Figure 8, which represents a data-set description containing only a single ambiguity. Both Figure 8 and Figure 9 exhibit maximal S-redundancy, but differ in R-redundancy. These illustrations demonstrate that any departure from the lowest entropy, maximal variety message for a data set (i.e. the most parsimonious cladogram) must result in lowered information content.

Figs. 4–9). Cladistic analysis thus conforms to the required information-theoretic criteria laid out in the introduction. It also satisfies the empirical and cybernetic criteria.

Given n terminal taxa, the number of branching diagrams possible for those taxa may be calculated a priori. The goal of a phylogeneticist is the elimination of all but one of those diagrams. The preferred cladogram becomes the only unrefuted or the least-refuted solution for the data set (Wiley, 1975), according to the minimum entropy calculations. Wiley (1975) first showed the compatibility of phylogenetic systematics and hypothetico-deductive approaches to science. Farris (1979b) showed the empirical relationships between information theory and phylogenetic systematics, confirming the empirical nature of cladistics. Bateson (1972) showed relationships among cybernetics, information theory and language. I have attempted to combine those views in a syllogism and show that there is a definite link among cybernetics, information theory and empirical studies in comparative biology. That link can best be demonstrated by treating classifications as languages. It should be no surprise, therefore, to find that professional linguists interested in comparative linguistics have utilized analytical methods compatible with cladistic analysis for some time (Platnick and Cameron, 1977).

Note that I have said nothing about phylogeny or evolution or even about any particular point of view regarding degrees of similarity among organisms. Platnick (1979) suggested,

> And if contemporary cladistics had to be summarized through three main principles . . . . They might run something like this: first, that nature is ordered in a single specifiable pattern which can be represented by a branching diagram or hierarchical classification; second, that the pattern can be estimated by sampling characters and finding replicated, internested sets of synapomorphies; and third, that our knowledge of evolutionary history, like our classifications, is derived from the hierarchic pattern thus hypothesized.

So, with apologies to paleontologists and population biologists who think differently, the only necessary and sufficient criterion for recognizing that evolution has occurred is the discovery of synapomorphies. As we have now seen, the discovery of synapomorphies is based not on any assumption about evolution but rather on an information theoretic criterion. If there has been evolution the minimum entropy classification will reflect that phenomenon because the evolutionary history will be the cause of the ordering of the data into a detectable pattern. If there have been cases of multiple rather than dichotomous speciation, the minimum entropy message will not allow attainment of a fully dichotomous branching diagram. The most important thing about all of this is that Hennigian systematics, being independent of any a priori assumptions about evolutionary mechanisms, allows for the first time truly scientific tests of the synthetic theory of evolution (neo-Darwinism). And the tests run so far do not support that theory particularly well. Rather, the picture of evolution which is emerging is one of relatively long periods of evolutionary stability broken up by fairly short periods of tremendous evolutionary change, usually correlated with major geological changes in earth history. Data supporting this view come from a variety of sources, including genetics (Hampton Carson's open and closed genome hypothesis), developmental biology (epigenetics research), biogeography (Leon Croizat's vicariance biogeography and subsequent analytical techniques), and paleontology (punctuated equilibrium model) as well as systematics. Riedl (1978) presented the first systems analysis of evolution, concluding that there are many aspects which cannot be represented adequately by a population genetical theory of evolution.

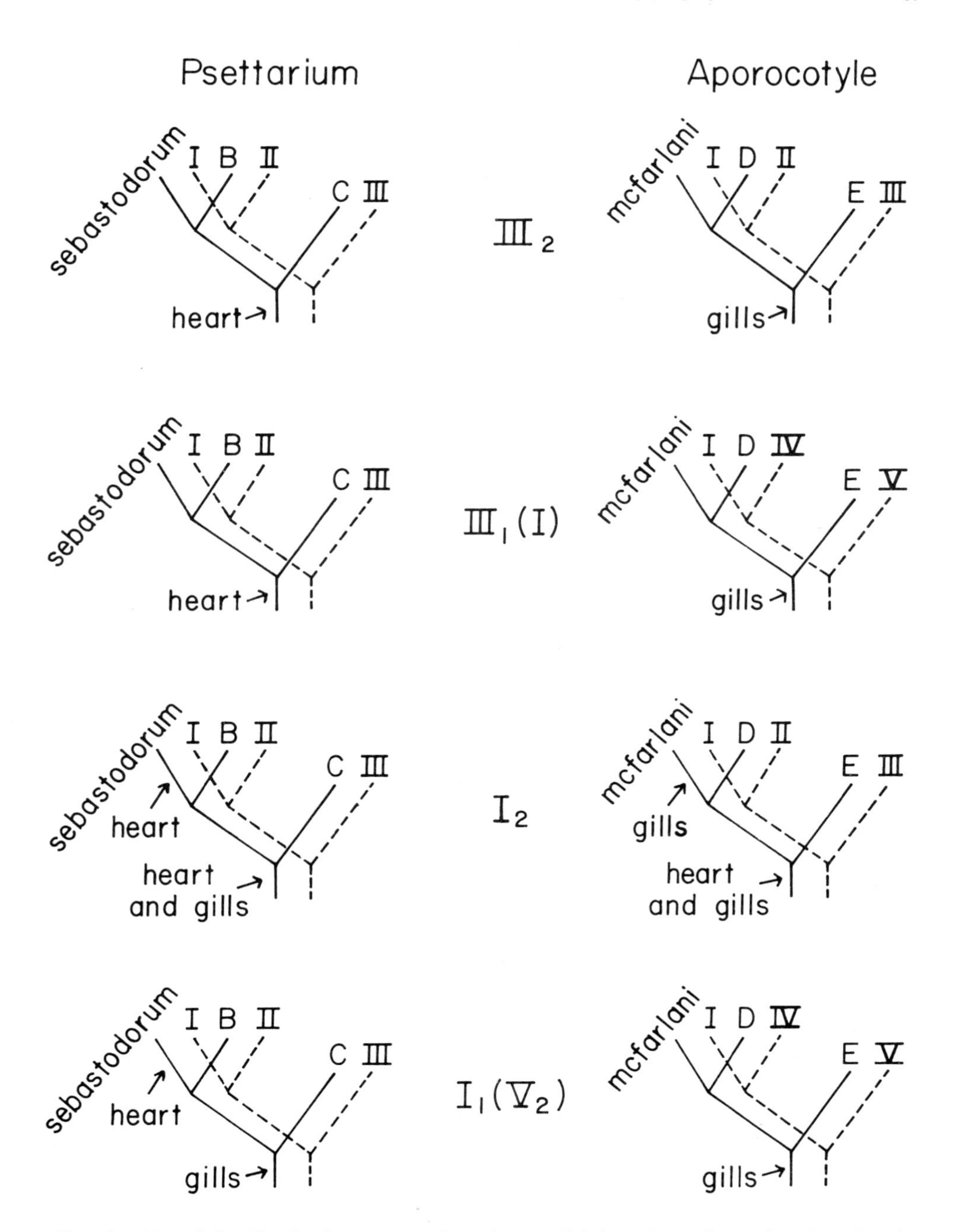

Fig. 10. Use of classification languages to formulate explicit hypotheses for testing, from Brooks (1980). In this instance, hypotheses concerning reasons for non-interactive site-selection by two species of blood flukes in the same host were compared.

Systematics is the general reference system for biology, and biology is a comparative science. For that reason alone, the utility of systematic methods in comparative studies should be closely examined and improved upon constantly. In addition, this Hennigian revolution in systematics provides a new dimension to comparative biology. Through the medium of an explicit language representing stored information maximally and also sensitive to potential information, biologists from varying disciplines can communicate precisely with each other and judge their differences about the same set of observations according to some objective criterion. Once formulated, these cladograms can thus serve as a *metalanguage* (Bateson, 1972) in comparative biology.

Earlier I cited a statement by Gatlin (1972) to the effect that all general situations or hypotheses can be expressed in the form of a language message. In this manner, any hypothesis in comparative biology may be formulated as a single cladogram or a particular set of cladograms exhibiting a given set of relationships to each other. Data may then be gathered, a minimum entropy message may be computed from the data, and the hypothesis which best agrees with the findings may be accepted as the best working hypothesis. Figure 10 depicts just such a set of hypotheses, taken from Brooks (1980). Hennigian systematics is not restricted by data type or any other a priorism. Physiologists can talk to ecologists; morphologists can talk to biochemists—all in precise and unambiguous terms.

There is one major drawback to accepting such an approach in one's science. This technique tosses out the possibility of relying on untested intuitive insight. It does not toss out intuitive insight, but it does not allow such insights to attain any scientific status until they have been tested.

Botanical Cladistics

Special Concerns in Estimating Plant Phylogenies

V. A. Funk

In general, botanists have not been as enthusiastic as zoologists concerning cladistics (=phylogenetic systematics of Hennig, 1966). This method of interpreting the pattern of evolution was begun by Hennig (1950, 1966) and developed by others (e.g., Brooks, 1979; Eldredge and Cracraft, 1980; Farris, 1979b; Gaffney, 1979; Nelson and Platnick, 1981; Platnick, 1979; Platnick and Nelson, 1978; Rosen, 1978; Wiley, 1979). It is true that some botanists, especially the pterydologists, have been using the less formalized (more generalized) Groundplan Divergence method of Wagner (e.g., Fryxell, 1971; Mickel, 1962, Wagner, 1961), but for the most part botanists in general have been reluctant to use cladistics. This reticence on the part of many of my colleagues is, I believe, a result of certain gaps that exist between conceptual and operational cladistics. These gaps, some valid and some imaginary, reflect problems that we tend to encounter that seem not as prevalent in many animal groups. The problems can be segregated into three areas: 1) hybridization, 2) parallel evolution, and 3) poorly understood suprageneric groups. Certain aspects of at least the first two areas have been addressed by others (e.g., Bremer and Wanntorp, 1979; Nelson and Platnick, 1980; Platnick, 1977; Wagner, 1980) and various solutions have been suggested. This paper discusses the problems, evaluates some proposed solutions, and offers some suggestions of its own.

Hybridization and Polyploidy are outstanding features of many plant groups. According to some authorities 30 to 80% of the species of angiosperms (Goldblatt, 1979; Lewis, 1979; Stebbins, 1974) and up to 95% of the ferns (Grant, 1971) are polyploid in origin, and these figures do not include hybridization at the diploid level. There are many facets of polyploidy and hybridization and these can affect cladograms in various ways. First, a brief diagrammatic explanation of the various types of hybridization and polyploidy is necessary (Fig. 1). Most workers in cytogenetics agree that confusion exists in the terminology (e.g., deWet, 1979; Grant, 1971; Stebbins, 1979). In particular, allopolyploidy and autopolyploidy have been variously defined. These two terms are here used according to the

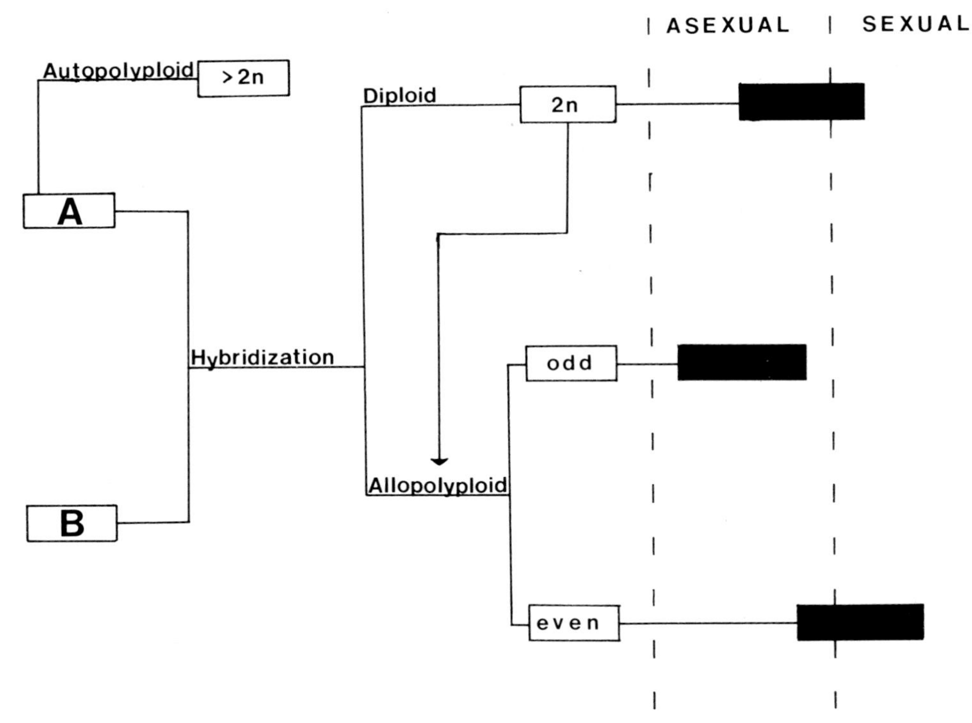

Fig. 1. Possible hybrid and polyploid relationships of two genomes (A and B) and their reproductive capabilities (see text for explanation).

strict definition (autopolyploidy involves only identical genomes). Segmental allopolyploidy is used to indicate genomes that are only partially different (Stebbins, 1950). Figure 1 illustrates the various means of hybridization at both the diploid and polyploid levels. A and B represent genomes that are at least partially different. Autopolyploidy can occur in either genome, but such individuals are similar to the parent individual and, therefore, present no problem in a cladistic analysis; the autopolyploid is either identical to, or the sister group of, the parent species. Autopolyploids are not discussed further, and unless otherwise specified in the remainder of the paper the term hybrids refers to diploids and allopolyploids.

Hybridization (Fig. 1) can occur at either the diploid or polyploid level, and the polyploids can be at either odd or even ploidy levels (e.g., odd = 3x, 5x; even = 4x, 6x). Some hybrids do not persist in nature, but a great number of them do. Even with reproductive irregularities many survive through asexual reproduction (Nygren, 1967). Indeed, nearly all gametophytic apomicts are polyploids (deWet, 1979). This enables the hybrids (both diploid and polyploid) to maintain themselves even if they cannot undergo normal meiosis. Sometimes the meiotic irregularities will eventually be eliminated and normal and sexual reproduction will follow (e.g., Wagner and Whitmire, 1957).

Diploid hybridization can result in either hybrid speciation or the formation of hybrid swarms. In hybrid speciation, such as is discussed by Grant (1971; Straw, 1956) in the genus *Penstemon,* two of the species that hybridize form an intermediate. *Penstemon centranthifolius* has red, trumpet-shaped hummingbird pol-

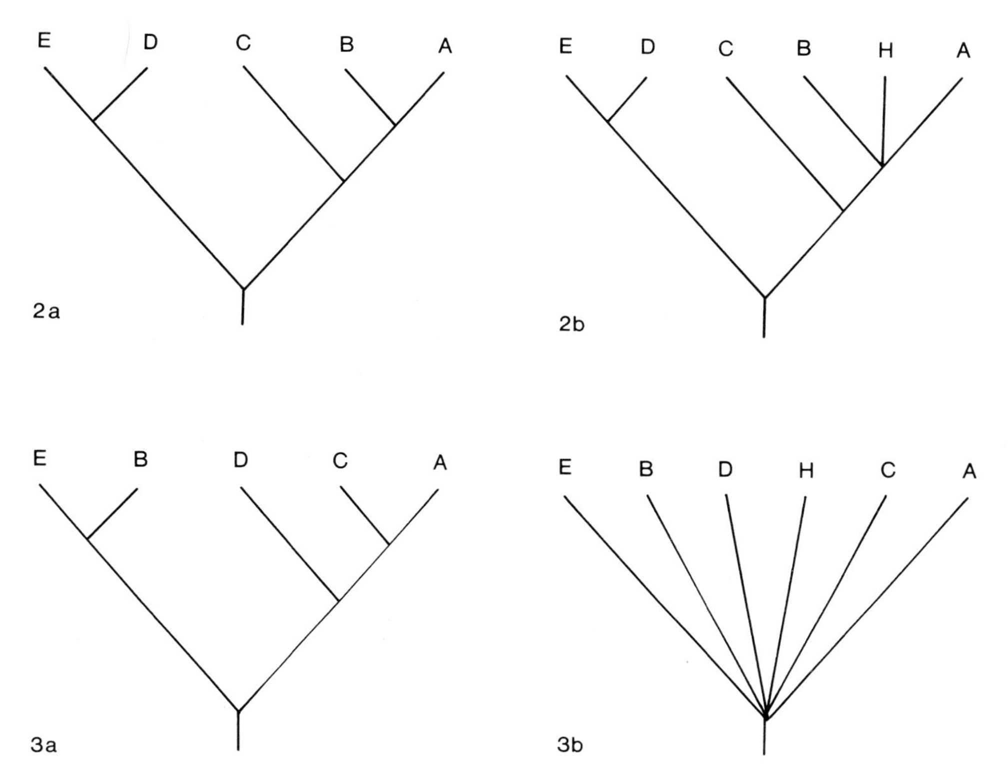

Figs. 2–3. Cladograms illustrating the result of hybridization between taxa A and B. In Figures 2a and 2b the parental taxa are sister taxa and the hybrid (H) causes a trichotomy. In Figures 3a and 3b the parental taxa are not closely related and the cladogram collapses to the first node that the two parents have in common.

linated flowers; *P. grinnellii* has flesh-colored, broad-throated, carpenter bee pollinated flowers; the diploid hybrid, *P. spectabilis,* is fertile, intermediate in color and shape, and is pollinated by wasps and anthophoid bees. Because this hybrid, like most others, tends to be intermediate between its parents, it has a predictable effect on the cladistic analysis. That effect is a function of the relationship among the three species. If the two parent species are sister taxa (sister taxa share a most recent common ancestor and are more closely related to one another than to any other taxon) there would be no problem in constructing the cladogram, for as Nelson and Platnick (1980) have pointed out it would result in a trichotomy. However, a trichotomy does not reflect the true pattern of evolution: a hybrid does not share an ancestor with the two parents. A hybrid between two non-sister species, on the other hand, would result in a collapse of the cladogram. For example, if the evolution of *Penstemon* species mentioned above looks like Figure 2a and the two parents (A and B) are sister taxa, the cladogram would look like Figure 2b. If the evolution is like Figure 3a and they are non-sister taxa, then it would look like Figure 3b.

In a diploid swarm such as is found in *Phlox* (Heiser, 1973; Levin, 1963; Levin and Smith, 1966), *Quercus* (Burger, 1975) and others, hybridization is followed by backcrossing with one or both parents resulting in a blurring of specific limits. These backcrosses have the potential to be fixed at any morphological stage,

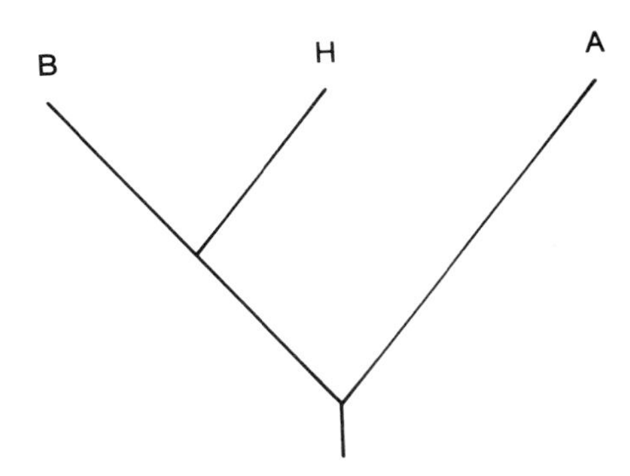

Fig. 4. Cladogram illustrating the result of the hybrid (H) segregating with one parent (B).

which means that the hybrid could segregate with either one of the parents and give a false representation of the pattern of evolution. One example is illustrated in Figure 4. The actual phylogeny involves A and B hybridizing and the hybrids backcrossing with B. The hybrid (H) eventually stabilized and had synapomorphies with only one parent.

These same problems exist when dealing with polyploid hybrids. True allopolyploids are usually intermediate in morphology between the two parents and they show little segregation. They are often fully fertile and constant and the backcrosses with either parent are usually sterile. These would affect a cladistic analysis in much the same way as diploid hybrid speciation. If the parents are sister species a trichotomy results (Fig. 2B) but if they are not, then the cladogram collapses to the point of the most recent common ancestor of the parental species (Fig. 3B).

Segmental allopolyploids form partially fertile hybrids through backcrossing from either or both parent species. The resulting problem of segregation from one parent or the other is the same as is found in diploid hybrid swarms (Fig. 4). The same problem occurs when dealing with auto-allopolyploids.

All of the types of hybridization are often complicated by two additional possibilities: 1) they can involve more than two parent species, and 2) the hybrids can then hybridize.

The multiple parent problem is well illustrated by *Gilia*. Grant and Day (e.g., Day, 1965; Grant and Grant, 1956; Grant, 1964) have shown that many of the diploids in *Gilia* can cross and produce fertile polyploids. These hybrids may be intermediate between two parents but the parents are not necessarily in the same monophyletic group. For instance, Day (1965) showed that two species from one group of diploids and one species from another could cross (Table I; *G. minor* × *G. clokeyi* and *G. minor* × *G. aliquanta*), and these are not the only species of *Gilia* that do this. A cladogram of all of the species in *Gilia* would most likely appear as a single polytomy.

The reticulate evolution that results from hybridization is best illustrated by Wagner's work on Appalachian *Asplenium* (1954). Nine taxa are the result of hybridization and polyploidy of three diploid species and their hybrids (Fig. 5). A cladogram of these nine taxa would be a nine branched polytomy and relatively uninformative.

The one type of allopolyploidy that can be reflected accurately in a cladogram is the condition termed "diploidized" polyploidy. Usually these are "old" polyploids and they have greatly diverged from the parent species which may or may

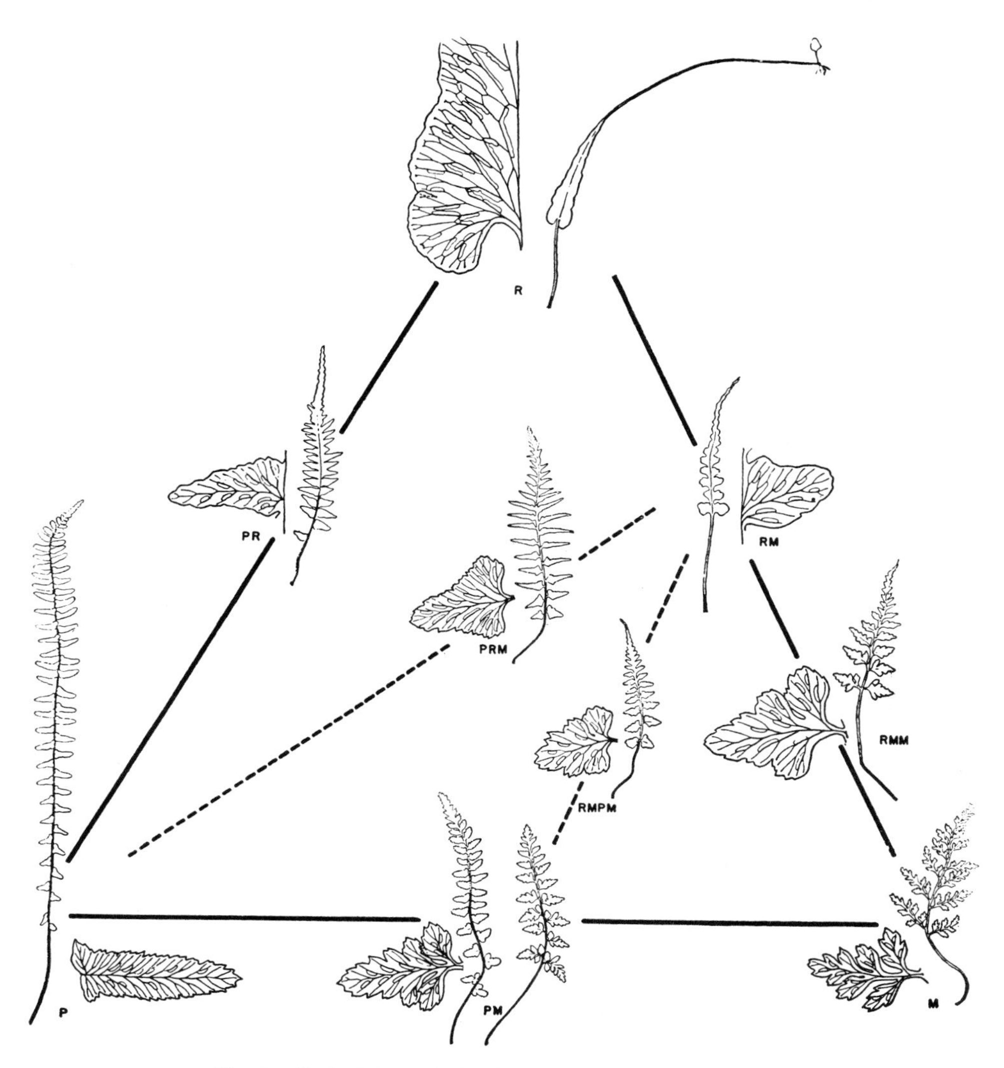

Fig. 5. Reticulate evolution in *Asplenium* (Wagner, 1954).

not still exist. These polyploids all undergo regular meiosis with bivalent pairing, the pollen is viable, and seeds are fertile; they can be found on all taxonomic levels. The genus *Montanoa* has three such polyploids. These three species share a number of features in common but are not closely related phylogenetically (Fig. 6). Such "old" polyploids are also found on higher taxonomic levels. Some families (e.g. Rosaceae, Oleaceae) contain entire tribes or subfamilies that are of allopolyploid origin. In addition, some families (e.g. Magnoliaceae) are believed to be of polyploid origin. However, the members of these taxa have long since stabilized. The task of dealing with these "old" polyploid taxa in a cladistic analysis is no more difficult than dealing with diploid taxa. The presence of such groups helps account for the large percentages that have been given for the numbers of species of higher plants that are polyploids (see introduction) and hence makes the task of treating angiosperm groups cladistically less formidable.

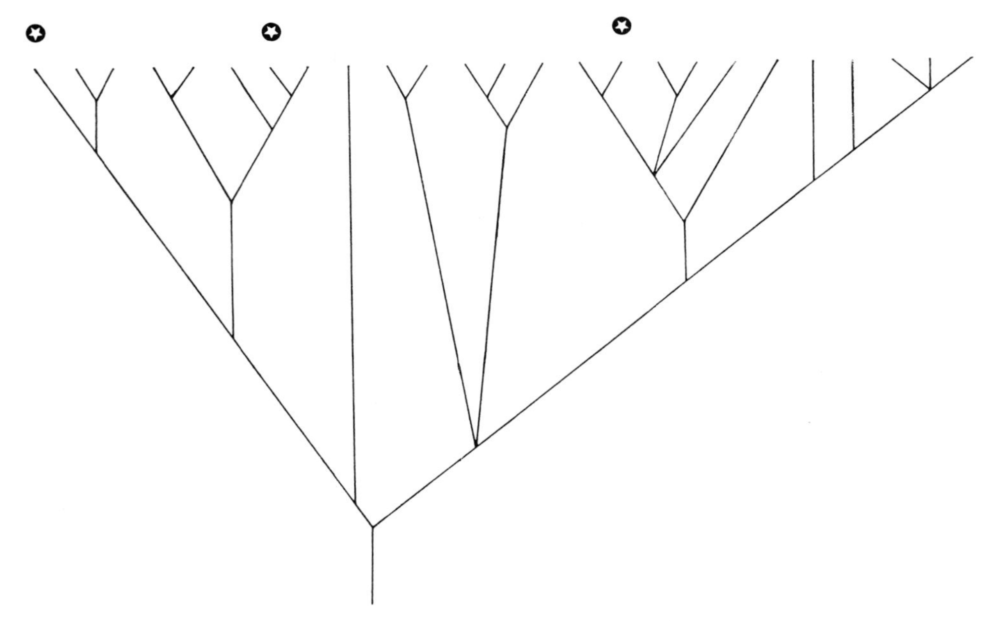

Fig. 6. Cladogram of *Montanoa* Cerv.; terminal taxa with stars are polyploids.

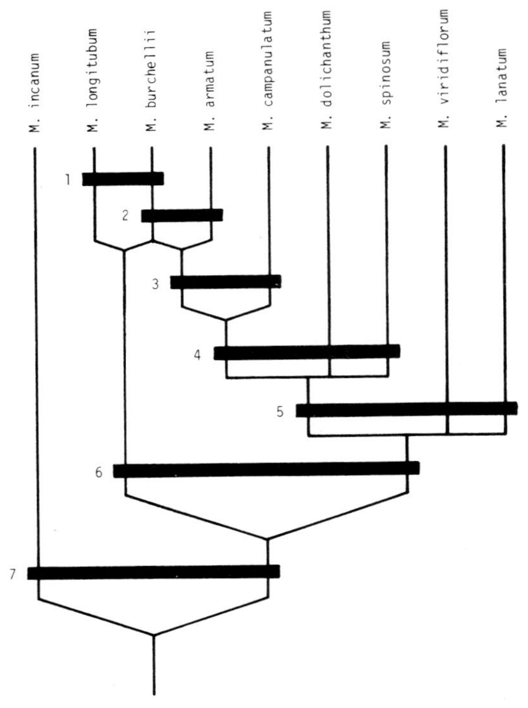

Fig. 7. Cladogram with a hybrid taxon (*M. burchellii*) shown as a trichotomy with the two parents (Bremer and Wanntorp, 1979).

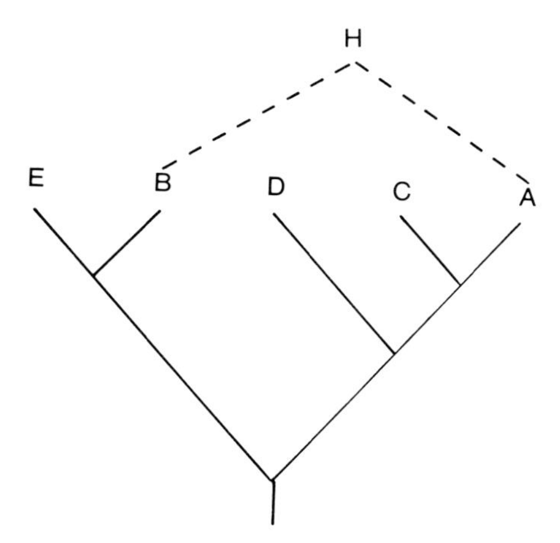

Fig. 8. Cladogram illustrating hybridization between taxa A and B (Fig. 3b) with the hybrid taxon (H) displayed above the terminal taxa.

All of these problems with hybridization, when combined with the large number of species that are of hybrid origin, help to explain part of the reluctance botanists have to using cladistics in groups with extensive hybridization. Although the plight of the botanists has been ignored for some time, recently a number of suggestions have been made on ways to deal with hybridization.

Bremer and Wanntorp (1979) recognized some of the problems with hybridization. They recommended leaving the hybrids in the analysis and illustrating their placement as in Figure 7. One of the significant features of cladistics is that the cladogram makes no statement about overall similarity and difference and because of this the cladogram is mobile, it can be rotated at any point, and, therefore, the distance between two terminal taxa is not meaningful. By putting the hybrids on the diagram as in Figure 7 the position of the terminal taxa is fixed and the mobility of the cladogram is lost. Also, you must have synapomorphies with both parents, and if more than one set of parents is involved, or if the parents are far apart, the cladogram becomes difficult to construct. Nelson and Platnick (1980) suggest leaving the polytomies that result from hybrids in the cladogram because this display reflects the actual character information. If the two parents are sister species and you have no objection to displaying the two parents and the hybrid as having a common ancestor then there is no problem. Because plants often hybridize with non-sister taxa, including them in the analysis will collapse the cladogram to the first common node (ancestor) between the parental species. This type of display actually obscures character information.

Nevertheless, there is a way around these problems. Some botanists (e.g., Stebbins, 1979; Wagner, 1970, 1980) feel that although polyploids, apomicts, inbreeders and hybrids are common today and probably have been throughout the history of vascular plants, the fact that "normal" species (diploid, sexual, outbreeding) still prevail today indicates that such plants have always been the primary thrust of evolution (Wagner, 1970, has referred to the others as evolutionary noise). The main pattern of evolution then is determined by the diploid species. I am here proposing, as has Wagner (1980), that the cladogram be constructed without the hybrids which can then be added after the basic framework of evolution within the genus has been determined. This could be accomplished by constructing an initial cladogram using all the taxa; polyploids and putative diploid

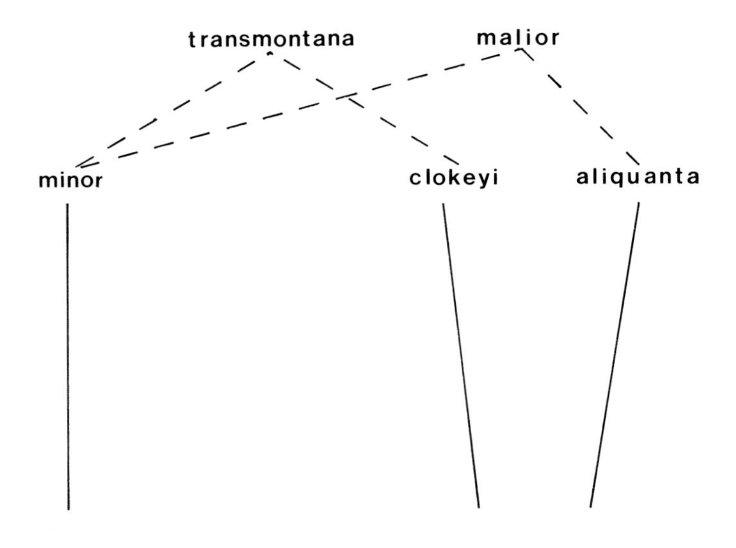

Fig. 9. Allopolyploid hybrids of *Gilia* and their parents (Table I).

hybrids involved in polytomies could then be removed and the cladogram further resolved. Figure 8 shows how this method would alter the cladogram in Figure 3b. This type of display will show clearly the branching pattern. Most hybrid character states tend to break down the distribution of synapomorphies and the branching pattern. A taxonomist familiar with the group can generally spot hybrids and these can often be substantiated by the pairing patterns at meiosis, chromosome number, distribution patterns and fertility level. Once the cladogram has been constructed one can add the hybrids at the top (Fig. 9 and Sanders, 1981, Fig. 12). This indicates accurately the hybrids as the combining of the genomes of two terminal taxa, and maintains the dichotomy and flexibility of the cladogram. This approach has been used by Humphries (1981) in his paper on the cytogenetics of *Anacyclus* (Fig. 10) in the preliminary construction of the cladogram, and also by Rosen (1978, 1979).

If the identification of a hybrid is in doubt, then it should be left in the analysis

Table I

System of classification of the diploid cobwebby *Gilia*s (Day, 1965)

I. *Gilia ochroleuca* group	II. *Gilia tenuiflora* group	III. *Gilia brecciarum* group
1. *Gilia ochroleuca*	7. *Gilia tenuiflora*	14. *Gilia becciarum*
2. *Gilia exilis*	8. *Gilia leptantha*	15. *Gilia diegensis*
3. *Gilia cana*	9. *Gilia latiflora*	
4. *Gilia mexicana*	10. *Gilia interior*	
5. *Gilia clokeyi*	11. *Gilia austrooccidentalis*	
6. *Gilia aliquanta*	12. *Gilia minor*	
	13. *Gilia jacens*	

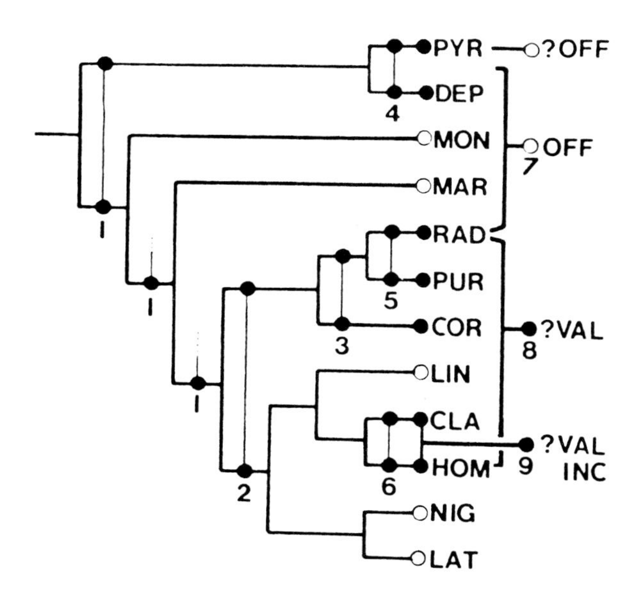

Fig. 10. Preliminary cladogram with the hybrid taxa placed above the terminal taxa (Humphries, 1981).

and the resulting polytomy may help identify the parents. The hybrid could then be removed from the cladogram and placed above it.

The allegation that hybrids are not the main line of evolution is supported in a number of ways. Stebbins (1979) has recently made some interesting observations that support the hypothesis that most polyploids are evolutionary noise [most diploid hybrids are relatively infertile (Fig. 1) and usually have to undergo ploidy increase before they can reproduce sexually and really be considered a possible candidate for evolutionary significance]. It is Stebbins' opinion that the buffering effects of ploidy reduce the effect on populations of individual mutations. This would reduce greatly the rate of evolution due to gene changes and would therefore tend to preserve archaic genotypes. He compared the relative number of derived characters a taxon had with its ploidy and its position in the hierarchy. Stebbins found two different levels of correlation that when compared lead to an interesting conclusion. The first level is within the genus and the second is above the genus. Within many genera the most specialized species are often polyploids (e.g., *Antennaria,* Stebbins, 1974; *Penstemon,* Stebbins, 1979). The reverse of this is often true at the second level (higher taxonomic levels). For instance, the true fern with the highest chromosome number is *Ophioglossum* ($x = 630$) and it has less specialized eusporangiate morphology, whereas *Hymenophyllum* with the lowest number ($x = 11$) is highly specialized. The same is true at the family level with the Annonaceae ($x = 7$, 8, and 9) being more specialized in Stebbins' opinion than the Magnoliaceae ($x = 19$) or the Degeneriaceae ($x = 12$). So it appears that after polyploids have been formed they evolve more slowly. Families are older than their modern genera so that during the course of evolution polyploids that, when first formed are among the most morphologically specialized representatives of their family, later become no more specialized than the majority of the diploid genera and species within the family, and finally become its least specialized members. These are, of course, only general tendencies and exceptions can be found.

Another aspect that should be considered is that since only the hybrids are removed and not the unknowns, no information is removed from the cladogram by formulating the branching pattern without the hybrids. Since the parents are included in the analysis, so is the information from the hybrids. Hybrids result in polytomies; the further apart the species are on the cladogram, the larger the polytomy. In the terms of Brooks (1981) and Gatlin (1972) polytomies lower the *S* redundancy (information density) of the cladogram; a fully dichotomous diagram has a high *S* redundancy. By analyzing the "normal species" first, and then adding on the hybrids you maintain the high *S* redundancy without removing any character information.

Hybrids usually do not reflect the main pattern of evolution and they are relatively easy to identify. Furthermore, they can be removed from the initial analysis of the data without removing any information, and the polytomies the hybrids create do not reflect the true pattern of evolution within the genus. The only logical way to analyze the data, therefore, is a priori exclusion of hybrids from the analysis and a posteriori inclusion on the cladogram.

Parallel Evolution is a problem in many vascular groups (convergent evolution[1] is not included here because it is more easily detected and "usually" does not present a problem in carefully studied taxa). Parallel evolution is often the result of plants developing similar character states as a result of similar life history strategies, pollinator syndromes, etc. They are in effect, false synapomorphies (Platnick and Nelson, 1981). In some instances this can be detected by a careful examination, but some groups have undergone true parallel evolution in that they have the same character developed from the same gene(s) independently so it cannot be detected by structure alone. True parallel evolution can only be identified by comparison with the pattern of evolution within the group. If the parallel evolution is extensive, then the pattern becomes progressively more difficult to determine. Grouping taxa based on these false synapomorphies leads to the recognition of artificial groups (members which do not share a common history).

In many Angiosperm families, especially the more recently and rapidly evolving ones, such as the Asteraceae, there are few characters, either above or below the generic level, that have not developed in parallel at least once. These give conflicting patterns of evolution depending on which characters are emphasized. In some groups such as the genus *Montanoa* Cerv. (Asteraceae) parallel evolution has reached staggering proportions. In this genus 80% of the non-autapomorphies (autapomorphy = apomorphy found in only one taxon) have evolved in parallel at least once (Fig. 11). This does not mean that a cladistic analysis is impossible, it simply means one must be suspicious of one character branches, and even more diligent in searching for additional characters. Also, the cladistic analysis must proceed somewhat slowly and as the parallel characters are identified they should not be relied upon for determining the branching pattern. The characters should not be removed from the analysis, however, because they are often useful in separating closely related taxa. Note that in Figure 11 the parallel characters are all clustered at the tips of the branches.

Wagner (1980) has indicated that if one does not understand the parallel and convergent evolution, mistakes will be made in determining or estimating the phylogeny of the group. While it is certainly necessary to have a thorough un-

[1] Parallelism is the independent development of similar apomorphies from the same plesiomorphic state; convergence is the development of apparently similar apomorphies from different plesiomorphous states (Bremer and Wanntorp, 1978).

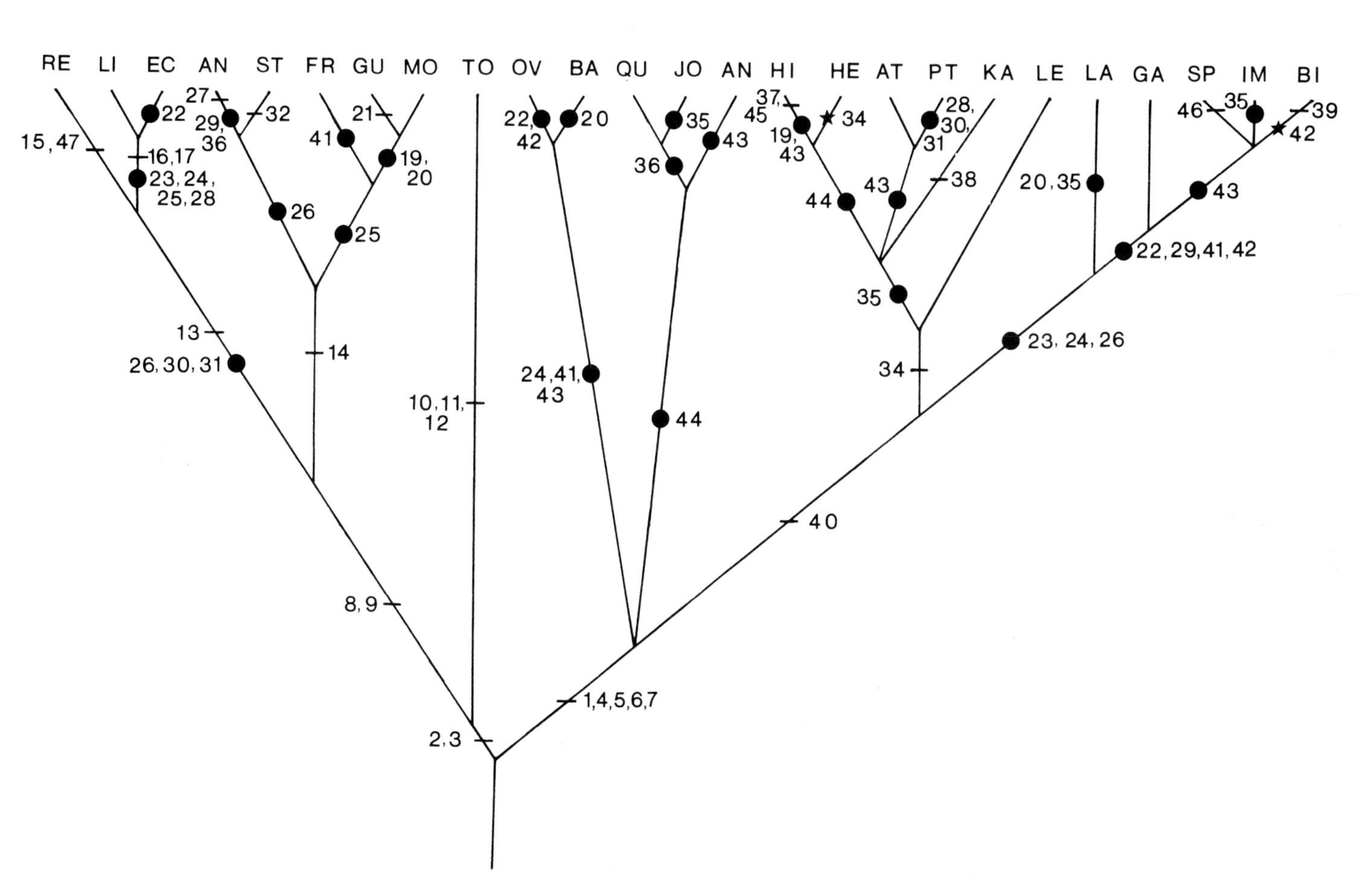

Fig. 11. Cladogram of *Montanoa* Cerv. (lines = apomorphies; dots = characters that have evolved in parallel; stars = reversals).

derstanding of the taxa, one really must construct the pattern (cladogram) before the parallel evolution can be identified. For instance, in *Montanoa* there has been an increase in head size three times in the genus. Previous treatments had grouped all of the plants with the same head size together (Characters 25 and 43, Fig. 11). The present analysis shows this not to be correct. This, as well as the other parallel characters, became evident as the cladogram was constructed. Ultimately, a better understanding of the parallel evolution within the genus was obtained by using cladistics. An analysis of a group with extensive parallel evolution requires a thorough examination of all characters and a careful evaluation of the resulting cladogram for possible alternative interpretations of characters that show more than one introduction on the cladogram.

Poorly Understood Suprageneric Groupings is the least easily solved of the three problem areas and has resulted in artificial categories, and lack of outgroups.

Many of the groups in vascular plants are non-monophyletic (lack a common history). For instance, the Gymnosperms are now believed to be polyphyletic (Delevoryas, 1979) and the families of the Pteridophytes are in a continual state of flux so that most of them are, at this point, artificial (sensu Hennig). Many of the families in the Angiosperms have problems with the generic boundaries. For instance, in the Asteraceae many of the genera and subtribes are just arbitrary divisions of almost continual variation (e.g., within the tribes Heliantheae, Senecioneae and Astereae). Also, recent studies in the Brassicaceae by Hauser and Crovello (1980) have shown most of the tribal delimitations to be artificial. Since a cladistic analysis involving groups that are non-monophyletic is futile (because the members possess no common history to be discovered) one should strive to define the groups by synapomorphies. In most cases if a group has some non-members a cladistic analysis will bring these taxa to light. As the analysis progresses the non-members will continually be shifted to the bottom divisions of the diagram and their relationship will eventually be evident. In a group that, in addition, is plagued by extensive parallel evolution, the lack of dependable characters combined with the artificial groups sometimes make the task extremely difficult. The solution to this combination of problems is usually to examine the characters that are appearing in several places on the cladogram more carefully in an effort to determine false synapomorphies, and to increase the data base so that new characters can help in determining the affinities. For groups where all of these efforts fail, a cladogram that maximizes *S* and *R* redundancy can be produced by rooting a network to give maximum parsimony (Brooks, 1981).

Another manifestation of the lack of information on the suprageneric level, is the determination of the polarity of the character states. Various papers have recently discussed a number of ways to determine polarity (Crisci and Stuessy, 1980; DeJong, 1980; Stevens, 1980). All of the various types can be put into three general categories: 1) outgroup comparison, 2) ontogeny, and 3) "common is primitive" (the most common state is the primitive state). Of these three, the arbitrary determination of "common is primitive" is not useful, resulting in balanced cladograms and unresolved trichotomies (e.g., Eldredge and Cracraft, 1980; Ross, 1974; Watrous and Wheeler, 1981; Wiley, 1980). Also, it is impossible to differentiate those cases where the common state really is the primitive one from those where it is not, without using outgroup comparisons. Both outgroup comparison and ontogeny are useful and can even be effectively combined (Funk, In press). Of the two, outgroup has greater efficacy because ontogeny is often difficult to analyze.

Watrous and Wheeler have a recent paper dealing with the ins and outs of

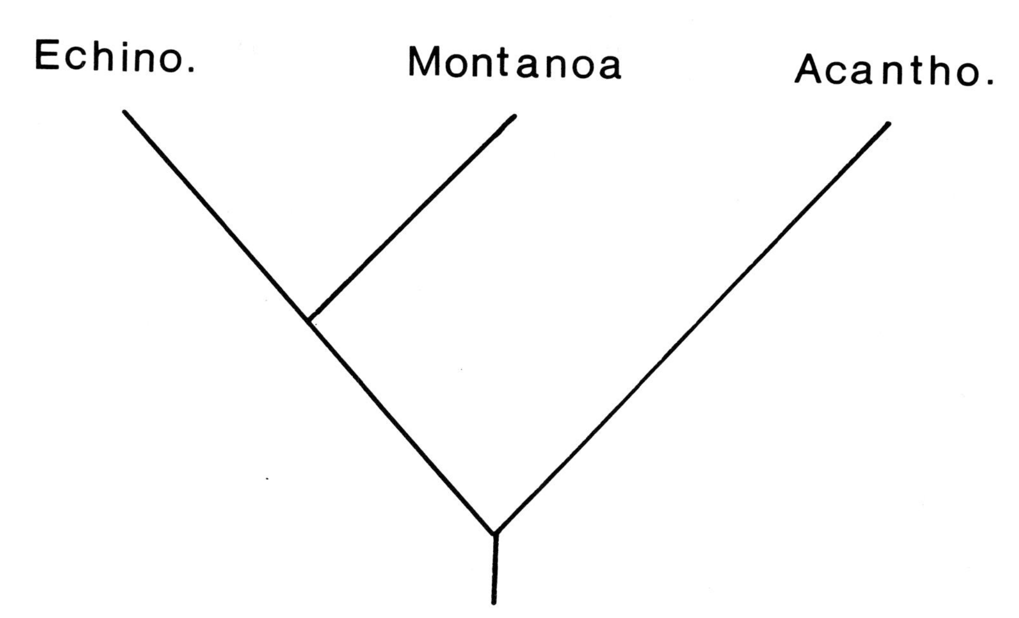

Fig. 12. Relationship of the three infrageneric groups of *Montanoa* Cerv.

outgroup comparison (Watrous and Wheeler, 1981).[2] Their operational definition of outgroup comparison is as follows:

> "For a given character with two or more states within a group, the state occurring in related groups is assumed to be the plesiomorphic state."

The number of times a character state occurs is not important, only the presence of that character state in both groups makes it plesiomorphic. It is because of the many poorly understood relationships that there is often confusion concerning the identification of the outgroup. Even if an outgroup is present, for instance, a sister genus, because of the extensive parallel evolution in many families both character states are sometimes present in both taxa. Rarely are plant suprageneric relationships worked out to the point that one can conveniently polarize the character states. Watrous and Wheeler (1981) have formalized a concept that aids in remedying some of these problems. They divide outgroup comparison into two types: ones performed with Taxonomic Outgroups (TOGs), and ones performed with Functional Outgroups (FOGs). Their contention is that there should be no constraints on taxonomic levels. It is only necessary to work with monophyletic groups; the taxonomic level is not important.

An excellent example of the usefulness of the Functional Outgroup concept is found in *Montanoa*. The genus is distinct in the tribe Heliantheae of the Asteraceae, so much so that it has been recognized as a monotypic subtribe (Robinson, 1978). The subtribal classification of the Heliantheae is in dispute and some of the groups are probably artificial; it therefore becomes difficult to determine an outgroup. In addition, the parallel evolution is so great in the tribe that most of

[2] This paper was presented by Watrous as part of this symposium but was already committed for publication elsewhere.

the characters cannot be polarized using the tribe as the outgroup. Therefore, Watrous and Wheeler's method of using FOGs was employed. There were 11 characters that could be polarized using ontogeny and outgroup comparison (using the tribe as the outgroup). These 11 characters divided the genus into three monophyletic groups (Fig. 12) which were then used as FOGs and the remaining characters were polarized. Referring to Figure 12, typical *Montanoa* was used as the outgroup of *Echinocephalae* and the plesiomorphic character states of these two sister groups were used as the outgroup of *Acanthocarphae*. The cladogram was then constructed (Fig. 11). Using this type of analysis no data were removed because each of the three FOGs had its own data set and was treated independently; all the characters were used each time. While the suggestions of Watrous and Wheeler (1981) do not eliminate all of the problems that botanists face in determining polarity, they certainly take care of quite a few, and they allow the use of outgroup comparison so one is not faced with the less useful method of "common is primitive."

Some of the suggestions in this paper may give the impression that they make the results of the cladogram more speculative. A cladogram is, however, an estimation of the pattern of evolution within a group. As such, it is a hypothesis and is subject to testing and falsification by the addition of new characters (e.g., Gaffney, 1979; Wiley, 1975). Nothing that is suggested here lessens the testability of the cladogram.

Phylogenetic analysis is only easy on unsuccessful groups like *Ginkgo*. In successful and rapidly evolving plant groups the task is much more difficult. This should not be interpreted as meaning that cladistics is not useful in the study of evolution in plants. On the contrary, it is the best method for addressing many of the problems (e.g., parallel evolution). It simply means that sometimes botanists are faced with special considerations in estimating plant phylogenies, and while these considerations may not be unique to plants they are certainly unusual in their magnitude. These special considerations require some modifications in the formulation of cladograms, but on the whole, serve to illustrate the utility of cladistics in the study of plant evolution.

Acknowledgements

I thank Drs. Daniel Brooks, Arthur Cronquist, Jack Maze, Scott Mori, Harold Robinson, G. Ledyard Stebbins and Warren H. Wagner for their comments on the manuscript. I am indebted to Gareth Nelson, Norman Platnick, Donn Rosen and Randall T. Schuh for the many discussions on the topics included herein. While due consideration was given to all suggestions, the final decisions were my own. Support for the study of *Montanoa* was obtained from NSF Grant DEB-77-14812.

The Cladistic Approach to Plant Classification

Kåre Bremer and Hans-Erik Wanntorp

Table of Contents

Introduction

Cladistically unsound taxa are found at all hierarchic levels and they are especially blatant when they occur at the higher levels, representing the basic groupings of the living world. We will give a few examples from the kingdom to the generic level, showing how a cladistic approach will change the existing traditional classification of plants.

Procaryotes and Eucaryotes

When the fundamental difference between the procaryotes and the eucaryotes was demonstrated, it was proposed that the organisms should be divided into two kingdoms corresponding to these groups (Traub, 1963, followed e.g. by Taylor, 1974, 1978, and Whittaker and Margulis, 1978). The relationship between these two groups is dependent on whether the eucaryotes arose gradually from procaryotes through advancing compartmentalization or by symbiotic associations of several different procaryotes (Margulis, 1970). The latter theory is gaining acceptance. Nevertheless, by excluding from consideration their probable endosymbionts the chloroplasts and the mitochondria, the eucaryotes may well be a cladistically acceptable monophyletic group, characterized by a number of shared derived characters—synapomorphies. By contrast, all defining characters of the procaryotes are thought to be primitive in comparison with those of the eucaryotes (most are in fact merely the lack of eucaryote characters) and there is not a single character that might be hypothesized to represent a synapomorphy. The sister group of the eucaryotes is most probably only a fraction of the procaryotes, which seem to be a heterogeneous assemblage of a number of components meriting the same or higher rank as the eucaryotes. A treatment of the procaryotes, demonstrating this point, was recently presented by Fox et al. (1980). The authors proposed that the procaryotes should be divided into at least two kingdoms representing the major branches of their alleged phylogenetic tree. Essentially a phenetic clustering method was used to demonstrate this classification but a cladistic analysis would also lead to several procaryote groups and in the future we will have a number of procaryotic kingdoms as well as the eucaryotic one.

Kingdoms of Organisms

In this context it is appropriate to comment on some other attempts at kingdom classification. A particular one that has become widespread in textbooks is that presented by Whittaker (1969). His kingdoms are based on a few criteria, levels of organization and modes of nutrition. Even with the later modifications in grouping (Whittaker and Margulis, 1978), which tend to obscure these simple criteria, these kingdoms are unnatural; the Monera and the Protista are paraphyletic and the Plantae and probably also the Fungi are polyphyletic. Probably because they used simple criteria the system has become widely accepted. In this respect it is comparable to the sexual system of Linnaeus and like that it is an artificial system grouping unrelated taxa. From a cladistic standpoint it is completely unacceptable.

An attempt to recognize kingdoms as monophyletic groups only was made by Leedale (1974). This led to the establishment of some 18 eucaryote kingdoms and most of these may in fact be monophyletic. Leedale gives no hypothesis as to the relationships of all these kingdoms but unites them in one gigantic polytomy (Leedale, 1974, Fig. 6). This system may be regarded as an advancement to that of Whittaker and Margulis but a very small one, since all his groups have been recognized before as natural entities but at lower taxonomic levels. In our opinion, but this is a matter of taste, it would be better to retain these groups at a lower level and accommodate the eucaryotes in a single kingdom.

The Green Plants

We will now consider the relationships within the green plants, i.e. the green algae and the higher plants (the land plants). During the past decade research in green algae has provided a wealth of new information, especially at the cytological level (e.g. Pickett-Heaps, 1975). Stewart and Mattox (1975) recognize at least two major groups of green algae, the chlorophytes and the charophytes. Within the charophytes there are a number of comparatively more derived algal groups which are more closely related to the higher plants than to other algae. We have summarized this information in the cladogram in Figure 1. It is not to be accepted as complete, as with increasing information on various green algae new branches will have to be added. The sister group relationship between *Klebsormidium* and the conjugates is inferred from Stewart and Mattox (1978, Fig. 3) but a synapomorphy has yet to be demonstrated. It is clear that the green algae is not a monophyletic group with the higher plants excluded. The sister group of the higher plants is a fraction of the green algae, viz. the Charales.

In traditional systems for the green plants the green algae and the higher plants are separated and the green algae are placed within the lower plants in a group called the Thallobionta (see e.g. Cronquist et al., 1966). The angiosperms have their sister group amongst some of the gymnosperms and consequently they

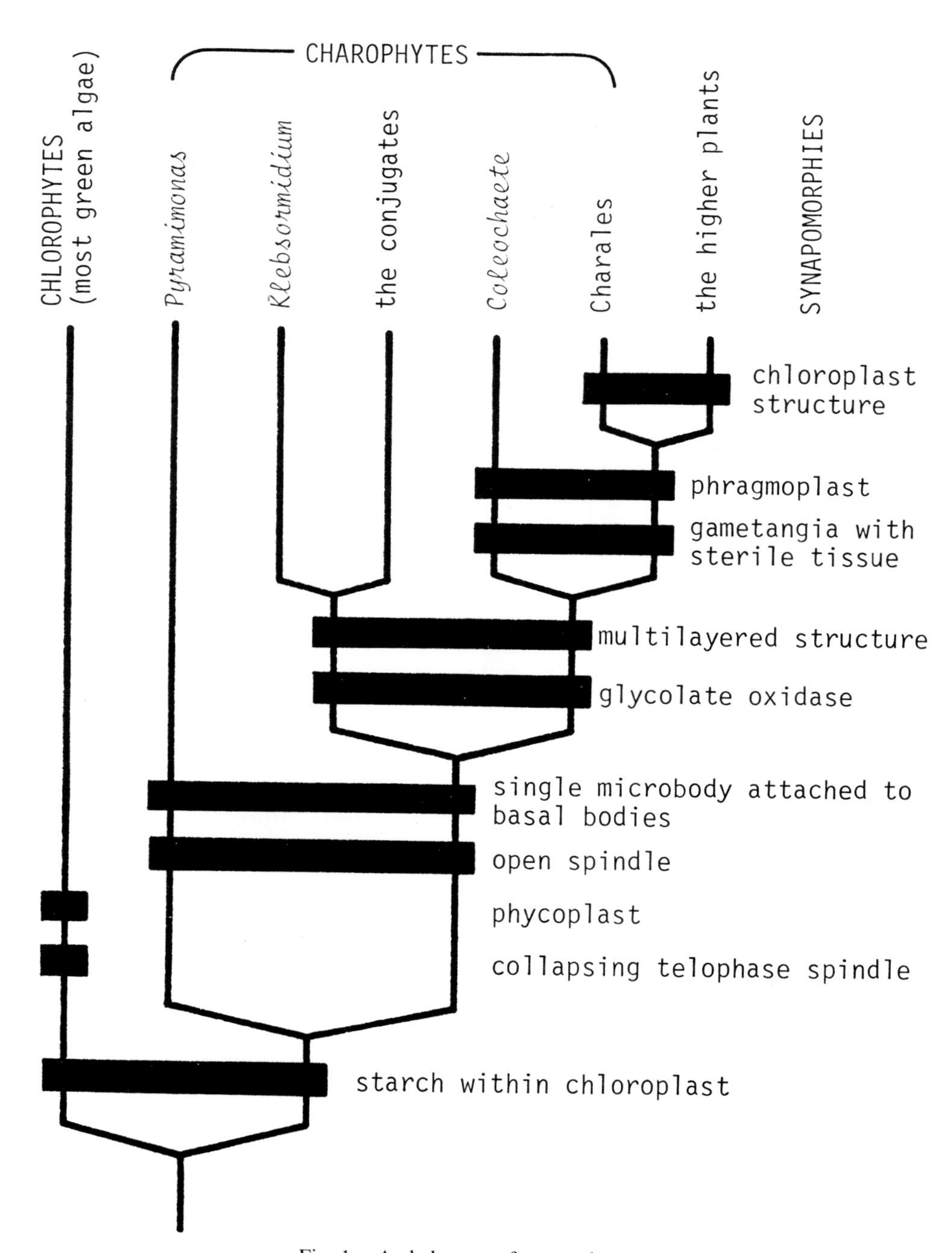

Fig. 1. A cladogram of green plants.

should be classified together. Nevertheless, in many systems the gymnosperms are still recognized as a taxon, the Pinophyta (Cronquist et al., 1966). The Thallobionta (the lower plants) and the Pinophyta (the gymnosperms) are paraphyletic groups, unacceptable in a cladistic classification. They are based on shared prim-

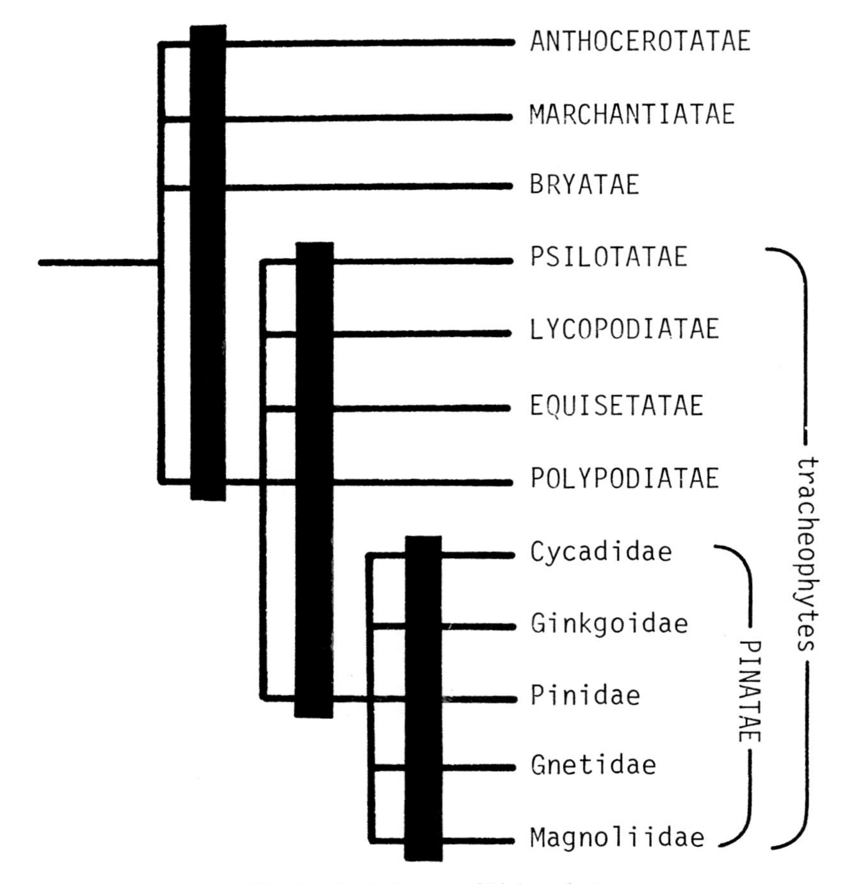

Fig. 2. A cladogram of higher plants.

itive characters or symplesiomorphies, such as thalloid habit and naked seeds, respectively. We have already proposed a cladistic classification (Bremer and Wanntorp, 1981), reproduced in Table I, where these paraphyletic groups are abandoned and the unity of green plants as well as the relationship of the higher plants to a part of the green algae is recognized. This classification reflects what is generally accepted about the cladistic relationships of the higher plants, shown in Figure 2. The groups included are considered to be monophyletic but some of them, e.g. Gnetidae, may have to be reevaluated.

Recently a more detailed cladogram of the land plants was presented by Parenti (1980) but several of her terminal taxa are not monophyletic groups, defined by synapomorphies. Examples of such groups are the Chlorophyceae (the green algae), the Lyginopteridopsida (the seed ferns), and the Magnoliopsida (the dicotyledons). Other, monophyletic groups of her cladogram have their sister group within these non-monophyletic groups and their inclusion in the cladogram will confuse rather than elucidate. Parenti has also taken the identification of synapomorphies much too lightly and thus made things look a lot easier than they really are. An example is the alleged synapomorphy of "seed and pollen cones." Those structures are clearly not homologous in the groups listed, e.g. the cycads and

Table I

A cladistic classification of green plants, after Bremer and Wanntorp, 1981

- Subkingdom **Chlorobionta** (green plants)
  - Division **Chlorophyta** (most green algae)
    - several subdivisions of green algae
  - Division **Streptophyta** (some green algae and higher plants)
    - several subdivisions of green algae, e.g.
    - Subdivision **Charophytina** (Charales)
    - Subdivision **Embryophytina** (higher plants)
      - Class **Anthocerotatae** (hornworts)
      - Class **Marchantiatae** (liverworts)
      - Class **Bryatae** (mosses)
      - Class **Psilotatae** (*Psilotum, Tmesipteris*)
      - Class **Lycopodiatae** (clubmosses)
      - Class **Equisetatae** (horsetails)
      - Class **Polypodiatae** (ferns)
      - Class **Pinatae** (seed plants)
        - Subclass **Cycadidae** (cycads)
        - Subclass **Ginkgoidae** (*Ginkgo*)
        - Subclass **Pinidae** (conifers)
        - Subclass **Gnetidae** (*Gnetum, Ephedra, Welwitschia*)
        - Subclass **Magnoliidae** (angiosperms)

the conifers, and thus this is not a synapomorphy but a case of convergence. Although Parenti's initiative is to be credited, her cladistic analysis contains several obvious mistakes and the result is unacceptable.

The tracheophytes are not recognized in our proposal but could be treated as a superclass. This would, however, necessitate the recognition of bryophyte superclasses as well (but see Farris, 1976, for an alternative viewpoint). This brings us to the question of complete cladistic classifications, where all branchings of the cladogram are represented by named taxa. This requires a great number of

Table II

A cladistic classification of higher plants, after Bremer and Wanntorp, 1981, and annotated as devised by Wiley, 1979

- Subdivision **Embryophytina**
  - Embryophytina, incertae sedis: "bryophytes"
    - Class **Anthocerotatae,** sedis mutabilis
    - Class **Marchantiatae,** sedis mutabilis
    - Class **Bryatae,** sedis mutabilis
  - Superclass of tracheophytes
    - Class **Psilotatae,** sedis mutabilis
    - Class **Lycopodiatae,** sedis mutabilis
    - Class **Equisetatae,** sedis mutabilis
    - Class **Polypodiatae,** sedis mutabilis
    - Class **Pinatae,** sedis mutabilis
      - Subclass **Cycadidae,** sedis mutabilis
      - Subclass **Ginkgoidae,** sedis mutabilis
      - Subclass **Pinidae,** sedis mutabilis
      - Subclass **Gnetidae,** sedis mutabilis
      - Subclass **Magnoliidae,** sedis mutabilis

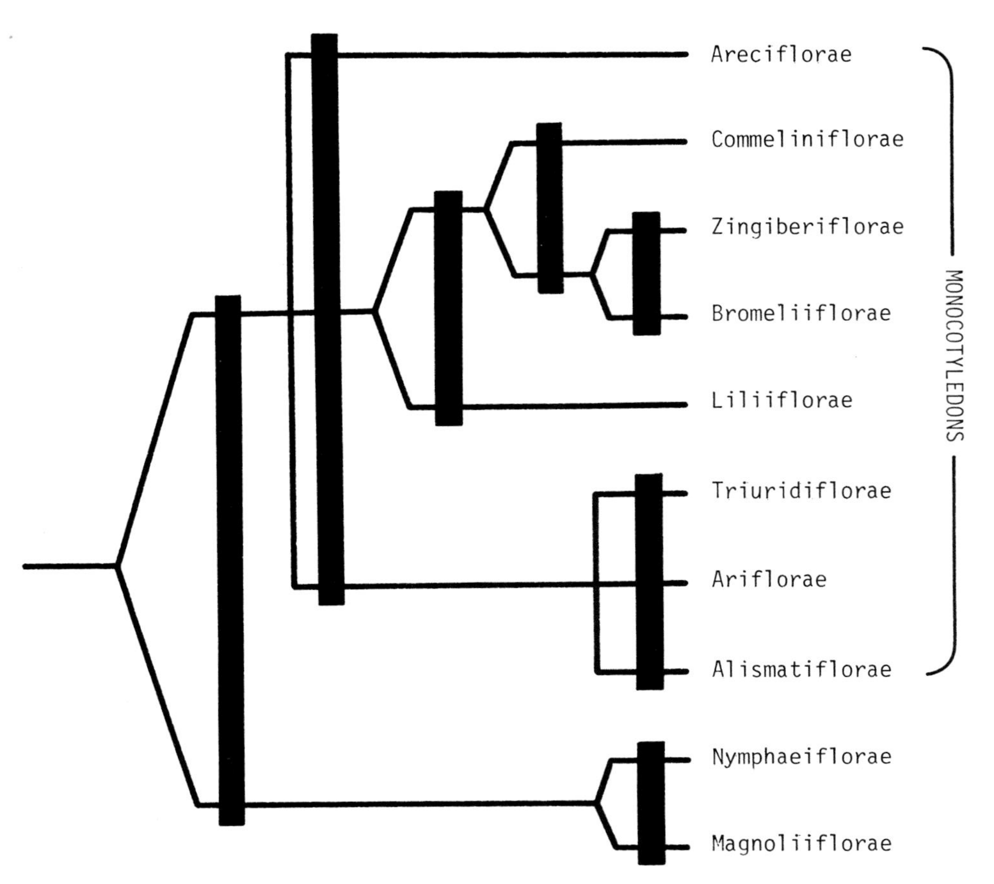

Fig. 3. A cladogram of monocotyledons, after Dahlgren (unpubl.).

categories in addition to those in general use and it is doubtful if such a complete classification would be useful. A way out of this quandary was proposed by Wiley (1979), who devised a system of annotations to the names of taxa at the same rank so that the cladistic relationships can be derived from the classification. In Table II Wiley's method is exemplified by the higher plants. The bryophytes are not necessarily a monophyletic group and it would be unwise to recognize them as a superclass. Thus they are put within shutter quotes and their unknown relationship to other higher plants is indicated by incertae sedis. The annotated system proposed by Wiley includes a sequencing convention to avoid unnecessary multiplication of ranks. This means that when an unannotated list of taxa of equal rank is encountered, "it is understood that the list forms a completely dichotomous sequence with the first taxon listed being the sister group of all subsequent taxa and so on down the list" (Wiley, 1979, p. 321). The mutual relationships within the bryophytes, the tracheophytes, and the seed plants (Pinatae) are unknown, however, and the sequencing of the classes and subclasses within these groups is thus arbitrary, indicated by sedis mutabilis.

Extinct groups are not included in this proposed classification but they can be included at the appropriate level, if they are considered monophyletic. Wiley (1979) also discussed the inclusion of extinct groups in cladistic classifications.

The Angiosperms

We will now move further down the taxonomic hierarchy and comment on the subclass Magnoliidae, the angiosperms. Little is known about the general relationships of these and there has been a number of different systems of angiosperms. Among the recent ones are those of Cronquist (1968), Takhtajan (1969), Thorne (1976), and Dahlgren (1980). These are not cladistic or phylogenetic systems in our sense, however. A case in point is that of the dicotyledons and the monocotyledons. The angiosperms have since long been divided into these two groups and the recent systems are no exception. The monocotyledons are certainly a monophyletic group, characterized by a number of synapomorphies such as their single cotyledon and unique sieve tube plastids (Behnke 1969; Dahlgren 1980). The dicotyledons are a typical example of a paraphyletic group, however. They are characterized by shared primitive features, symplesiomorphies, such as two cotyledons and the absence of the advanced features of the monocotyledons. We have already stressed this point in an earlier paper (Bremer and Wanntorp, 1978; see also Parenti, 1980). It becomes even more obvious in view of attempted reconstructions of angiosperm phylogeny. Dahlgren (unpubl.) made a cladistic analysis of major monocotyledon groups. His cladogram, shown in Figure 3, clearly shows that the sister group of the monocotyledons is not the dicotyledons but a fraction of them, probably the two superorders Magnoliiflorae and Nymphaeiflorae. Since the dicotyledons are not monophyletic, it follows that we should abandon the taxonomic division of angiosperms into dicotyledons and monocotyledons. These are merely levels of advancement, analogous to spore plants/seed plants and gymnosperms/angiosperms. In the latter cases systematists are abandoning spore plants and gymnosperms as taxonomic units and this will certainly be the fate of the dicotyledons, too. Dicotyledons and monocotyledons are still useful concepts but they are not units in a cladistic classification.

If we wish to recognize the monophyletic group of the monocotyledons as a taxon, their sister group the Magnoliiflorae + Nymphaeiflorae should be established as a taxon of the same rank. The remaining dicotyledons are not necessarily monophyletic, since there are no synapomorphies to unite them. Thus, a further splitting of the dicotyledons will probably have to be done, although this is hardly possible at the present state of knowledge. It seems best to regard the angiosperms as a subclass (see above) with a large number of superorders (Dahlgren, 1980, recognized 31 superorders). The question of ranking was further discussed in our earlier paper (Bremer and Wanntorp, 1981). If the superorders can be defined as monophyletic units, characterized by synapomorphies, a cladogram comprising all superorders of angiosperms could be constructed. Eventually it will provide a better classification of the angiosperms than the present non-cladistic dichotomy of the dicotyledons and the monocotyledons.

Conclusion

The cladistic approach to plant classification is now increasingly being employed by botanists working at the everyday taxonomic level of families, genera, and species. An early example of a cladistic plant classification at the generic level was provided by Koponen (1968). He revised the bryophyte family Mniaceae and presented a cladogram of its genera. Three of these, *Rhizomnium, Plagiomnium,* and *Pseudobryum,* are new genera separated from *Mnium* in order to achieve monophyletic taxa. The resulting classification has been largely accepted by bryologists. We think that cladistic classification will be increasingly used in botany and the bryophyte example is illustrative of this point. Cladistic classification may seem revolutionary but a cladistic plant classification, although significantly different, need not be drastically different from the traditional one; we hope that it will be considered an improvement by the majority of botanists.

Acknowledgements

We thank Rolf Dahlgren for putting his cladogram of the monocotyledons at our disposal. This study was supported by travel grants from the Swedish Natural Science Research Council (grant R-RA 3296–102) and the Netzel fund, University of Stockholm.

Cladistic Analysis of *Agastache* (Lamiaceae)

Roger W. Sanders

Table of Contents

Introduction

Agastache, a genus of odorous herbaceous mints, is the North American representative of the predominantly Eurasian catnip tribe, the Nepeteae. The entire genus was last revised by Lint and Epling (1945) who intuitively recognized two sections. *Agastache* sect. *Agastache* extends across boreal North America with one species in eastern Asia. *Agastache* sect. *Brittonastrum* is confined to the southern cordilleras of North America (Fig. 1).

Having recently revised sect. *Brittonastrum* through morphological, chromosomal and chemical studies (Sanders, 1979, In press), I would like to propose a predictive, cladistic phylogeny for this group. Two major techniques are available—parsimony and compatibility, the latter of which is rather controversial. For proponents, see Estabrook and Anderson (1978) and Estabrook (1978); for a critique see Farris and Kluge (1979). The parsimony methods of Hennig (1966) which are employed here are based on the principles of grouping by shared derived characters and determination of primitive character-states by outgroup comparison. The Groundplan/Divergence parsimonious method (Wagner, 1980) has

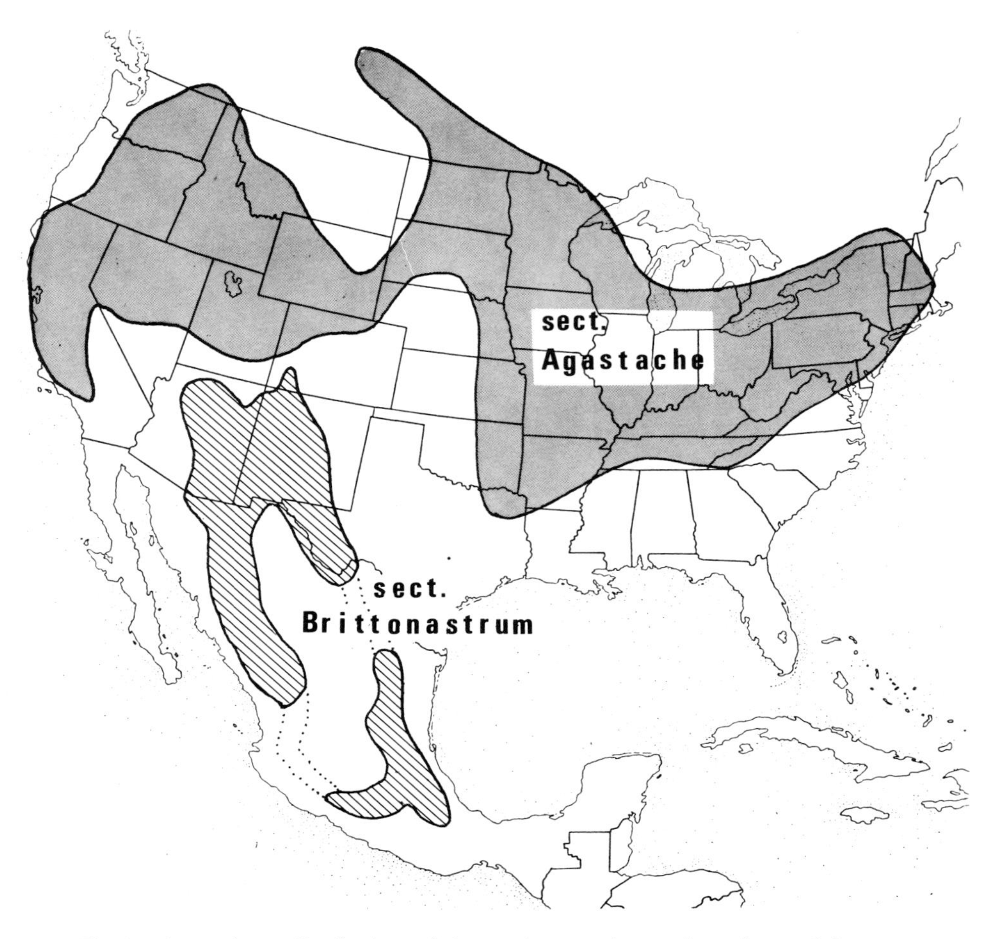

Fig. 1. Approximate distributions of *Agastache* sect. *Agastache* and sect. *Brittonastrum*.

been widely applied in botany, but it implicitly relies on the criterion that within a group the common condition is the primitive condition. This concept has been discussed and rejected: Eldredge and Cracraft (1980) and Watrous and Wheeler (1981) have shown that, within a lineage, all advanced states which group a plurality of taxa will always be misinterpreted as being primitive when this criterion is used.

Several assumptions are inherent in Hennig's methods: (*1*) The group must be monophyletic, (*2*) An outgroup must be known, (*3*) Characters in which the advanced or apomorphic state is homologous throughout the group must outnumber characters showing homoplasy, and (*4*) The pattern of clades and terminal taxa is one of strictly divergent branching. Because these assumptions are often more difficult to verify or are more frequently invalid in plant groups than in animal groups (Funk, 1981), many botanists are reluctant to attempt Hennigian cladistics.

This paper presents a case study for dealing with these problems as an example for other botanists. Even though reticulate evolution has been treated and discussed in the context of Groundplan/Divergence methods (Wagner, 1980), this paper represents one of the few studies in which a plant group with numerous hybrids is treated by Hennigian cladistics; others include: Bremer and Wanntorp, (1979) and Humphries (1981).

Procedures

The steps to producing a manually drawn cladogram can be summarized as follows: (*1*) Select a monophyletic group, (*2*) Select an outgroup, (*3*) Determine the evolutionary polarities of the character-states, (*4*) List the distribution of character-states in the taxa, (*5*) Group taxa by shared derived character-states.

Monophyly and Outgroups

In *Agastache,* I approached the problem of selecting a monophyletic group by looking for internested groups, each of which is defined by unique characters. In all species of *Agastache* there is the development of transverse intercostal veins over most of the calyx length. This feature is not found in other members of the hypothetically next inclusive group, the tribe Nepeteae, and the occurrences appear to be homologous as evidenced by correlated associated features and the strong similarities of the calyces in these species. Therefore, *Agastache* should be monophyletic. The Nepeteae in turn is characterized by a feature unique in the Lamiaceae. The upper pair of stamens is longer than the lower pair; the reverse is true in all other didynamous labiates. The Nepeteae can be grouped with three other tribes, all of which show a pollen-type, a megagametophyte-type, and an embryo-type which are unique in the Lamiaceae and Verbenaceae. Thus by forming these internested sets of characters and groups, the argument that *Agastache* forms a monophyletic unit is strengthened.

In this study, no unique characters were discovered in sect. *Brittonastrum* which separate it as a unit from sect. *Agastache*. This presents a dilemma as to what should be chosen as the outgroup—sect. *Agastache* or the whole Nepeteae. If sect. *Agastache* is chosen, the close morphological sequence of the extant species and strong similarity in character associations greatly strengthen the treatment of homologies. The small number also allows for a detailed study of character-distributions. However, if the exclusion of sect. *Agastache* from sect. *Brittonastrum* makes the latter paraphyletic rather than monophyletic, then the polarities of the characters may be erroneous. On the other hand, if the Nepeteae is adopted, the study is further beset with two very serious problems. First the number of taxa increases to above 200, making an in-depth analysis of character-distributions time consuming. The second point is crucial and is exactly why many botanists are reluctant to attempt Hennigian cladistics (Funk, 1981). Because of the simpler growth form and development in plants, homoplasious character-states become increasingly common with comparisons of species which are more and more distantly related beyond the limits of single genera. That is, the presumption of homology becomes increasingly tenuous. Therefore, in this study a compromise was followed. Section *Agastache* was treated as a provisional outgroup and polarities were assigned. The polarities of these characters were then evaluated through a survey of the Nepeteae representing most genera and about one-tenth of the species of the tribe.

Criteria for Determining Polarities of Character-States

Outgroup comparisons support most decisions of apomorphy (Eldredge and Cracraft, 1980; Watrous and Wheeler, 1981). Morphological measurements were made on rehydrated dried specimens. Thirty-four characters were chosen for their intraspecific invariant nature. Twenty-four of these were uniform in sect. *Agastache* and were assigned polarities by comparison with species of sect. *Agastache* and a representative sample of the species of the Nepeteae. These 24 characters were used in the construction of a primary cladogram.

The ten remaining characters varied within both *Agastache* sect. *Agastache* and the Nepeteae at large, yet were consistent within species-groups as delimited in the primary cladogram. Functional outgroups (Watrous and Wheeler, 1981) were designated within sect. *Brittonastrum,* such that a lineage invariant in the character was chosen as the outgroup of a sister-lineage in which the character did vary.

Ontogeny was also used as a criterion in determining the polarity of leaf characters. The earlier stages in heteroblastic sequences were accepted as the plesiomorphic (primitive) states. Because of the difficulty of determining an outgroup for sect. *Brittonastrum,* ontogenetic analysis, which can often be conducted independently of outgroups (Crisci and Stuessy, 1980), would be desirable, especially for certain problematic characters which are discussed below. However, there has been insufficient time for a detailed developmental anatomical study in *Agastache*.

The characters, their states and polarities are provided in Appendix I.

Taxa of Hybrid Origin

Before compiling the data, it is necessary to consider the cladistic assumption of divergence. Strong morphological, chromosomal, and chemical evidence suggest that hybridization between sympatric species has occurred and that selection of some of the hybrids on the diploid level has produced distinct taxa of hybrid origin (Sanders, 1979). Because cladistic analysis depicts strictly divergent evolution, the taxa of putatively hybrid origin were excluded from the cladistic analysis but were inserted on the final phylogenetic diagram as reticulations. Wagner (1980) has also accepted this approach, and Funk (1981) discusses the theoretical basis of this technique more fully.

Lest it seem that I am imagining hybrids, my operational species concept should be stated explicitly. For practical reasons the biological species concept is untenable. Rather, I accept a species as a population or series of populations that are morphologically discontinuous with other populations but of which the constituent elements are morphologically continuous. This is based conceptually on genetic discontinuities between species. Occasional natural hybrids are common in plants (Grant, 1971) but do not obscure these discontinuities. Even though species of hybrid origin are intermediate between other species, their own populations vary continuously amongst themselves and discontinuously with the parental species. For example, *Agastache breviflora* appears to have been selected from hybrids between *A. mearnsii* and *A. wrightii.* Natural hybrids between it and the putative parent, *A. mearnsii,* have been found but are sterile.

The major evidence of hybrid origin of these various taxa is morphological intermediacy. Lacking direct evidence of hybridity by re-creation, I also used the following evidence: the highly derived nature of much of the intermediacy, greater intrapopulational variation, greater interpopulational segregation, transgressive and developmentally abnormal characters (genetic disharmony), chemical intermediacy and intermediate and transgressive types of habitats. Experimental studies have confirmed that these conditions accompany hybridization (Grant, 1953, 1963, 1971; Lewis and Epling, 1959; Wagner, 1969).

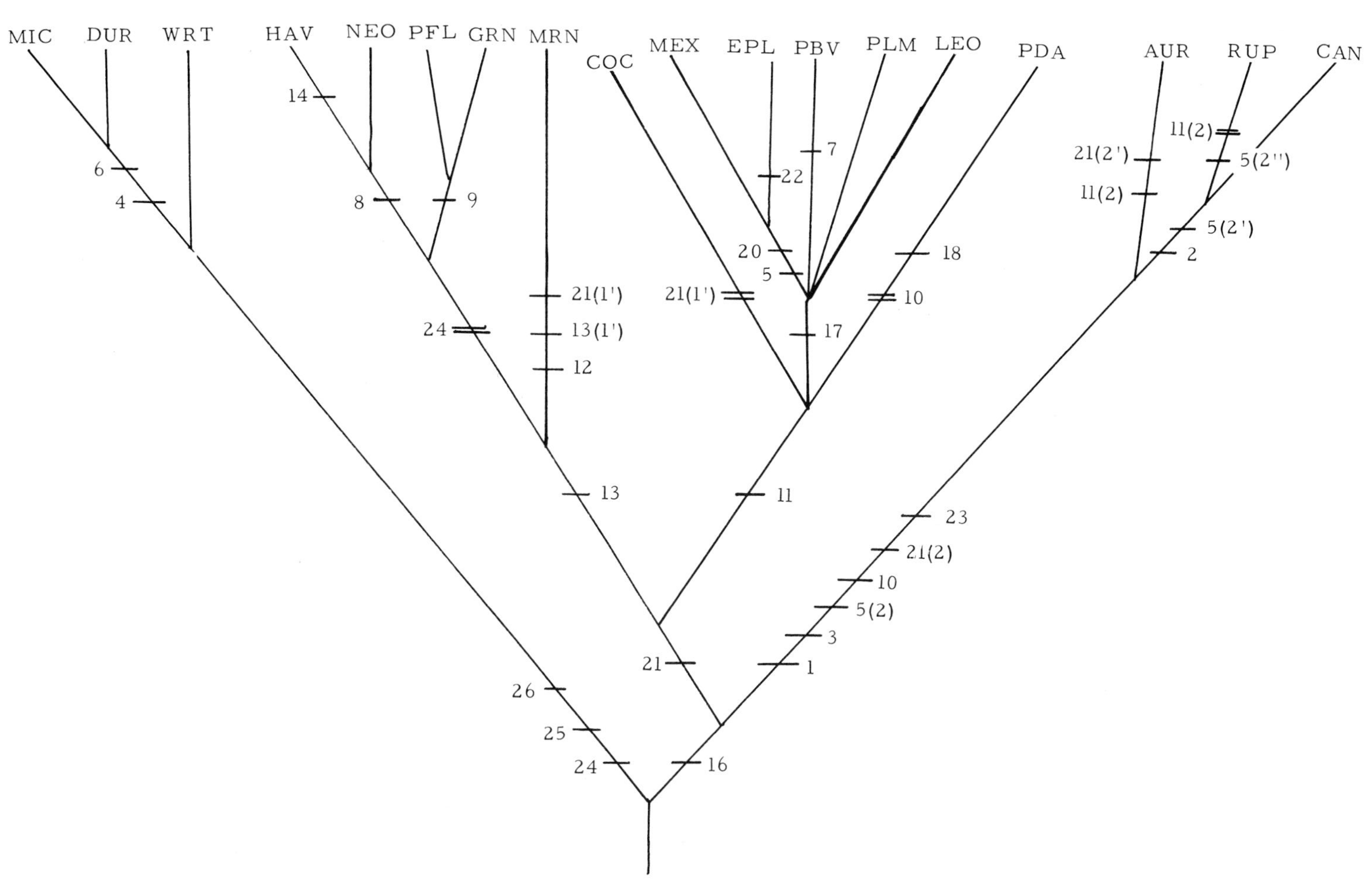

Fig. 2. Primary cladogram. Single bars indicate apomorphies; parallel bars indicate parallelisms.

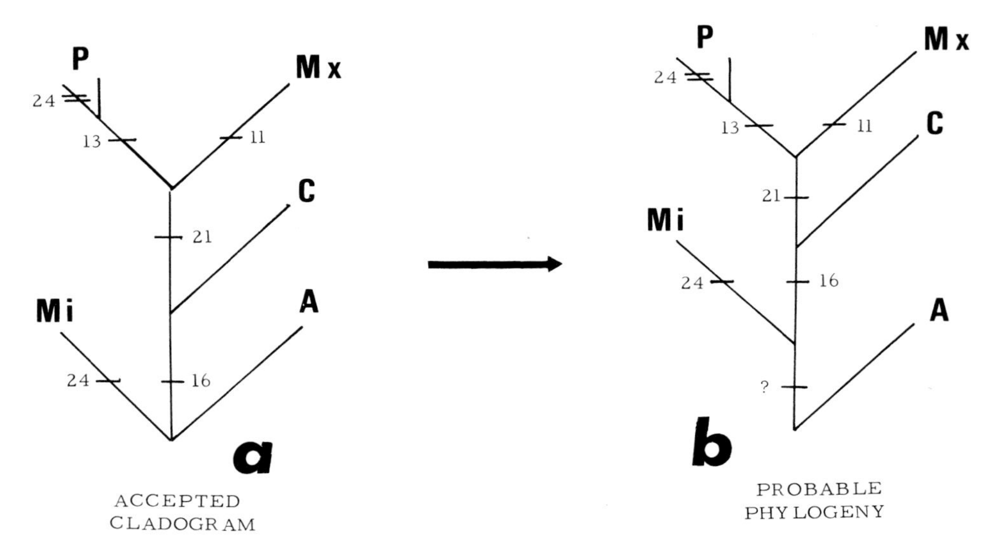

Fig. 3. Simplification of Figure 2 with insertion of sect. *Agastache*. Indicated characters are those whose polarities are suspect. A = sect. *Agastache*, C = *cana*-complex, Mi = *micrantha*-complex, Mx = *mexicana*-complex, P = *pallidiflora*-complex. a. Most highly resolved cladogram for these groups. b. Probable phylogeny of the genus based on auxiliary evidence of geography and hybridization.

The names and acronymic codes of the divergent and hybrid taxa are listed in Appendix II.

Construction of the Cladogram

Distributions of the character-states in the taxa are tabulated in Appendix III. Character-states are assigned integral values; character-states with different integral codes are considered independently evolved. Those with a primed (′) value are considered to form transformation series. From the compiled data matrix, taxa with the largest number of synapomorphies (shared derived character-states) were grouped manually. In the data set, a new character distribution (of their hypothetical ancestor) of only those states shared by the two were substituted for each pair. The grouping and drawing of ancestral character distributions were repeated until all branches were joined.

Results

Primary Resolution of Taxa

Using the 24 characters which were assigned polarities based on sect. *Agastache* as the outgroup (Nos. 1–26 Nos. 15 and 19 were excluded late in the study because of their nearly continuous ranges of states and ambiguous distributions in the taxa), the cladogram in Figure 2 was obtained. Four major clades or lineages appear and are tentatively designated as the "*micrantha*-," "*pallidiflora*-," "*mexicana*-" and "*cana*-complexes," respectively. However, for a balanced evaluation of this hypothesis, it should be remembered that the known cladistic characters in the genus do not resolve sect. *Agastache* and sect. *Brittonastrum* into unequivocal sister sections (Fig. 3a). We must consider the effect on the

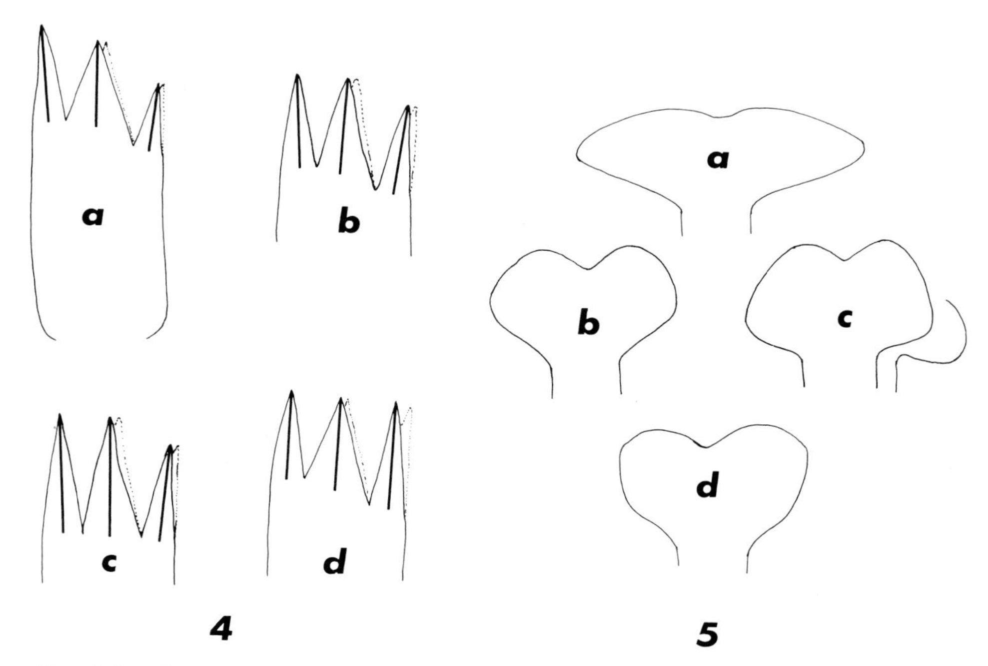

Figs. 4–5. Illustration of controversial characters. 4. Character 16, ratio of length of upper calyx teeth (3) to length of lower calyx teeth (2). Bars indicate relative lengths of calyx teeth. a. State in *micrantha*-complex and in sect. *Agastache*. b. State in *cana*-, *mexicana*-, and *pallidiflora*-complexes. c. State in sect. *Agastache*. d. State in *pallidiflora*-complex. 5. Character 21, shape of median lobe of corolla. a. State in sect. *Agastache*. b. State in *micrantha*-complex. c. State in *mexicana*- and *pallidiflora*-complexes. d. State in *cana*-complex.

cladogram of re-evaluating the character-state polarities with the survey of the Nepeteae. The previously determined plesiomorphic states for most characters occur uniformly in the Nepeteae. Apomorphic states of characters appear sporadically in other genera besides *Agastache,* but based on comparative morphological studies these multiple expressions clearly are not homologous. On the other hand, characters 11, 13, 16, 21, and 24, which are plotted in Figure 3a, appear repeatedly in the other genera. More detailed comparisons only weakly support reversing the directions of characters 11, 13, and 24. Even if the polarities of the latter three are reversed, the branching relationships are only slightly affected.

Much more critical are characters 16 and 21, whose positions affect all major branches of the cladogram and of which the justification for directionality is controversial. Character 16 is the ratio of the length of the lower calyx teeth to the upper calyx teeth (or upper calyx lobes). It is consistently less that 1.0 in sect. *Agastache* and in the *micrantha*-complex and is equal to or greater than 1.0 in the *pallidiflora*-, *cana*- and *mexicana*-complexes (Fig. 4). This character should not be assumed to be homologous throughout the tribe for the following reason. It is variable within the remaining genera of the tribe. As an expression of the degree of fusion of the ancestral calyx lobes which were presumably free, it is most probably subject to simple morphogenetic alteration.

Character 21, the shape of the limb of the medial corolla lobe, is assumed to be primitively transversely rhombic or elliptic as in the *micrantha*-complex (Fig.

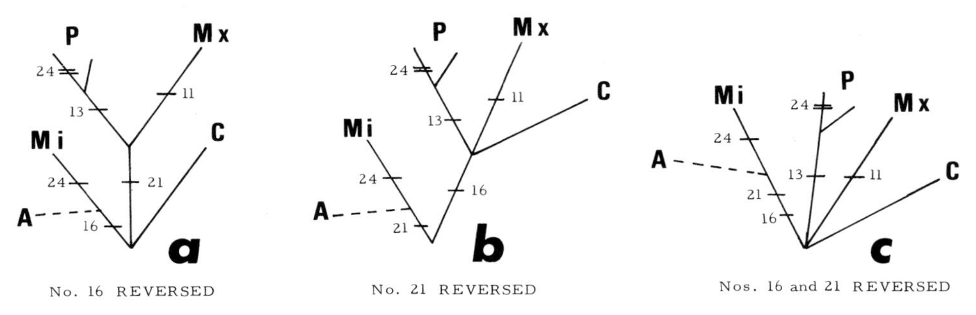

Fig. 6. Alternate cladograms based on reversal of characters. a. Character 16 reversed. b. Character 21 reversed. c. Characters 16 and 21 reversed.

5). Essentially all members of sect. *Agastache* are very broadly transversely elliptic, and this is interpreted as a sequential transformation of the former less extreme character-state. However, transversely elliptic lobes are otherwise rare in the Nepeteae, while the typical shape is semicircular or trapezoidal and elongately two-lobed as in the *pallidiflora*- and *mexicana*-complexes.

The reduced cladogram (Fig. 3a) is altered in Figure 6 by reversing the polarities of characters 16 and 21 individually, 16 and 21 jointly, and in Figure 7 by reversing all five equivocal characters. The result of reversing the trends of characters 16 and 21 is that all or most of the species-complexes arise simultaneously in unresolved tri- or tetrachotomies. Figure 7 consists of equally parsimonious cladograms both of which exhibit two parallelisms instead of one as in Figures 3a and 6. Moreover, they depict an unlikely history in that the *micrantha*- and *cana*-complexes, the two most divergent and dissimilar groups, arise from a common ancestor separate from the more generalized *pallidiflora*- and *mexicana*-com-

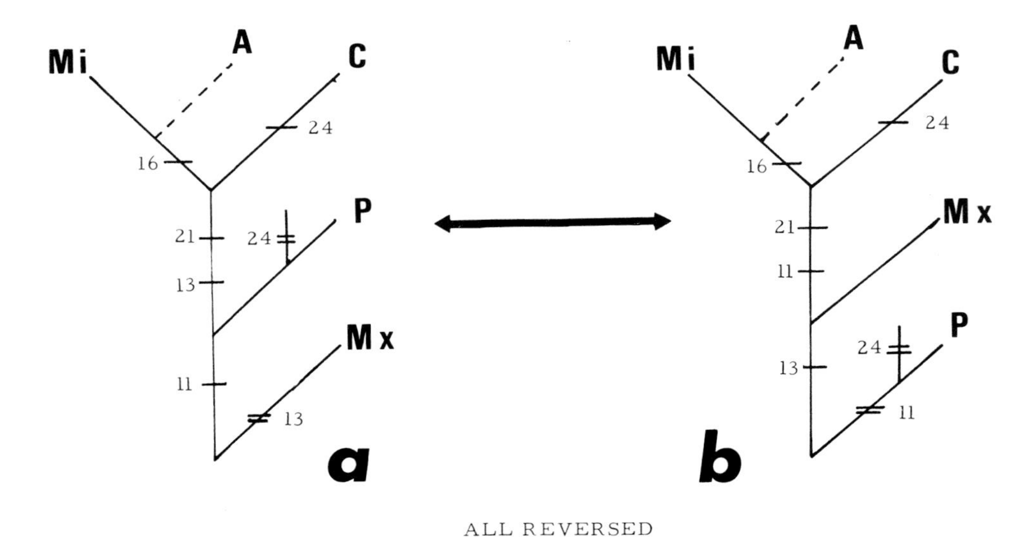

Fig. 7. Alternate cladograms based on reversal of characters 11, 13, 16, 21, and 24. a. Rooted with *mexicana*-complex. b. Rooted with *pallidiflora*-complex.

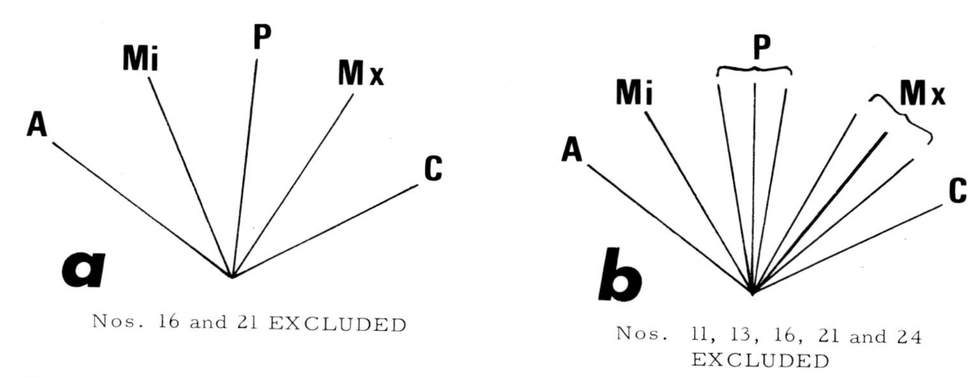

Fig. 8. Alternate cladograms in which polarity of characters is based exclusively on Nepeteae as outgroup. a. Only characters 16 and 21 excluded. b. Characters 11, 13, 16, 21, and 24 excluded.

plexes. When sect. *Agastache* is interpolated onto the diagrams, it always arises as a sister-group to the *micrantha*-complex within sect. *Brittonastrum*.

If the more extreme view is taken that the Nepeteae should be adopted directly as the outgroup, then characters 16 and 21 would certainly have to be excluded from consideration due to their wide variation throughout the tribe. Characters 11, 13, and 24 might be dropped on similar grounds. The results of these approaches are shown in Figure 8a and b. One could argue that Figure 8a provides the objectively most highly resolved cladogram, but Figure 8b is essentially unresolved and contains so little information as to be almost worthless.

At this point, let us consider certain extraneous evidence which suggests that Figure 3a is the most highly resolved cladogram. First, the allopatry of the two sections suggests distinct histories from the time of divergence (Fig. 1). This does not give us any absolute method of choosing between the alternative cladistic placements of sect. *Agastache*; however, the chances that this distribution would be produced by the history described above are much greater than if the *micrantha*-complex gave rise to sect. *Agastache*. Moreover, the *micrantha*-complex is adapted for a specialized oak-savannah habitat in the ecotone between the subtropical desert and the dry montane forests. Section *Agastache*, on the other hand, occupies a range of habitats in the temperate zone, including montane conifer-deciduous forests, coastal rainforests and the Eastern deciduous forest. These habitats approach those which are generalized for sect. *Brittonastrum*, namely montane conifer and conifer-deciduous forests. Secondly, the genomes of the two sections seem to have diverged. Herbalists and nurserymen have produced intrasectional hybrids in both sections but have produced no documented hybrids between the sections (Sanders, pers. obs.). Within sect. *Brittonastrum* sympatric contact of any two species has usually resulted in, at least, sporadic hybrids despite the morphological differentiation between some of these species. One could argue that sect. *Agastache* originated from the *micrantha*-complex and diverged at a much greater rate, especially in evolving breeding barriers. But this argument disagrees with the observation that between closely related allopatric taxa, ethological barriers are rarely or only poorly developed (Grant, 1971, Ch. 10).

Because the analysis is based both on the presumed sister-group (which might possibly be an ingroup) and on representatives of the absolute outgroup, the directionalities of all characters except 16 and 21 are considered to be strongly supported. The additional biological evidence suggests as well that the direction-

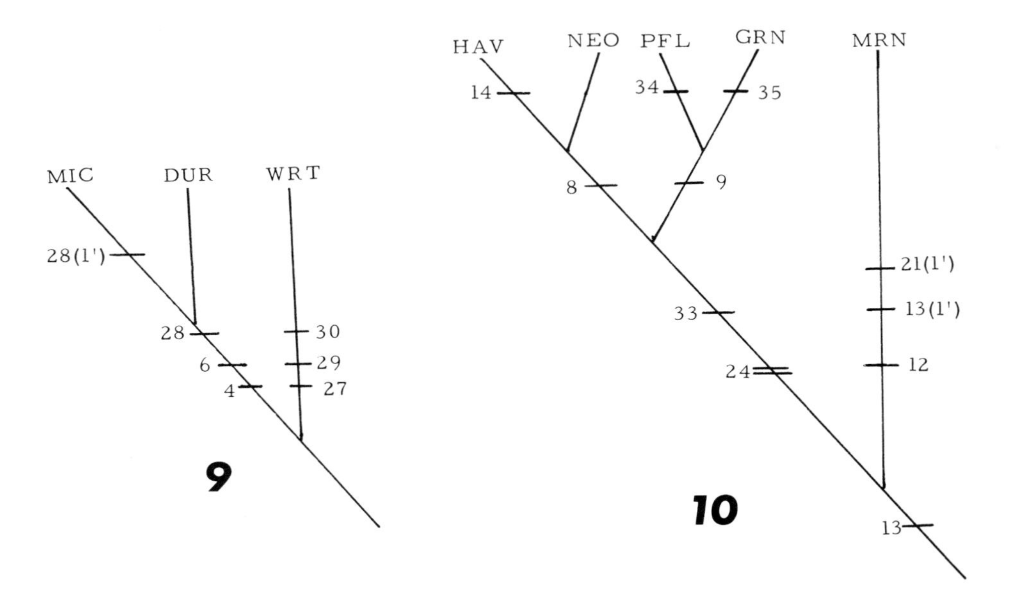

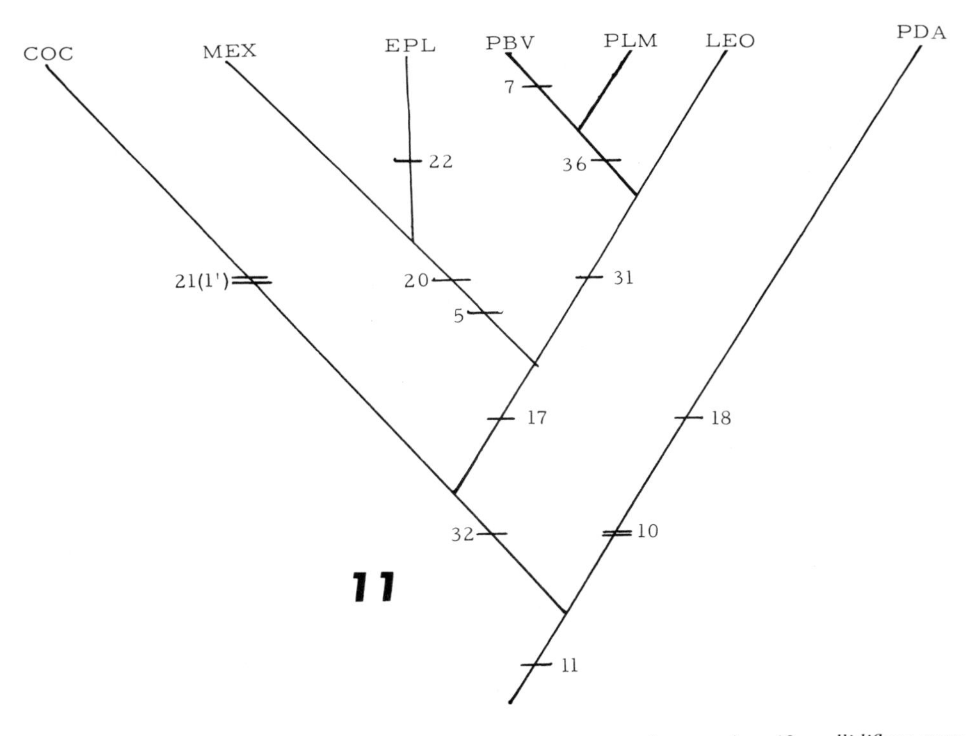

Figs. 9–11. Secondary resolution of the complexes. 9. *micrantha*-complex. 10. *pallidiflora*-complex. 11. *mexicana*-complex.

alities of the remaining two characters are correctly hypothesized. Therefore, Figure 3a is considered sufficiently strongly supported to be the most highly resolved cladogram. Furthermore, Figure 3b represents a probable phylogeny which can be deduced from the cladistic study, the geography and genetics and represents hypothetical chronistic, as well as cladistic relationships. An in-depth study of technical characters in the Nepeteae is planned and it is hoped that one or more synapomorphies uniting the lineages of sect. *Brittonastrum* will be discovered.

Ultimate Resolution

To determine the polarity of characters 27–36, appropriate outgroups were selected from within the primary cladogram. A strict application of the outgroup criterion was sufficient to polarize all characters except three. Character 28, flower size, has not been employed throughout sect. *Brittonastrum* due to extreme plasticity in certain of the large-flowered species (Sanders, 1979). In the taxa of the *micrantha*-complex, corolla length is stable and is shorter than in any other taxon of the genus. Since in this complex there is a range of rather small to very small or minute flowers, this is interpreted as a transformation series from an ancestor with moderately or "average-"sized flowers.

Character 29, corolla color, has not been employed at large because differences in tint are often subtle (e.g., yellow-orange vs. red-orange vs. carmine, etc.) and because the chemical basis of these differences is unknown. In the *micrantha*-complex, only white and blue or blue-violet floral colors occur. White is simply the loss of pigment (blockage of sequential enzymatic reactions), and double occurrences can't be judged for homology. In the outgroup, blue doesn't occur; white occurs only sporadically in *A. pallidiflora*. Harborne (1977) argues that the magenta or purple (cyanidin) is more primitive, being found throughout many families. Blue tints result from the complexing of metal ions or flavone moieties with the anthocyanin and are thus more advanced. Therefore, white is grouped with purplish as plesiomorphic to blue-violet.

Character 36, hair length on the undersurface of leaves of *Agastache palmeri*, separates *A. palmeri* var. *palmeri* and var. *breviflora* from var. *leonensis*. The apomorphic state of this character was the only one discovered which groups any two of the three varieties. The shorter length is judged to be plesiomorphic because the state in the outgroup (*Agastache mexicana* and *A. eplingiana*) is shorter than any state in *A. palmeri*.

The final cladogram incorporates the independent resolution of the *micrantha*-complex (Fig. 9), the *pallidiflora*-complex (Fig. 10) and the *mexicana*-complex (Fig. 11), as well as the taxa of hybrid origin, which are added as reticulations (see Fig. 12).

It can be argued that because we lack positive proof of the hybrid nature of these taxa, they should be left in the original analysis (Parenti, 1980; Nelson and Platnick, 1980). This has been done in Figure 13. These interclade-hybrids cannot form terminal trichotomies as predicted by these authors, but they do increase disproportionally the number of homoplasies. The suggestion that these truly are parallelisms and reversals counters the principle of parsimony and disregards the patterns of hybridity which are so well documented in many plant groups. Thus, I reject the notion that the information of Figure 13 equals or exceeds that in Figure 12.

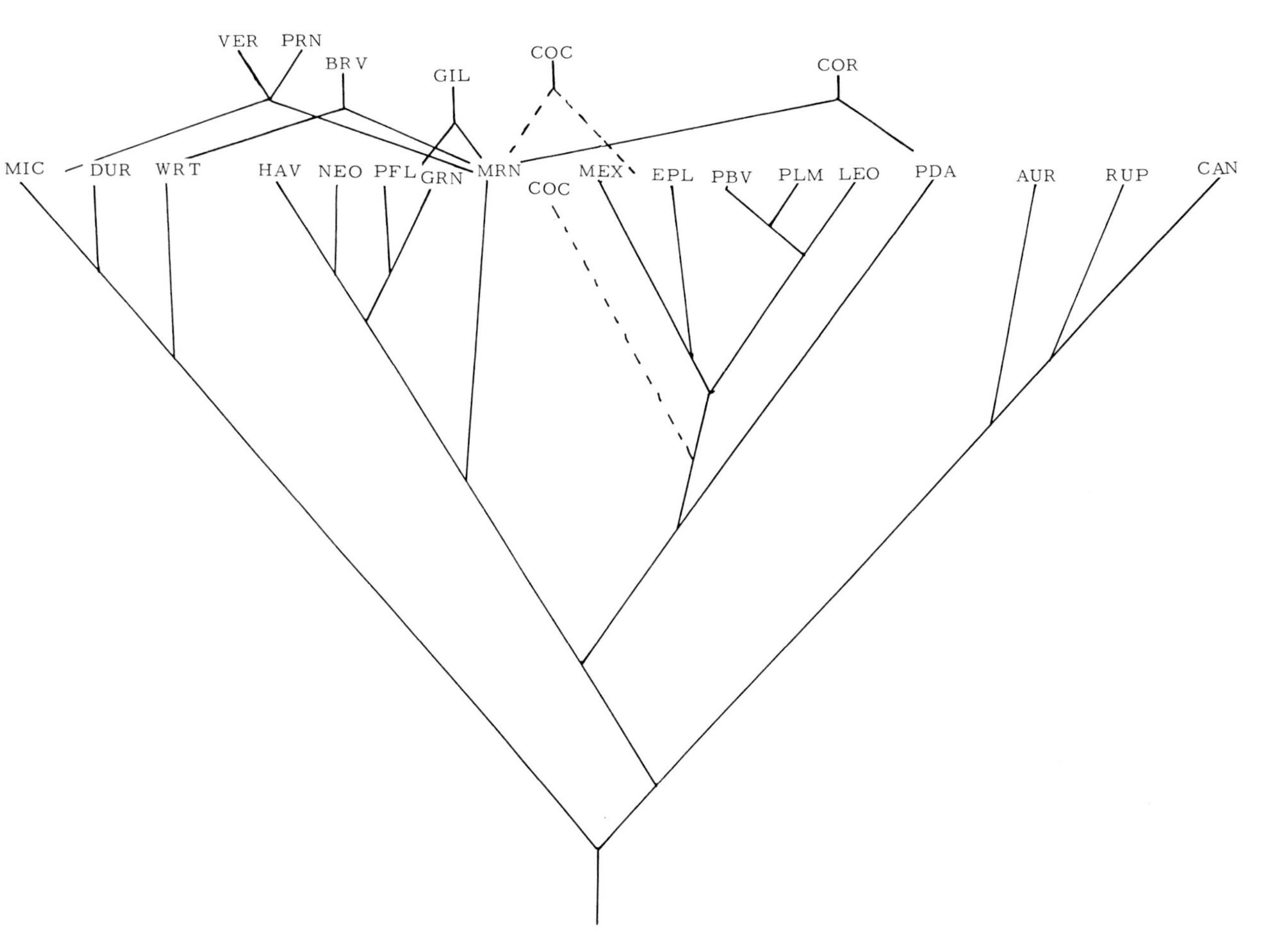

Fig. 12. Final cladogram. Dashed line indicates uncertain placement. Anastomoses indicate taxa of hybrid origin.

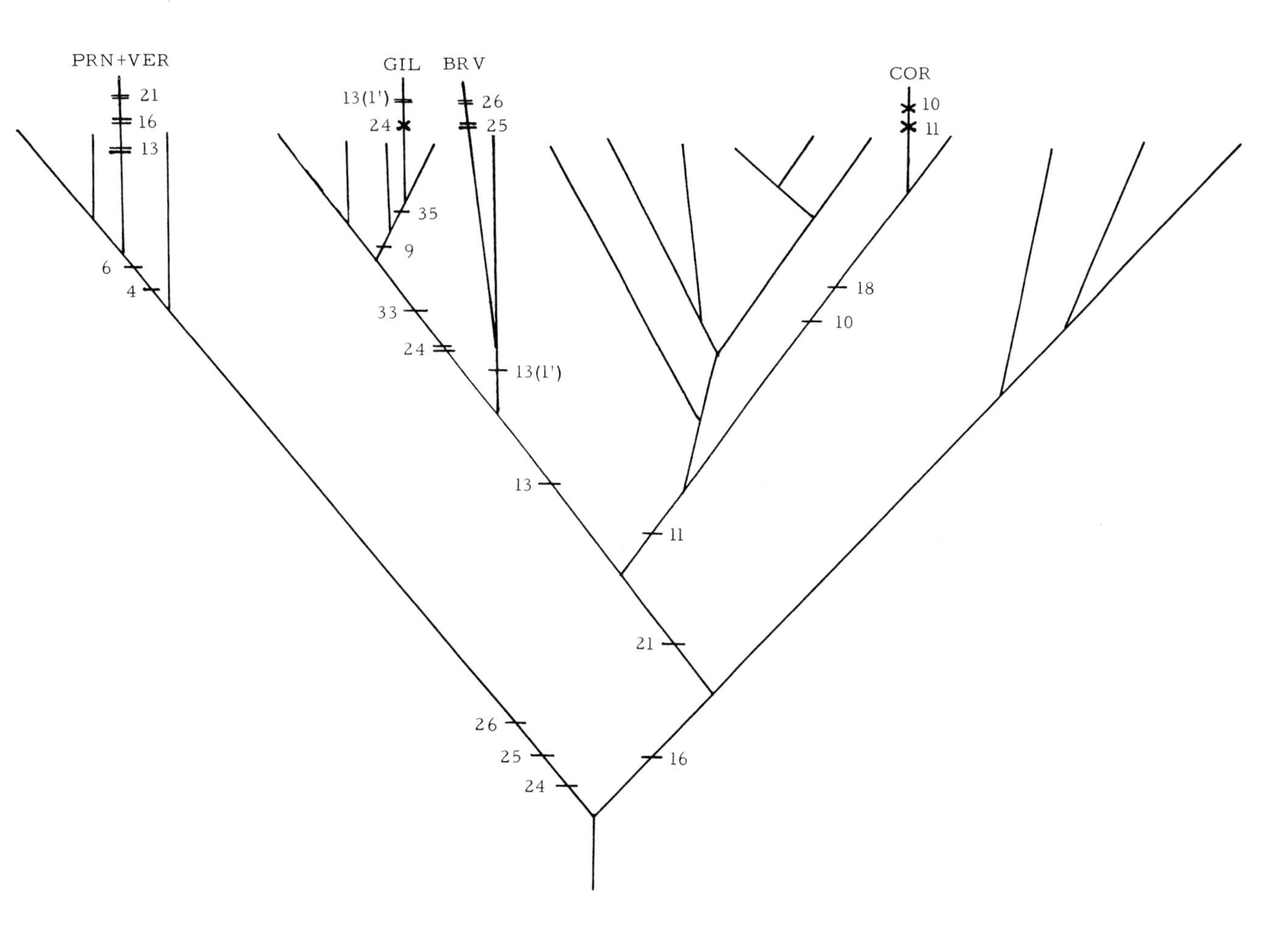

Fig. 13. Final cladogram with cladistic insertion of taxa of hybrid origin in the most parsimonious positions. Crosses indicate reversals.

Discussion

Taxonomic Conclusions

Cladistic analysis clearly shows there to be four independent or nearly independent lineages in sect. *Brittonastrum*. However, the recognition of these as formal taxonomic series is complicated by the taxa of hybrid origin between lineages. There exist three possible treatments which are artificial to various degrees.

Each hybrid could be recognized as a distinct series since each has its own distinct origin. However, the parental and hybrid series would be nonequivalent, the former could be interpreted as paraphyletic, and some hybrids are not distinct at the species level. This would require elevating parts of species to the series level.

Series need not be recognized since the lineages could be interpreted as being genetically too similar. However, this would obscure the distinctiveness and extreme morphological divergences exhibited by different lineages.

In the process of selection and speciation, *Agastache pringlei* and *A. breviflora* each have segregated toward a parental species, such that they are not perfectly intermediate. Moreover *Agastache pallida* and *A. pallidiflora* each possess varieties (geographic races) which are also of hybrid origin. Therefore, the hybrids may be grouped taxonomically with the lineage to which they bear the greatest similarity and for which a coherent taxonomic description can be made to include both divergent and hybrid taxa. It may be argued that the recognized taxonomic series are not strictly monophyletic; but neither are they truly para- or polyphyletic. This solution is practical, comes closest of the three to depicting the situation occurring in nature and was adopted in a taxonomic treatment of *Agastache* sect. *Brittonastrum* (Sanders, 1979).

One species, *Agastache coccinea,* remains problematic. Several of its characters including cladistic character No. 21 (1′), occurs in parallel with *A. mearnsii*. Other characters (not included in the data set) suggest its intermediacy between *A. mearnsii* and the *mexicana*-complex; *A. coccinea* is also geographically intermediate. Yet it possesses several distinctive apomorphies (also not included in the data set). Whether this species is exhibiting parallelisms or is representing an ancient hybrid can't be answered by the available data. Both possibilities are represented in the final cladogram.

Congruence of Data

The greatest power of cladistic analysis is not that it is operational but that it provides us with a predictive hypothesis. If characters and polarities are chosen with discretion, the cladogram should be a very good approximation of the evolutionary history of the group. New systematic data should also show the same phylogenetic patterns and should show complete or nearly complete congruence with the cladogram. Indeed, we can predict unknown distributions of data.

As an example, the taxonomic distribution of the flavonoids of *Agastache* sect. *Brittonastrum* has been determined (Sanders, 1979). Three drawbacks in determining polarities of the flavonoid data prevented the use of these data in the original cladistic analysis. The chemistry of the outgroups, sect. *Agastache* and the tribe Nepeteae, is too poorly known. The biochemical pathways of the flavonoids in these plants are unknown. Moreover, most of the taxa differ from the basic profile by the absence of one or more typical compounds; the absence of data cannot be tested for homology. However, by plotting the flavonoid data on the cladogram, a phylogenetically congruent pattern is obtained (Sanders, 1980).

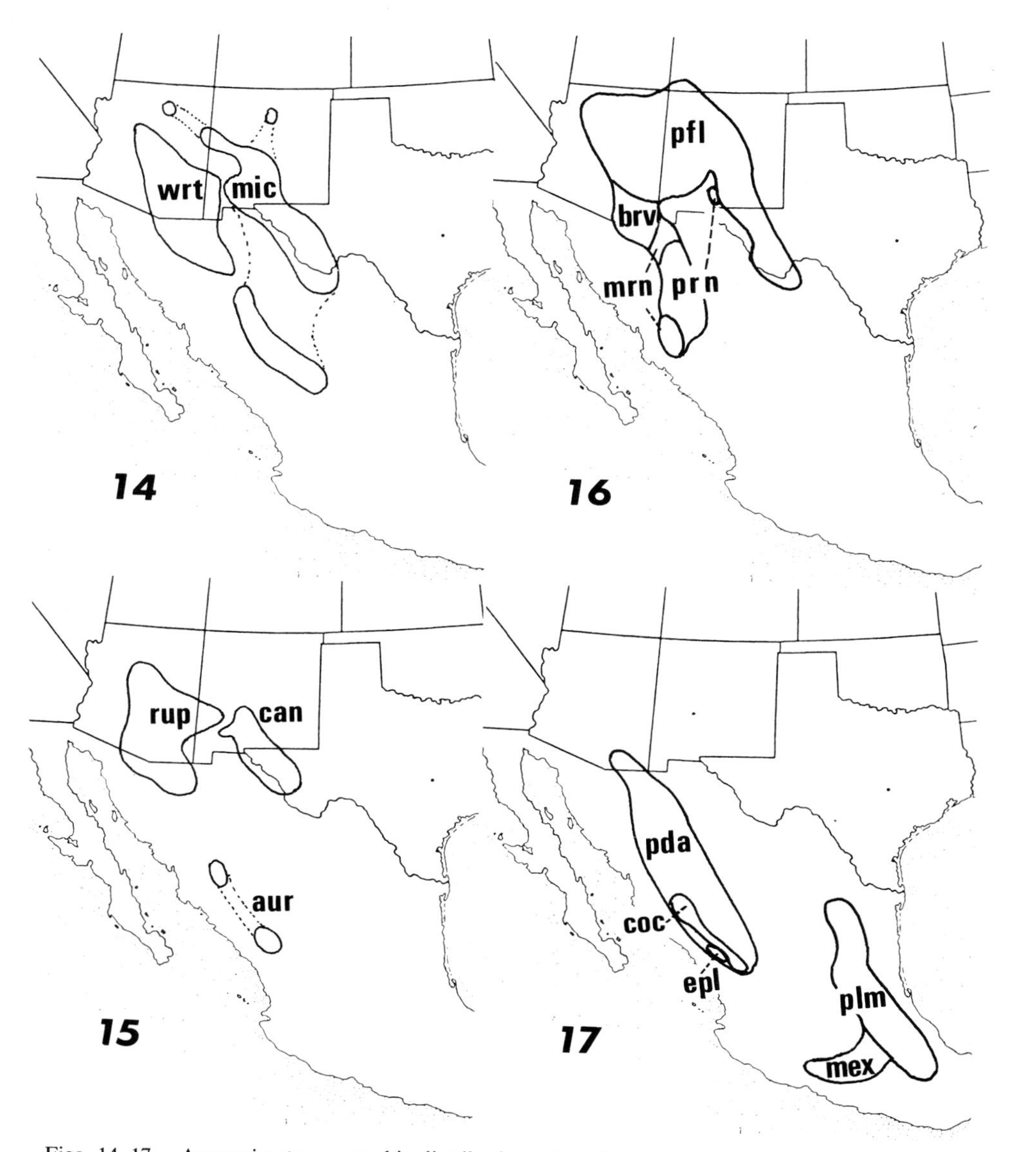

Figs. 14–17. Approximate geographic distributions of clades and associated taxa of hybrid origin. Only the nominal acronym is given for species with infraspecific taxa. 14. *Micrantha*-complex. 15. *Cana*-complex. 16. *Pallidiflora*-complex. 17. *Mexicana*-complex.

The geographic distribution of the clades (and associated taxonomic series) is coherent and congruent with the cladogram (Figs. 14–17). The first two dichotomies of the cladogram produce three clades which are geographically sympatric, but which occupy basically different ecological zones (Sanders, 1979). The third dichotomy produces a clade (*mexicana*-complex) which is largely allopatric but ecologically equivalent to its sister group.

Cladistic analysis also gives us insights into possible modes of speciation of certain species-pairs. The terminal species-pairs of both the *micrantha*- and *cana*-

complexes are geographically distributed in parallel; i.e., *A. micrantha* is sympatric with *A. cana* in the Chihuahuan Desert Region and *A. wrightii* is sympatric with *A. rupestris* in the Sonoran Desert Region (Figs. 14 and 15). Speciation in both pairs probably was vicariant having been governed by the same geologic and climatic factors.

Overall Conclusions

As numerous zoological papers have shown, Hennigian cladistics can resolve many phylogenetic and taxonomic problems. Botanists have been slow to follow this lead due to their understanding that the morphology and development of plants are under simpler genetic controls than in most animal groups. Thus, monophyletic groups and outgroups are harder to delimit. Often a hierarchically much more inclusive outgroup must be used to prevent using a mistakenly chosen ingroup as an outgroup. However, at this level the homology of character-states is more frequently invalid. Moreover, hybrids and taxa of hybrid origin more commonly persist in plants.

As seen in this paper, a modification of techniques can allow one to circumvent these problems. Here, hybrid elements were not allowed to produce a myriad of polytomies or homoplasious apomorphies. In the resulting phylogenetic reconstruction, the monophyletic species complexes are evident, and the relationships among all the divergent taxa are clarified. Furthermore, the distributions of other characters, especially chemical and geographic, corroborate the cladistic analysis and support its validity.

Acknowledgements

This research was supported by the University of Texas Graduate School and the National Science Foundation [Dissertation Improvement Grant DEB 78-18895 (to R. W. S.) and Grant DEB 76-09320 (to T. J. Mabry and B. L. Turner)]. The manuscript was improved through discussions with Robert Jansen, Vicki Funk, Daniel Brooks, and Larry Watrous. Collecting was permitted by the Victorio Co., Phoenix, Arizona; Fort Bliss, Texas; and the Big Bend and Guadalupe National Parks.

Note added in proof: In continued studies an additional charcter, the relative fusion of the corolla lobes, has been discovered which possibly could serve as a synapomorphy for the species in sect. *Brittonastrum*. This will be discussed in a revision of *Agastache* sect. *Brittonastrum*.

Appendix I. Characters and their states used in the cladistic analysis

Code	Character	Plesiomorphic	Apomorphic
		(Sect. *Agastache* as outgroup)	
1	Habit	herb	suffrutex
2	Stem insertion	simple	strongly fascicled
3	Branching pattern	strict or fastigiate	divergent
4	Leaf texture	herbaceous	membranous
5	Leaf type	ovate, crenate, hirtellous	(1) lanceolate, basally crenate, apically entire, hirtellous (2) lance-oblong, basally crenate, apically entire, tomentose (2′) ovate, basally hastately toothed to entire, finely incanous (2″) linear, entire, incanous
6	Leaf teeth, shape	antrorsely bluntly crenate	acutely retrorsely crenate
7	Inflorescence internode length	= or > than calyx length	⩽ calyx length
8	Bract size	= or > than calyx, evident	< calyx, inconspicuous
9	Development of outer bracteoles	similar to inner bracteoles	those of outer 1–3 nodes much larger than inner ones
10	Ratio bracteole length : cyme-internode length	1.5 or greater	1.0 or less
11	Calyx tube shape	campanulate	(1) tubular (2) barrel-shaped
12	Course of calyx costae	curving or flexuous	rigid, straight, giving fluted aspect to calyx
13	Prominence of costae	thin, not prominent	(1) thick, prominent (1′) cartilaginous, differentiated from lamina
14	Calyx coloration	green or tinged magenta	bluish or blue-lavender
15	Omitted		
16	Ratio upper calyx teeth : lower calyx teeth length	ca. 0.7	1.0–1.5
17	Upper calyx teeth, degree of connation	<40°	>40°
18	Calyx teeth apex	bifacial	unifacial
19	Omitted		
20	Corolla, lateral lobe shape	as broad or broader than long, obtuse	longer than broad, acute
21	Corolla, median lobe shape	transversely rhombic or elliptic	(1) semicircular, transversely ovate or reniform (1′) trapezoidal, obtrapezoidal (2) transversely obovate (2′) strongly bifurcate

Appendix I. Continued

Code	Character	Plesiomorphic	Apomorphic
22	Corolla, median lobe margin	entire to moderately toothed	lacerate
23	Ratio style-branch length : anther length	1.0–1.5	ca. 0.7
24	Ratio thickness of calyx intercostal veins : thickness calyx costae	1/4	ca. 1/2
25	Calyx teeth, intramarginal veins, development	well developed, at least in lower teeth	absent or only as thin, short free-ends
26	Upper stamens, filament length	exserted beyond upper corolla lip	included under upper corolla lip
	(Characters in *micrantha*-complex, remainder of sect. *Brittonastrum* as outgroup)		
27	Calyx teeth pigmentation	chlorophyllous	scarious, achlorophyllous
28	Flower size	moderate	(1) small (1′) very small
29	Corolla color	white or purple	blue-violet
30	Calyx teeth shape	triangular	oblong-acute
	(*Pallidiflora*-complex, COC and PDA as outgroup to MEX, EPL, PLM, PBV, & LEO)		
31	Upper calyx teeth, degree of overtopping lower calyx teeth	weakly or moderately	strongly, far exceeding
	(*Cana*- or also *micrantha*-complex as outgroup to *mexicana*- and *pallidiflora*-complexes)		
32	Calyx texture, in fruit	indurate	thin, herbaceous or membranous
33	Calyx symmetry	symmetrical	oblique-asymmetrical
	(*Mexicana*-complex as outgroup to *pallidiflora*-complex)		
34	Internodes of cymal branches	developed	reduced, obsolete
35	Upper calyx lip shape	flat	incurved, convex
	(MEX and EPL as outgroup to PLM, PBV, & LEO)		
36	Hairs, leaf undersurface, length	<0.15 mm	>0.15 mm

Appendix II. Species and varieties of *Agastache* sect. *Brittonastrum*

Taxa	Codes
Taxa of presumed divergent origin	
A. aurantiaca (A. Gray) Lint & Epling	AUR
A. cana (W. J. Hooker) Wooton & Standley	CAN
A. coccinea (Greene) Lint & Epling	COC
A. eplingiana R. Sanders	EPL
A. mearnsii Wooton & Standley	MRN
A. mexicana (HBK) Lint & Epling	MEX
A. micrantha (A. Gray) Wooton & Standley	
var. *micrantha*	MIC
var. *durangensis* R. Sanders	DUR
A. pallida (Lindley) Cory	
var. *pallida*	PDA
A. pallidiflora (Heller) Rydberg	
ssp. *pallidiflora* var. *pallidiflora*	PFL
ssp. *pallidiflora* var. *greenei* (Briquet) R. Sanders	GRN
ssp. *neomexicana* (Briquet) Lint & Epling	
var. *neomexicana* (Briquet) R. Sanders	NEO
ssp. *neomexicana* var. *havardii* (A. Gray) R. Sanders	HAV
A. palmeri (B. L. Robinson) Lint & Epling	
var. *palmeri*	PLM
var. *breviflora* (Regel) R. Sanders	PBV
var. *leonensis* R. Sanders	LEO
A. rupestris (Greene) Standley	RUP
A. wrightii (Greenman) Wooton & Standley	WRT
Taxa of presumed hybrid origin	
A. breviflora (A. Gray) Epling	BRV
A. pallida var. *coriacea* R. Sanders	COR
A. pallidiflora ssp. *pallidiflora*	
var. *gilensis* R. Sanders	GIL
A. pringlei (Briquet) Lint & Epling	
var. *pringlei*	PRN
var. *verticillata* (Wooton & Standley) R. Sanders	VER

Appendix III. Basic data matrix

Taxa	Characters																																
	1	2	3	4	5	6	7	8	9	10	11	12	13	14	16	17	18	20	21	22	23	24	25	26	27	28	29	30	31	32	33	34	35	36
AUR	1	0	1	0	2	0	0	0	0	1	2	0	0	0	1	0	0	0	2′	0	1	0	0	0										
CAN	1	1	1	0	2′	0	0	0	0	1	0	0	0	0	1	0	0	0	2	0	1	0	0	0										
RUP	1	1	1	0	2″	0	0	0	0	1	2	0	0	0	1	0	0	0	2	0	1	0	0	0										
WRT	0	0	0	0	0	0	0	0	0	0	0	0	0	0	0	0	0	0	0	0	0	1	1	1	1	0	1	1						
MIC	0	0	0	1	0	1	0	0	0	0	0	0	0	0	0	0	0	0	0	0	0	1	1	1	0	1′	0	0						
DUR	0	0	0	1	0	1	0	0	0	0	0	0	0	0	0	0	0	0	0	0	0	1	1	1	0	1	0	0						
PDA	0	0	0	0	0	0	0	0	0	1	1	0	0	0	1	0	1	0	1	0	0	0	0	0						0	0			
COC	0	0	0	0	0	0	0	0	0	0	1	0	0	0	1	0	0	0	1′	0	0	0	0	0						1	0			
MEX	0	0	0	0	1	0	0	0	0	0	1	0	0	0	1	1	0	1	1	0	0	0	0	0					0	1	0			
EPL	0	0	0	0	1	0	0	0	0	0	1	0	0	0	1	1	0	1	1	1	0	0	0	0					0	1	0			
LEO	0	0	0	0	0	0	0	0	0	0	1	0	0	0	1	1	0	0	1	0	0	0	0	0					1	1	0			0
PLM	0	0	0	0	0	0	0	0	0	0	1	0	0	0	1	1	0	0	1	0	0	0	0	0					1	1	0			1
PBV	0	0	0	0	0	0	1	0	0	0	1	0	0	0	1	1	0	0	1	0	0	0	0	0					1	1	0			1
MRN	0	0	0	0	0	0	0	0	0	0	0	1	1	0	1	0	0	0	1′	0	0	0	0	0						0	0	0	0	
PFL	0	0	0	0	0	0	0	0	1	0	0	0	1	0	1	0	0	0	1	0	0	1	0	0						0	1	1	0	
GRN	0	0	0	0	0	0	0	0	1	0	0	0	1	0	1	0	0	0	1	0	0	1	0	0						0	1	0	1	
NEO	0	0	0	0	0	0	0	1	0	0	0	0	1	0	1	0	0	0	1	0	0	1	0	0						0	1	0	0	
HAV	0	0	0	0	0	0	0	1	0	0	0	0	1	1	1	0	0	0	1	0	0	1	0	0						0	1	0	0	
PRN	0	0	0	1	0	1	0	0	0	0	0/1	0	1	0	1	0	0	0	1	0	0	1	1	0/1	0	0	0	0		0	0	0	0	
VER	0	0	0	1	0	1	0	0	0	0	0/1	0	1	0	1	0	0	0	1	0	0	1	1	0/1	0	0	0	0		0	0	0	0	
BRV	0	0	0	0	0	0	0	0	0	0	0/1	1	1	0	1	0	0	0	1	0	0	0	1	1	0/1	0	0	0		0	0	0	0	
GIL	0	0	0	0	0	0	0	0	1	0	0	0	1′	0	1	0	0	0	1	0	0	0	0	0						0	1	0	1	
COR	0	0	0	0	0	0	0	0	0	0	0	0	0	0	1	0	1	0	1	0	0	0	0	0						0	0	0	0	

A Cladistic Analysis of *Salmea* DC. (Compositae-Heliantheae)

M. R. Bolick

Table of Contents

Introduction

Salmea DC. (Compositae-Heliantheae-Verbesininae) is a small genus of neotropical shrubs traditionally allied with *Salmeopsis* Benth., *Otopappus* Benth., *Notoptera* Urb., and *Spilanthes* Jacq. (Blake, 1915). Recent studies suggest that *Notoptera* should be recombined with *Otopappus* (Anderson et al., 1979; Hartman and Stuessy, 1980). Work by Jansen on *Spilanthes* indicates that this taxon should be split into *Spilanthes* (sensu stricto) and *Acmella* Rich. in Pers. (Jansen, 1979; Jansen and Stuessy, 1980).

A cladistic analysis of *Salmea* using Farris's Wagner '78 algorithm is presented here as part of a revisionary study; the taxonomy will be presented elsewhere (Bolick, In prep.). The major nomenclatural changes given in Bolick and Jansen (1981) are briefly summarized as follows: *Salmea pauciceps* Griseb. is returned to *Salmea* from its placement in *Spilanthes* by Blake (1915); two other Cuban *Spilanthes* are also transferred to *Salmea, S. insipida* (Jacq.) Bolick and Jansen and *S. montana* (Britton and Blake) Bolick and Jansen. The genus *Salmeopsis* is not recognized, its one species *S. claussenii* Benth. being only doubtfully distinct from the widespread *Salmea scandens* (L.) DC.

Intuitively, the species of *Salmea* considered here seem to fall into three groups. The first group consists of *S. scandens,* which is found from northeast Mexico to Argentina, *S. orthocephala* Standl. and Steyerm. found in the Central Depression of Chiapas and in adjacent areas of Guatemala to Honduras, *S. oligocephala* Hemsl. found on the edges of the Central Plateau of Mexico, and *S.*

Table I

Characters of *Salmea pauciceps, S. insipida, S. montana* and *Spilanthes urens*

	Taxa			
Characters	*pauciceps*	*insipida*	*montana*	*urens*
1. Leaves with gold resinous dots	present	present	absent	absent
2. Upper leaf surface always glabrous	yes	no	yes	no
3. Number of capitula	3–6	1–3	1	1
4. Leaf shape	deltoid to lance-ovate	spatulate to lance-ovate	narrowly lanceolate	spatulate to lance-ovate
5. Leaf margin	dentate	dentate to entire	entire	entire
6. Petiolate or sessile	distinct petiole	petiole with decurrent wing	distinct short petiole	sessile, with three veins of blade to stem
7. Internodes	(2.5) 4–5 (12) cm	(0.5) 2–3 (7) cm	0.4–1.0 cm	(0.5) 2–3 cm
8. Bracts on peduncle	present	sometimes present	present	absent
9. Large-tipped glands on corolla, pales, anther appendages	present	present	present	absent
10. Habit	shrub, woody	small sub-shrub, ± woody	shrub, woody	herb
11. Habitat	limestone rocks	forests, sometimes serpentine	serpentine pinelands	wet areas
12. Distribution	Las Villas, Pinar del Río	Las Villas, Matanzas, Habana, Pinar del Río	Pinar del Río	Oriente, Camagüey, Las Villas, Matanzas, Habana, Pinar del Río

palmeri Wats. of Jalisco and Guerrero. The second group consists of the Bahamian endemic, *S. petrobioides* Griseb. The third group is endemic to Cuba: *S. caleoides* Griseb., *S. glaberrima* C. Wright in Griseb., *S. pauciceps* Griseb., *S. insipida* (Jacq.) Bolick and Jansen, and *S. montana* (Britton and Blake) Bolick and Jansen. A cladistic analysis seemed to be the most reasonable way to corroborate or refute the intuitive groupings.

Within the groups of species, I suspect that both hybridization and convergent evolution have played a role in the evolution of the Cuban Salmeas. *Salmea montana,* known only from the type collection, is the most distinctive of the Cuban taxa, showing many features considered typical of serpentine plants. This xeric morphology is shared to some extent with *S. caleoides*. *Salmea caleoides* also shares morphological features with *S. glaberrima*. In some specimens, *S. caleoides* and *S. glaberrima* appear to intergrade while in others they differ considerably in leaf shape and pubescence. The two always differ in flowering time and habitat preference, *S. caleoides* preferring the basal areas of the mogotes where serpentine outcrops and *S. glaberrima* preferring the upper limestone of the mogotes. *Salmea pauciceps* has a leaf shape like that of the non-caleoid specimens of *S. glaberrima* but differs from both *S. glaberrima* and *S. caleoides* in the capitulescence shape and in the presence of gold resinous dots on the leaf surface. The highly variable group of plants called *S. insipida* shares many characters with *S. pauciceps* in combination with characters of a plant like *S. montana* or *Spilanthes urens* Jacq. (Table I).

In the case of *Salmea insipida*, the suspicion of a hybrid origin for the taxon was strengthened by the data from pollen stainability. *Salmea insipida* is represented by very few collections in U.S. herbaria; there are even fewer specimens with sufficient material to permit sampling for pollen stainability. Nine collections were deemed to have enough material so that a few florets could be removed without rendering the specimen useless to future workers. The pollen was stained in cotton-blue-lactophenol; this only crudely approximates true viability but the error is one of underestimating the number of non-viable grains. Five of the nine plants had highly stainable pollen (between 80% and 100%); two plants had somewhat lower stainability (62% and 74%) while the last two specimens showed very few stainable grains and an abundance of aborted tetrads. *Salmea insipida* seems to resemble *S. pauciceps* most closely; *Spilanthes urens* seems to be closer in morphology to a hypothetical other parent than does *Salmea montana*.

To further complicate this picture, the species of *Salmea* and of the other genera in this group of the Verbesininae show few consistent morphological differences (Tables II, III, IV). Even the "microcharacters" of foliar and floral cellular type and structure, seen only with high (400×) magnifications, reveal a remarkable uniformity. Additional data from cytology and phytochemistry are needed to help sort and define the taxa in this group. I hope this preliminary herbarium study and cladistic analysis will encourage someone with free access to Cuba to pursue field studies in the group.

Methods

I found the arguments of Farris and Kluge (1979) in favor of parsimony methods most convincing and have chosen Farris's Wagner '78 algorithm for use. The analysis was performed on an IBM 360 computer.

The degree of variation shown by the specimens of *Salmea insipida* made it difficult to enter these plants onto the data matrix. The specimens were treated as a group showing the maximum divergence from *S. pauciceps* rather than coding each plant separately or emphasizing the similarities to *S. pauciceps*.

One problem that I was unable to resolve arose in trying to use parameters that serve well to distinguish the species but that do not seem amenable to determinations of polarity. Differences in range of elevation and in season of flowering are often directly related to species isolation and separate several species of *Salmea* (a difference in elevational range separates *S. scandens* from *S. orthocephala*; a difference in flowering season separates *S. scandens* from *S. orthocephala*, *S. oligocephala* from *S. palmeri*, and *S. caleoides* from *S. glaberrima*). If only apomorphies (shared derived characters) define taxa, then these characters of reproductive isolation cannot be used. I cannot honestly claim to know what the ancestral elevational range or flowering season was, or that I think outgroup comparison will be of use in this instance.

Twenty-four remaining characters were used in the analysis (Tables II, III, IV). The polarities of these characters were determined where possible, by outgroup

Table II

Outgroup comparison of characters used in analysis

Characters	Taxa					Primitive state
	Otopappus	*Salmea*	*Chryso-salmea*	*Spilanthes*	*Acmella*	
1. Number of flowers per capitulum	few–several–many	few–several	few–several–many	several	several	?[a]
2. Number of capitula per capitulescence	one–few–many	few–many	one–few–many	one	one	?[a]
3. Leaves mesomorphic (+) or leaves xeromorphic (−)	+	+ (−)[b]	+ (−)	+	+	+
4. Plants glabrous (−) or hirsute (+)	+	+ (−)	+ (−)	+ (−)	+ (−)	+
5. Found on karst hills of W. Cuba (+)	−	−	±	−	−	−
6. Serpentine soil	−	−	(+) −	−	−	−
7. Marginal cells on angles of achene	+	−	+	+	+	+
8. Large-tipped glands on corollas	−	−	+	−	−	−
9. Apex of pale flaring	−	−	+	+	−	−
10. Glands on anther appendage	−	−	+	−	−	−
11. Large-tipped glands on pale	−	−	+	−	−	−
12. Apex of pale with an acute petaloid appendage	−	±	−	−	−	−
13. Plants erect (−) or scandent (+)	±	(+) −	−	−	−	−
14. Multiseriate glandular trichomes on corolla (+)	−	(+) −	−	−	−	−
15. Inner phyllaries and pales with an obtuse, white, scarious margin (+)	−	(+) −	−	−	−	−
16. Phyllaries strongly indurate basally with an herbaceous apex (+)	−	(+) −	−	−	−	−
17. Pales densely hirsute with wide uniseriate hairs on upper ¼ (+)	−	(+) −	−	−	−	−

Table II

Continued

Characters	Taxa					Primitive state
	Otopappus	*Salmea*	*Chryso-salmea*	*Spilanthes*	*Acmella*	
18. Pales with a green ligulate apex (+)	(+) –	(+) –	–	–	–	–
19. Leaves folding along midrib (+)	–	(+) –	–	–	–	–
20. Corolla lobes not papillose on innerside	–	(+) –	–	–	–	–
21. Leaves cuneate (+)	–	(+) –	–	–	–	–
22. Gold resinous dots on upper leaf (+)	–	–	(+) –	–	–	–
23. Hairs on upper leaf surface (+)	+	–	(+) –	±	±	?[a]
24. Leaf base straight (+)	–	–	(+) –	–	–	–

[a] Character with alternative states used as ancestral.
[b] denotes rare occurrence.

Table III

Data matrix

Character	scan	orth	olig	palm	petr	cale	glab	mont	insi	pauc
1	0	0	1	0	0	1	0	3	2	2
2	0	0	0	0	0	0	0	2	2	1
3	0	0	0	0	1	0	0	1	0	0
4	0	0	0	0	1	0	1	1	0	0
5	0	0	0	0	0	2	2	0	0	1
6	0	0	0	0	0	1	0	1	0	0
7	1	1	1	1	1	0	0	0	0	0
8	0	0	0	0	0	1	1	1	1	1
9	0	0	0	0	0	1	1	1	1	1
10	0	0	0	0	0	1	1	1	1	1
11	0	0	0	0	0	1	1	1	1	1
12	1	1	2	2	0	0	0	0	0	0
13	2	1	0	0	0	0	0	0	0	0
14	1	0	1	0	0	0	0	0	0	0
15	0	1	0	0	0	0	0	0	0	0
16	0	0	1	1	0	0	0	0	0	0
17	0	0	1	1	0	0	0	0	0	0
18	0	0	1	0	0	0	0	0	0	0
19	0	0	0	1	0	0	0	0	0	0
20	0	0	0	0	1	0	0	0	0	0
21	0	0	0	0	1	0	0	0	0	0
22	0	0	0	0	0	0	0	0	1	1
23	0	0	0	0	0	0	0	0	1	0
24	0	0	0	0	0	0	1	0	0	1

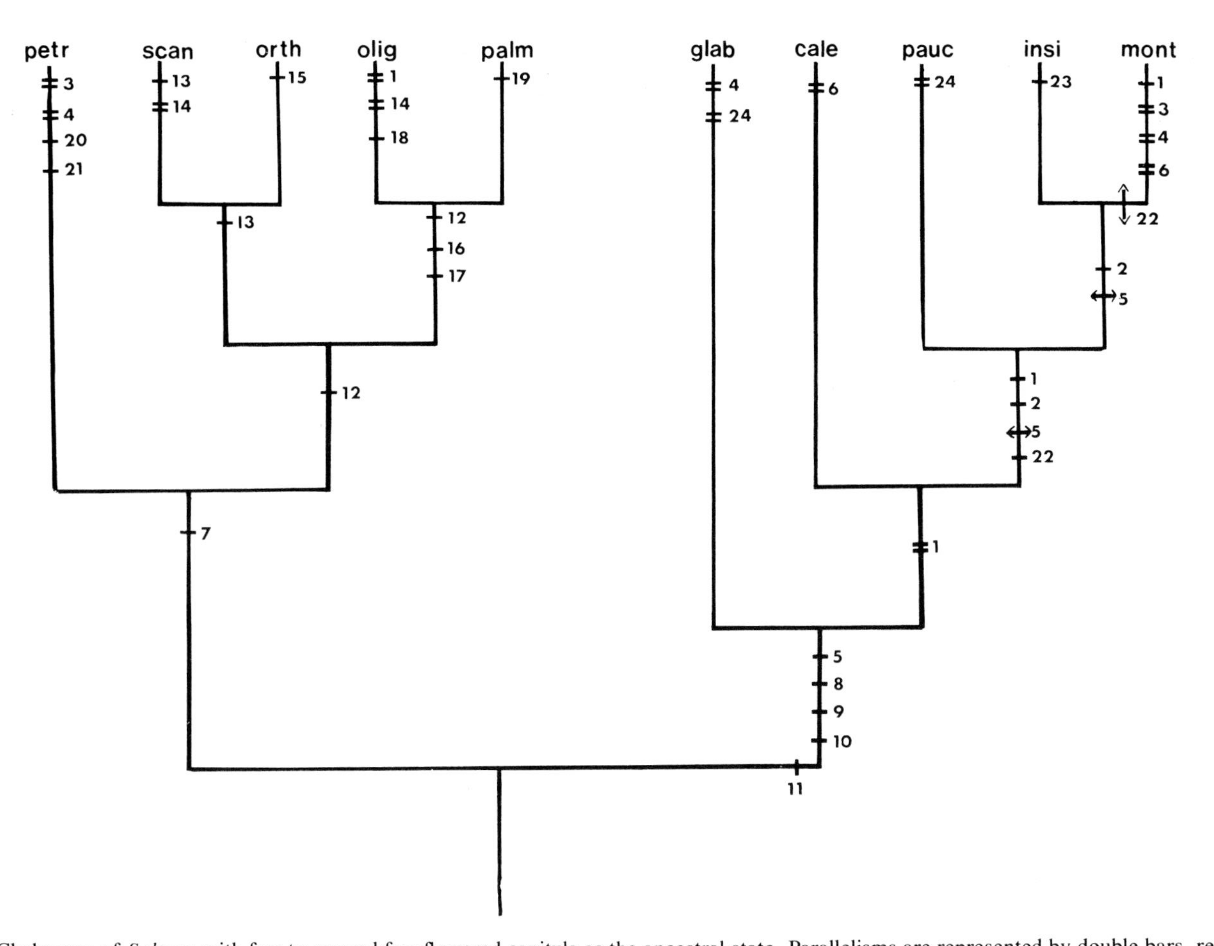

Fig. 1. Cladogram of *Salmea* with few to several few flowered capitula as the ancestral state. Parallelisms are represented by double bars, reversals by arrows. The numbers refer to the characters given in Table I. Each species is abbreviated to the first four letters.

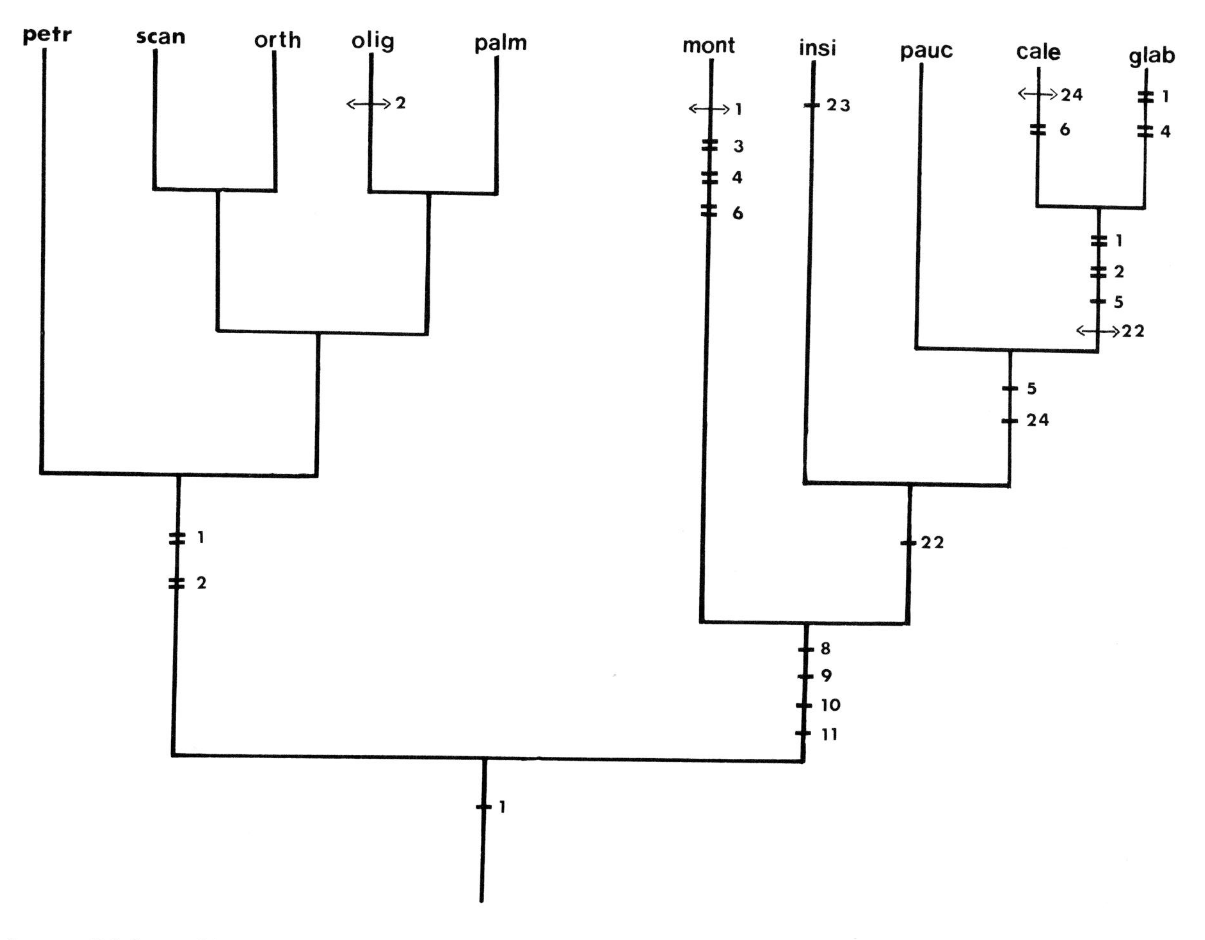

Fig. 2. Cladogram of *Salmea* with solitary many flowered capitula as ancestral. Only character changes differing from Figure 1 are shown. Abbreviations, numbers, and symbols as in Figure 1.

Table IV

Outgroup comparison of characters separating outgroup genera

Character	Outgroup	Primitive state	Taxa				
			Otop	Salm	Chry	Spil	Acme
1. Habit shrubby (+) or herbaceous (−)	+	+	+	+	+	−	−
2. Leaves petiolate (+) or sessile (−)	+	+	+	+	+	−	+
3. Bracts on peduncle present (+) or bracts absent (−)	±	+?	+	+	+	−	−
4. Number of capitula	few–solitary	?	few–many–solitary	few–many	few–many–solitary	solitary	solitary
5. Ray flowers present (+) or absent (−)	+	+	±	−	−	−	±
6. Corollas yellow-orange (+) or white (−)	+	+	±	−	−	−	±
7. Receptacle flat to convex (+) or long conic (−)	+	+	±	−	−	−	−
8. Pale apex nonpapillose (+) or flared & papillose (−)	+	+	+	+	−	+	+
9. Glands on pales, corollas, anther appendages missing (+) or present (−)	+ (−)	+?	+	+	−	+	+
10. Corollas not exserted beyond pales by ⅓–½ length (+) or strongly exserted (−)	±	+?	−	+	+	+	+
11. Corolla gradually expands to throat (+) or abruptly expands (−)	+ (−)	+?	+	+	+	−	+

Table IV

Continued

Character	Outgroup	Primitive state	Taxa: Otop	Salm	Chry	Spil	Acme
12. Anthers with only appendage exserted (+) or anthers exserted most of their length (−)	±	+?	−	+	+	+	+
13. Achenes dimorphic (+) or monomorphic (−)	+	+	+	−	−	−	+
14. Achenes non-winged (+) or winged (−)	±	+?	−	+	+	+	+
15. Callose margin of achenes narrow (+) or wide (−)	+	+?	No callose	No callose	−	−	+
16. Achene margin ciliate (+) or non-ciliate (−)	±	+?	−	+	+	+	+
17. Pappus of awns or sturdy bristles (+) or pappus of slender bristles or none (−)	+ (−)	+?	+	+ (−)	+	+	−
18. Chromosome numbers	n = 11, 12, 14, 15, 16, 17, 18, 19, 21, 30, 34	?	n = 16	n = 18, ca. 18, 32–33, 32 + frag.	?	n = 16	n = 12, 13, 24, 26, 39

comparison to *Otopappus, Spilanthes* and *Acmella*. In twenty-one cases, the presumed ancestral state seemed clear after the outgroup comparison. For characters 1 and 2, comparison to *Otopappus* suggested one ancestral state while comparison to *Spilanthes* and *Acmella* suggested the opposite.

A cladistic analysis of the larger group was done in an attempt to determine which of the outgroup genera was the sister group of *Salmea*. The characters in Table IV were used, where possible. It was assumed that the loss of ray flowers and dimorphic achenes was not reversible. This analysis gave several equally parsimonious cladograms. The autapomorphies of *Salmea* were the loss of ray flowers, dimorphic achenes and yellow flower color. These losses were paralleled in *Otopappus* and *Acmella*. *Spilanthes* alone, *Spilanthes* and *Acmella* together, and all three outgroup genera were the sister group of *Salmea* in at least one cladogram. Since this failed to clarify the polarities, another method was tried. No species has many capitula with many flowers or few capitula with few flowers; with these character states eliminated, characters 1 and 2 were paired and run on the computer with solitary capitula of many flowers coded as both a primitive and as an advanced state.

Upper leaf pubescence (23) shows no clear pattern in the outgroup. Its presence is evenly distributed in the three genera. It was also coded in alternate states, with both presence and absence as derived.

These characters, two paired and one singly, gave four combinations of characters to be run on the computer. In succeeding runs, the character combinations giving the most parsimonious cladogram were used.

In addition to possible homoplasy in characters 1, 2, and 23, homoplasy was suggested for characters 3, 4, 5, 6, 14, and 24 in the first computer runs. These characters were removed from the data matrix and the resulting data set was run on the computer again.

Results

The alternate computer runs with characters 1 and 2 gave cladograms with differing topologies. Assuming that the ancestral state for 1 was few flowers and that for 2 was few to several capitula gave the cladogram shown in Figure 1. Changing the polarities of 1 and 2 resulted in the cladogram of Figure 2. The two cladograms differ in the ordering of the Cuban taxa. The first computer run (Fig. 1) gave a more parsimonious cladogram so the corresponding character states for 1 and 2 were used in succeeding variations of the analysis.

In the computer runs for character 23, the topology of the cladogram remained the same. With the presence of upper leaf pubescence as the ancestral state, the loss of pubescence occurred before the first node of the cladogram, reversing itself in *S. insipida*. Coding the ancestral state as "no upper leaf pubescence" resulted in a more parsimonious cladogram, so this was used in further analysis.

The topology of the cladogram of the Cuban species changed after the removal of the homoplasic characters. With the removal of character 1, a trifurcation (*glaberrima, caleoides,* and the stem of the three remaining species) results. If character 1 is returned to the data matrix, *S. pauciceps* becomes the sister group to *S. insipida* instead of *S. montana*.

Conclusions

The cladistic analysis of *Salmea* corroborates the intuitive species groupings. The genus as now defined has two distinct groups, the Cuban species and the species centered around *S. scandens* with *S. petrobioides* distantly related to the latter group.

Salmea scandens, S. orthocephala, S. oligocephala, S. palmeri, and *S. petrobioides* are separated from each other by non-homoplasic derived characters. The topology of this cladogram is stable. The Cuban species are distinguished by homoplasic characters and include a possible hybrid species. This combination of hybridity and homoplasy precludes a consistent cladogram topology with the data now available.

The analysis of the outgroup genera re-emphasized the difficulties that may arise when taxa are separated by characters showing homoplasy. This analysis does suggest that the generic limits in this group should be re-examined when more information is at hand. Although two of the equally parsimonious cladograms support the current nomenclature, two suggest that the Cuban species be raised to generic rank as a sister group to *Salmea* and *Spilanthes*. Two other cladograms suggest all three genera be recombined under *Spilanthes*.

Acknowledgements

I thank R. L. Hartman, R. K. Jansen and T. F. Stuessy for information on *Otopappus, Spilanthes* and *Acmella*; J. D. Lynch and B. C. Ratcliffe for helpful discussions on cladistics; J. S. Farris for adapting Wagner '78 for use at the University of Nebraska; G. A. Littrell and M. A. Marcuson for technical assistance; and two reviewers for helpful comments.

A Phylogenetic Analysis, Classification and Synopsis of the Genera of the Grimmiaceae (Musci)

Steven P. Churchill

Table of Contents

Introduction

The Grimmiaceae is a large, mostly xerophytic moss family of some 150 to 200 species. The family distribution is worldwide: it is common in the Northern Hemisphere and frequent in the Southern Hemisphere where, in the tropics, it is confined to mountainous regions. This family of mosses is almost totally confined to rocky substrates (saxicolous), forming cushions or tufts that are usually dark green. Some species are specific as to whether the substrate is metamorphic, igneous, or sedimentary; other species are generalists that occupy a variety of rock types. Members of the Grimmiaceae are rarely found on soil or on trees (corticolous).

Traditionally the Grimmiaceae has been assigned to the subclass Haplolepideae. These are mosses that are characterized by having a peristome derived from the primary and inner peristomal layer with a unique 2:3 series (Edwards, 1979) and by having an acrocarpous habit, i.e. they have a sporophyte produced terminally on an usually upright stem. In addition to the Grimmiaceae, some 9 to 20 other moss families have been included as members of the Haplolepideae (Figs. 1–5). Several recent papers have suggested that the haplolepidous mosses are derived from an acrocarpous, diplolepidous ancestor (Buck, 1980; Crosby,

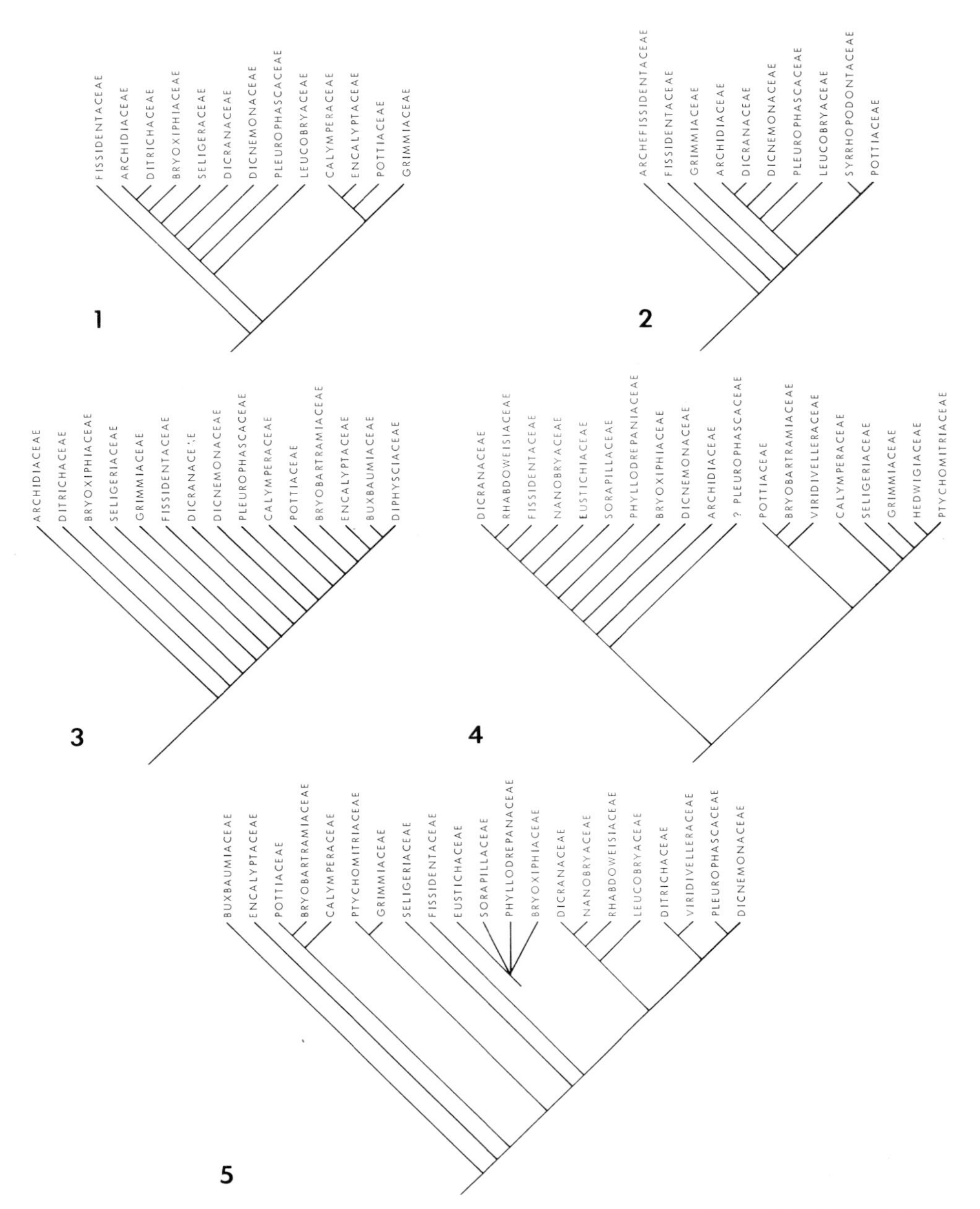

Figs. 1–5. Various hypothesized relationships for the moss families of the Haplolepideae. 1. Brotherus, 1924. 2. Dixon, 1932. 3. Robinson, 1971. 4. Crosby, 1980. 5. Vitt, 1980 and personal communication.

1980; Edwards, 1979; Vitt, 1980). No phylogenetic analysis has yet been offered to corroborate such phyletic conclusions.

The Grimmiaceae are characterized by the following features: small to rather robust plants that often form small to moderately extensive dark green, brown, or black cushions, tufts or mats. Stems usually ascending; in cross-section with a differentiated epidermis, cortex and central strand, rarely one or more regions lacking. Leaves hygroscopic, appressed or crisped about the stem when dry, patent to spreading when wet; lamina shape lanceolate, occasionally ovate-lanceolate, ovate or linear; costa strong, often extending beyond the lamina into a hyaline awn and costa in cross-section with three distinct differentiated regions (or in more derived forms, with a loss or gain of one or more regions); margins plane or recurved, rarely incurved or enrolled; upper laminal cells smooth, sometimes papillose, quadrate and incrassate, basal cells mostly rectangular; leaf cells occasionally sinuose or nodose; lamina unistratose or variously bi- or multistratose. Autoicous, cryptoicous or dioicous. Perichaetial leaves similar to stem leaves but larger; perigonial leaves differentiated from stem leaves, smaller, usually ovate-oval with obtuse or acute apex. Seta long or sometimes short, erect or occasionally curved or sigmoid. Capsule exserted above or contained among the leaves, ovoid, obloid or cylindrical; stomata confined to the base of urn; peristome with 16 teeth, these either whole (occasionally divided above) with distinct horizontal bars and with the outer (dorsal) layer of the tooth thicker than the inner (ventral) layer, or teeth perforated or divided or divided nearly to the base, without distinct horizontal bars, and with equal to subequal inner and outer layer thickening; operculum mamillate to long rostrate. Calyptra mitrate, rarely cucullate, barely covering the operculum to completely covering the urn, smooth to plicate, base entire, erose or lobate. Spores small and variously sculptured. Chromosome number commonly 12, 13 or 14.

In studying this large, diverse family, I have employed the methods of phylogenetic systematics to aid in elucidating relationships at the generic level. The goal of systematics is to determine the actual genealogical relationships that are shared between various monophyletic taxa. In this investigation of the Grimmiaceae, I have proposed a phylogenetic hypothesis that attempts to reflect the genealogy of this moss family. I have derived a classification for the Grimmiaceae directly from the phylogenetic analysis.

The view presented here is one in which the process of "reciprocal illumination" sensu Hennig (1966) is employed. This process involves two reciprocating steps. First, a hypothesis is proposed based upon known attributes of the organisms. The hypothesis is then tested by further studies of the same attributes (but in greater detail and often with a different viewpoint) and of previously undiscovered attributes; more related taxa are also often investigated. In this way the initial hypothesis will either be refuted or corroborated. The initial phylogenetic hypothesis merely provides the framework from which questions can be raised concerning the evolution of a particular group of organisms. That is what this study establishes for the moss family Grimmiaceae. Various hypotheses proposed here may or may not undergo modification in the future; however the important point is that one can test these hypotheses and propose alternative hypotheses if necessary.

Phylogenetic Analysis

Comparative biology generally operates under the evolutionary paradigm. If this paradigm is valid, then phylogeny is the focal point in biological systematics. The methods we choose to employ in the study of phylogeny will greatly affect most fields of biological investigation. The methods and principles of phylogenetic systematics allow us to analyze organisms in a scientific manner. Unlike the various systematic schools such as phenetics (i.e. Sneath and Sokal, 1973) or evolutionary systematics (i.e. Mayr, 1969; Stebbins, 1974), phylogenetic systematics is fully consistent with the axioms of evolution. Furthermore, the methods of phylogenetic systematics allow the formulation of hypotheses which are open to testing and allow the proposition of alternative hypotheses from the original hypothesis (Wiley, 1975, 1981). This study uses the methods of phylogenetic systematics (Hennig, 1966; for a brief summary of concepts and terms see Wiley, 1980) to develop a phylogenetic hypothesis for the Grimmiaceae. A classification for the Grimmiaceae is then possible by deriving it directly from the phylogenetic hypothesis.

Character analysis must be the basis for developing any phylogeny because phylogenies cannot be directly observed. The general plesiomorphic and apomorphic conditions for the Grimmiaceae, at the generic level, are listed in Table I. For the purposes of this investigation, the sister group is considered to be the Pottiaceae s.l. (including the Trichostomaceae), but the Haplolepideae was also studied so as to establish transformations. Unfortunately no testable phylogeny exists for the Haplolepideae (or any other higher moss taxa); however, there have been enough classifications and/or "phylogenies" proposed to gain some estimate of relationship. Five hypothetical relationships for the Haplolepideae are shown in Figures 1–5. For the most part, these hypotheses have been translated from linearly sequenced published classifications (a phylogenetic tree was presented by Vitt, 1980, and pers. comm.). Several of these classifications allude to a shared relationship between the Grimmiaceae and the Pottiaceae s.l. (Figs. 1, 4 and 5). The classification by Dixon (1932) (Fig. 2), and the somewhat similar one by Robinson (1971) (Fig. 3), propose fundamentally different relationships. Little or no justification, i.e. synapomorphies (shared derived characters), was given by the authors for these various classifications. What these resulting classifications actually represent are relationships based on intuitive, "overall similarity" (=the sum of plesiomorphic and apomorphic features). As such, one assumes that taxa above and below any particular branch point are similar and show "relationship" of some nebulous sort. Nevertheless, one can use these to gauge a rough estimate of relationship (Eldredge and Cracraft, 1980, Chapter 5).

The family Grimmiaceae can be defined by several synapomorphies: a midrib described by Kawai (1968) as a C-type (A-, B-, and E-types are derived forms) (Fig. 6, character 1), a mitrate calyptra (cucullate forms are derived) (Fig. 6, character 2), type of peristome ontogeny and morphology (Edwards, 1979) (Fig. 6, character 3), and a *Macromitrium*-type sporeling (Nishida, 1978) (Fig. 6, character 4). Possibly the saxicolous feature associated with a xerophytic environment is also a synapomorphy. Due to the shortage of detailed comparative character studies like those by Kawai (1965, 1968) or Nishida (1978), character distributions for most families are poorly known, so that synapomorphies at the family level may appear weak. This problem is compounded by authors' not using methods that employ the axioms of evolution and that will allow us to check their conclusions. The resulting relationships have led to ad hoc classifications. However, as we now address the various character distributions with methodologies such as phylogenetic systematics, these relationships should become better defined and understood.

Three subfamilies are recognized for the Grimmiaceae: Grimmioideae, Coscinodontoideae, and Ptychomitrioideae. Subfamily Grimmioideae is considered here to be of earliest origin; Coscinodontoideae and Ptychomitrioideae are derived.

Several features suggest a common ancestor for the four genera of the subfamily Grimmioideae: the short conic mitrate calyptra, and a peristome that in cross-section is composed of a thick outer layer of deposited material and a thin inner

Table I
General plesiomorphic and apomorphic features at the generic level for the Grimmiaceae

Character	Plesiomorphic	Apomorphic	Synapomorphy number (Fig. 6)
Gametophyte			
Sporeling type	*Bryum*-type (sensu Nishida, 1978)	*Macromitrium*-type	(4)
Leaf shape	lanceolate	modified, ovate or oblong	(15)
Lanceolate form	short lanceolate, acute apex	long lanceolate, acuminate apex	(7)
Lamina differentia	undifferentiated	differentiated, modified lower & upper lamina	(16)
Costa anatomy	E-type (sensu Kawai, 1968)	C-type	(1)
Sexual attributes			
Sexual condition	dioicous	cryptoicous	(21)
Sporophyte			
Seta length	long (longer than capsule)	short	(10)
Capsule shape	ovoid or short cylindrical	semi-spherical	(8)
Capsule mouth	narrow	wide	(9)
Basal membrane	absent	present	(19)
Peristome shape	linear or lanceolate	triangular	(3)
Peristome thickness	inner layer thicker than outer	outer layer thicker than inner	(6)
Peristome form	whole, horizontal bars present, papillae lacking or if present small	reduced & modified, bars absent, papillae dense & gemmate	(11)
Peristome outline	entire	perforated or divided to near the base	(13, 18)
Peristome thickness	outer layer thickest	outer & inner layer equal thickness	(12)
Calyptra type	cucullate	mitrate	(2)
Calyptra shape	conic	campanulate	(14)
Calyptra envelopment	operculum and upper third of urn	operculum & upper most of urn	(5)
Calyptra base	entire, erose	lobate	(17)

layer (Fig. 6, characters 5 and 6). The informal tribe "Guembelieae" consists of "*Guembelia*" and "*Rhabdogrimmia*" (use of quotes indicates taxa not given formal nomenclatural status in this study) and exhibits a synapomorphy of a long lanceolate leaf shape and an apex that is acuminate, usually with a hyaline awn (Fig. 6, character 7). "*Guembelia*" has a leaf margin that is plane or cucullate. Another feature in several species of "*Guembelia*" is the cucullate calyptra which may be plesiomorphic or apomorphic. If "*Guembelia*" is of earliest origin within the Grimmiaceae (and this may be the case), and those species exhibiting a cu-

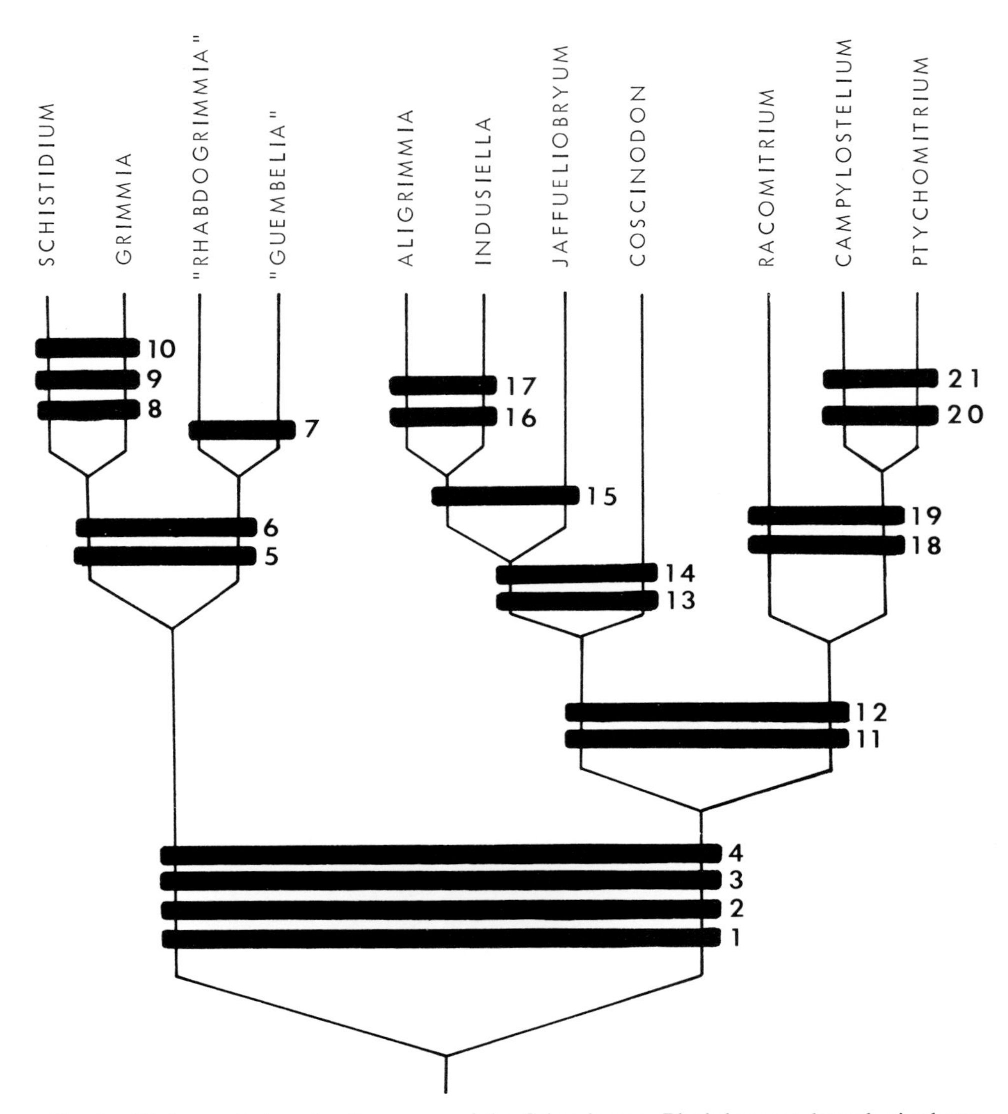

Fig. 6. Phylogenetic tree for the genera of the Grimmiaceae. Black bars are hypothesized synapomorphies. Synapomorphic characters: **1.** C-type midrib, **2.** mitrate calyptra, **3.** peristome shape, **4.** *Macromitrium*-type sporeling, **5.** low conic calyptra, **6.** outer peristome layer thicker than inner, **7.** long lanceolate leaf, **8.** short semi-spherical capsule, **9.** wide capsule mouth, **10.** short seta, **11.** reduced or modified peristome form, **12.** peristome with outer layer equal or subequal to inner, **13.** perforated peristome, **14.** campanulate calyptra, **15.** modified lamina, **16.** differentiated lower and upper lamina, **17.** lobate calyptra, **18.** divided peristome, **19.** basal membrane, **20.** E-type midrib, **21.** cryptoicous. Character descriptions have been abbreviated here, see text for further discussion. Synapomorphies (unique characters) for each of the eleven genera given in the text.

cullate calyptra are of earliest derivation within "*Guembelia*," then the cucullate calyptra may be in fact plesiomorphic, otherwise it is apomorphic. In most cases within the Grimmiaceae, the cucullate calyptra is considered uniquely derived from a mitrate calyptra, therefore this feature represents a convergence found between this family and other haplolepidous families. "*Rhabdogrimmia*" exhibits two synapomorphies: a curved or cygneous seta, and a ribbed capsule; possibly the capacity to produce propagula is another synapomorphy. Within the Grimmiaceae the ability to produce gemmae is mostly confined to this genus, and almost solely to this subfamily, although Deguchi (1977b) reported propagula for a species of *Racomitrium*. The two remaining genera, *Schistidium* and *Grimmia*, of the tribe "Grimmieae" exhibit three synapomorphies: a short, semi-spherical capsule, a wide capsule mouth, and a short seta (Fig. 6, characters 8, 9 and 10). *Grimmia* has an urn that is gibbous, and a seta that is sigmoid. Within *Grimmia*, the apomorphic features involve a bistratose lamina and/or a loss of a peristome. Several synapomorphies suggest that the species of *Schistidium* shared a common ancestor: a columella that remains attached to the operculum, the loss of an annulus, and no air space between the capsule wall and the spore sac (see Bremer, 1981, for a detailed phylogenetic systematic study of this genus).

Hydrogrimmia is placed incertae sedis within the Grimmiaceae. This genus exhibits features possibly shared with "*Guembelia*" such as having plane leaf margins and an erect, somewhat long seta (longer than the capsule), and a cucullate calyptra. Features unique for *Hydrogrimmia* are soft leaves, poorly defined epidermal and cortical regions of the stem, and thin cell walls. Further work is needed, particularly among the "Guembelieae" taxa, to determine the proper placement of *Hydrogrimmia*. It is likely to be a derived form within a genus, possibly "*Guembelia*," and not one that shared a common ancestor with any of the four genera of this sub-family.

The Coscinodontoideae and Ptychomitrioideae are thought to have shared a common ancestor based on the synapomorphies of a reduced or modified peristome form, and a peristome in which the outer and inner layers are subequal to equal in thickness (Fig. 6, character 11—see Figs. 9 and 10, and character 12—see Fig. 8). The characters which unite the Coscinodontoideae are the perforated peristome (Fig. 6, character 13—see Fig. 9), and a plicate, campanulate calyptra which covers most of the urn (Fig. 6, character 14). *Coscinodon* is the earliest derived member, exhibiting lanceolate, ±plicate leaves. Two important transformations are noted within *Coscinodon*: the reduction of seta length and the development of differentiated cells, rectangular rather than quadrate, along the lamina plication.

Jaffueliobryum, *Indusiella*, and *Aligrimmia* share a modified lamina, distinct from the plesiomorphic lanceolate form (Fig. 6, character 15). *Jaffueliobryum* has broadly ovate leaves that are unistratose. A short, straight seta immersed among the leaves is a plesiomorphic feature of *Jaffueliobryum*, while an elongated seta is apomorphic. This demonstrates the principle that what may be considered plesiomorphic at one level (a long seta at the generic level) may be apomorphic at another level (a long seta at the species level of analysis). One level of analysis cannot be extrapolated to the next lower or higher level; synapomorphic analysis with outgroup comparison must be conducted for each level and group.

Aligrimmia and *Indusiella* exhibit synapomorphies involving a differentiated lower and upper lamina, and a calyptra that is lobate (Fig. 6, characters 16 and 17). This lobate calyptra is found as a convergence between the above discussed genera and genera of the Ptychomitrioideae (some members of *Racomitrium* and all members of *Ptychomitrium*) as well as in a member of the genus *Jaffueliobryum* (*J. arsenei*). It is also found in a few members of the Grimmioideae. The differentiated lamina consists of a broad and somewhat sheathing ovate base with the upper lamina becoming narrow and incurved or enrolled.

Coscinodontella (Williams, 1927) is included in *Indusiella* based on the follow-

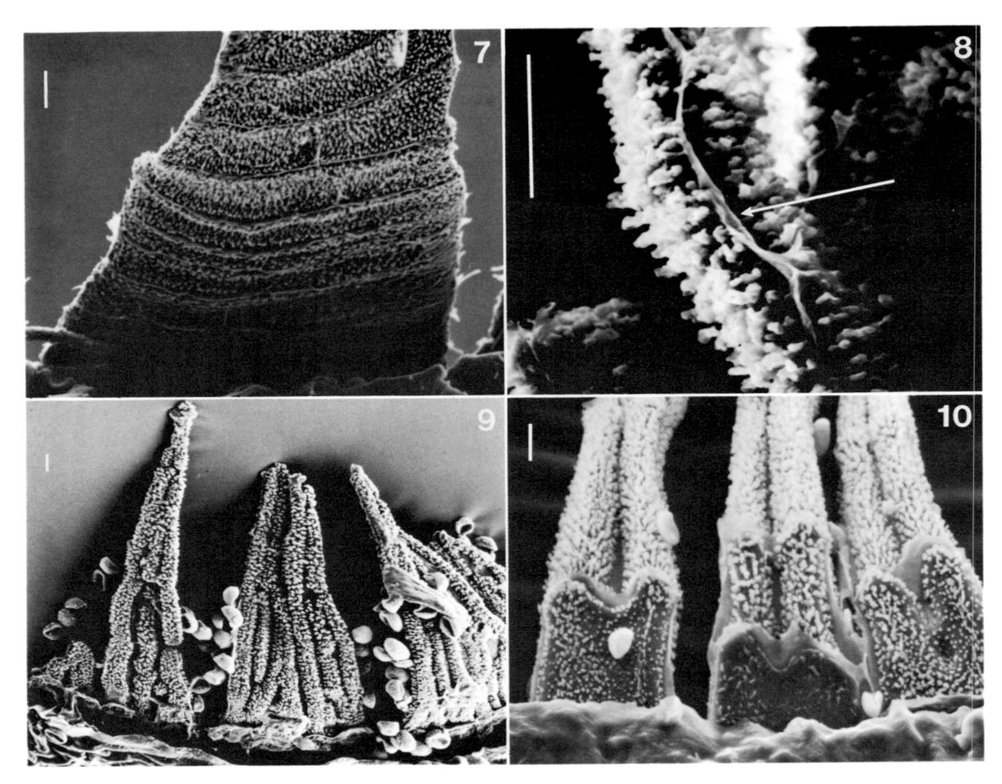

Figs. 7–10. Peristome types in the Grimmiaceae. 7. Whole peristome of *Schistidium*; *Schofield & Zales 47246* (MO). 8. Peristome side view of *Jaffueliobryum*, arrow indicates cell wall with equal material deposited on either side; *Harvill 2772* (MO). 9. Perforated peristome of *Jaffueliobryum*; *Mahler 5784* (SMU). 10. Divided peristome of *Ptychomitrium*, *Pringle 10418* (MO). Figures 7, 9, and 10, bar = 10 μm; Figure 8, bar = 1 μm.

ing shared features: a central strand that is reduced or absent, an upper lamina that is incurved to enrolled, a bistratose upper lamina, a costa cross-section exhibiting identical cell type distribution, a relatively wide capsule mouth, and the presence of a persistent annulus. *Aligrimmia* exhibits adaxial lamellae, the only member of the Grimmiaceae to display this feature. The presence of adaxial lamellae is a classic example of homoplasy, i.e. convergence, in the mosses; it is found as a synapomorphy in the large order Polytrichales uniting both Dawsoniaceae and Polytrichaceae, and it also occurs in several genera of the Pottiaceae (s. s.).

Members of the subfamily Ptychomitrioideae exhibit synapomorphies of a peristome divided nearly to the base and the presence of a basal membrane (Fig. 6, character 18—see Fig. 10, character 19). *Campylostelium* and *Ptychomitrium* share two synapomorphies: similar midrib anatomy (E-type of Kawai, 1968) and a sexual condition described by Deguchi (1977a) as cryptoicous (Fig. 6, characters 20 and 21). Cryptoicy condition involves a perigonium that is actually attached to the vaginula and therefore contained within the perichaetial leaves. *Campylostelium* exhibits a cygneous seta and linear leaves from an ovate base. *Ptychomitrium* has a calyptra that is plicate and deeply lobed at the base and perichaetial leaves that are similar in size or somewhat smaller than the stem leaves. Dixon's *Ptychomitriopsis* (1931) is considered here to be a derived form within the small species of *Ptychomitrium* (Brotherus' Section *Notarisia*, 1925) which have smooth leaf margins and a single sporophyte. While it has several autapomorphies (uniquely derived characters), including the very small size of the plants and cells that are large, distinct and pellucid, it is clearly a member within the genus *Ptychomitrium*. *Racomitrium* displays several synapomorphies: no central strand, a B-type costa anatomy (Kawai, 1968), cells that are sinuose, dioicous sexual condition, perichaetium and perigonium both on terminal and lateral branches, and vaginule epidermal cell walls that are thick and pitted.

Inclusion of the genera *Campylostelium* and *Ptychomitrium* in the Grimmiaceae has not always been followed (see section on classification). The view here is that the Grimmiaceae would be a paraphyletic group if these taxa were excluded. Character evidence from both chromosomal (Smith and Newton, 1968) and morphological analyses suggests a natural relationship of these genera to the Grimmiaceae and, in particular, *Racomitrium*.

An alternative hypothesis of *Racomitrium* relationship may involve a shared ancestor with "Guembelieae" or with one of the genera within the tribe, i.e. "*Guembelia*" or "*Rhabdogrimmia*." Characters which seem to suggest this relationship include symplesiomorphies (shared primitive characters) such as the general aspect of the plants, lanceolate leaf shape and an elongated seta. However, only the feature of nodulose basal cells found primarily in the tribe "Guembelieae" resembles the sinuose basal cells in *Racomitrium* (see Deguchi, 1979). The nodulose feature is only sporadic in "Guembelieae" and most likely represents a homoplasy rather than homologue (e.g. in a transformation series). It is suggested here that it is more parsimonious to accept a relationship between *Racomitrium* and a shared ancestor with *Campylostelium* and *Ptychomitrium* than to suggest a relationship with "Guembelieae" or any of the genera within the tribe.

Pseudohyophila is placed incertae sedis within the family Grimmiaceae until further investigations can establish its proper position within the family or if in fact the genus belongs to the Pottiaceae (s.l.). Williams (1903) described this originally as a species of *Hyophila*, but later Hilpert (1933) transferred this taxon from the Pottiaceae to the Grimmiaceae and placed it in its own genus—*Pseudohyophila*. Features mentioned by Williams (1903) for this taxon include: leaves crispate when dry, flat or slightly recurved margin, peristome lacking, operculum obliquely rostrate, and a cucullate calyptra.

The genus *Scouleria* traditionally has been placed in the family Grimmiaceae since it was first described by Hooker in 1829. In fact, *Scouleria* was treated as a section of the genus *Grimmia* by Müller (1851). This classification was followed for a short time by various bryologists in the last century including Mitten (1869), and Lesquereux and James (1884). *Scouleria* is certainly monophyletic and features several synapomorphies which are, even within the Haplolepideae, unique. These characters include rhizoids present on not only the stem but also on the abaxial leaf surface, lamina cells that are essentially undifferentiated above and below but with a submarginal or marginal band of elongate-rectangular cells that extend up from the base to half the lamina length, a spherical capsule which when mature shrinks proximally with the columella becoming exserted, columella vertically ribbed, annulus lacking, peristome of 32 teeth or absent, an unusual peristome formula (see Edwards, 1979), prostome present, and large spores (30–55 μm) which have gemmate papillae.

In light of these features there is little to suggest that *Scouleria* belongs to the

Grimmiaceae. The general habit aspect of the plant as well as an operculum that remains attached to the columella probably contributed to the concept that *Scouleria* bore some relationship to the Grimmiaceae, particularly the genus *Schistidium*. However, the recent investigation by Edwards (1979) demonstrated not only that the peristome formula pattern for *Scouleria* was unique for the Haplolepideae, but that the peristome in cross-section resembles the *Seligeria* type. *Scouleria* is here established in its own family, the Scouleriaceae. Its relationship among the Haplolepideae families as well as the status of the species in this genus are currently under investigation by the author.

The genus *Glyphomitrium* is excluded from the Grimmiaceae in this study. *Glyphomitrium,* a small genus of ten species, has had a varied history as to its proper familial placement. In this century, *Glyphomitrium* was placed in the Orthotrichaceae by Brotherus (1909) as a synonym of *Aulacomitrium*. Later Brotherus (1925) placed *Glyphomitrium* in the Ptychomitriaceae within the order Isobryales. In the expanded version of the family Grimmiaceae for North America, Jones (1933) placed *Glyphomitrium* not in the subfamily Ptychomitrioideae but in the subfamily Grimmioideae. Noguchi (1952) treated *Glyphomitrium* as a member of the Erpodiaceae, but Crum (1972b) excluded the genus in his taxonomic account of the family. I doubt that *Glyphomitrium* belongs to the Grimmiaceae, although Crum (1972a) presented an alternative view. Recently Edwards (1979) suggested that *Glyphomitrium* may be close to the Seligeriaceae. Alternatively, *Glyphomitrium* may be related to members of the Orthotrichaceae as indicated by Brotherus (1909). While several morphological features of *Glyphomitrium* are also exhibited in the Orthotrichaceae, it is the peristome that strongly suggests this relationship. The peristome teeth of *Glyphomitrium* are united in pairs and reflexed—features commonly found in the Orthotrichaceae. Detailed work on the peristome of the Orthotrichaceae is needed to clarify the evolutionary trends.

The notion of a relationship between the Grimmiaceae and the Orthotrichaceae has been persistent in the literature and classifications for many years, and may have important implications on our present view of the natural groups of mosses. Such a relationship is based on the scenario involving a peristomal transformation from an Orthotrichaceae-like ancestor which was diplolepidous and by reduction led to a Grimmiaceae-like ancestor which was haplolepidous. While I tend to agree with those such as Edwards (1979), Crosby (1980) and Vitt (1980) that the Haplolepideae is a monophyletic group, we should all be aware that there has been little research into this aspect of phylogeny. Whatever evidence comes to light, there exists an alternative hypothesis which suggests that the Haplolepideae are not monophyletic (and thus the Diplolepideae are paraphyletic).

Classification

Any consideration of the classification of the Grimmiaceae must start with Brotherus' major studies of the moss genera (1909, 1924, 1925). The original classification of the genera of Grimmiaceae proposed by Brotherus in 1909 underwent several major changes in the later 1924–1925 classification (Table II). The 1909 classification defined and included the subfamily Ptychomitreae in the Grimmiaceae (a view held here) whereas the later classification placed the Ptychomitreae genera in their own separate family, the Ptychomitriaceae. Further, the Ptychomitriaceae was removed from the Haplolepideae to the diplolepidous, acrocarpous Isobryales suborder Orthotrichineae. Also, *Glyphomitrium* in the 1909 classification was equal to *Ptychomitrium* as we now understand it, while those

Table II

Comparison of two classifications proposed by Brotherus for the Grimmiaceae (1909, and 1924, 1925)

1909 classification	1924–1925 classification
Order Grimmiales	Order Grimmiales
Family Grimmiaceae	Family Grimmiaceae
Subfamily Ptychomitreae	Subfamily Scoulerioideae
Genus *Glyphomitrium*	Genus *Scouleria*
Subgenus *Ptychomitrium*	Subfamily Grimmioideae
Subgenus *Brachysteleum*	Genus *Coscinodon*
Subgenus *Notarisia*	Genus *Indusiella*
Subgenus *Euglyphomitrium*	Genus *Aligrimmia*
Genus *Campylostelium*	Genus *Grimmia*
Subfamily Scoulerieae	Subgenus *Litoneurum*
Genus *Scouleria*	Subgenus *Gumbelia*
Subfamily Grimmieae	Subgenus *Hydrogrimmia*
Genus *Coscinodon*	Subgenus *Rhabdogrimmia*
Genus *Indusiella*	Subgenus *Schistidium*
Genus *Grimmia*	Subgenus *Gastrogrimmia*
Subgenus *Schistidium*	Subgenus *Streptocolea*
Subgenus *Gasterogrimmia*	Genus *Rhacomitrium*
Subgenus *Grimmia*	Order Isobryales
Subgenus *Rhabdogrimmia*	Family Ptychomitriaceae
Subgenus *Gumbelia*	Genus *Campylostelium*
Genus *Rhacomitrium*	Genus *Ptychomitrium*
Subgenus *Dryptodon*	Genus *Glyphomitrium*
Subgenus *Rhacomitrium*	

taxa now called *Glyphomitrium* were placed in *Aulacomitrium* in the family Orthotrichaceae.

Few modifications have been made to the 1924–1925 classification proposed by Brotherus. However, since the time of this work, several other genera have been proposed for inclusion in the Grimmiaceae: *Bucklandiella* by Roivainen in 1972 (originally *Bucklandia* Roiv. nom. illeg., 1955b), *Coscinodontella* by Williams in 1927, *Jaffueliobryum* by Thériot in 1928, and *Pseudohyophila* by Hilpert in 1933. Also, several recent authors (e.g. Deguchi, 1979; Smith, 1978) consider some sections of the genus *Grimmia* to be distinct genera: *Coscinodon, Dryptodon, Hydrogrimmia,* and *Schistidium*. Likewise, the family Ptychomitriaceae has often been placed close to or within the Grimmiaceae. Further consideration of previous classifications such as that of Hagen (1909), while important now in a historical context, is deferred for future work (see Deguchi, 1979, for a brief outline of the subject).

Neither Brotherus nor subsequent workers (except Loeske, 1913, 1930) gave much discussion of the possible relationships among the genera which might substantiate their classifications. Admittedly the purpose of regional floristic or monographic studies is not detailed treatments of either relationships or classification but resolving alpha-taxonomic problems that exist among species. Recently several have provided important character analyses and discussions that pertain to the classification of the Grimmiaceae (Bremer, 1981; Deguchi, 1979; Edwards, 1979; Hirohama, 1978; Kawai, 1965, 1968; Nishida, 1978). These kinds of studies provide the needed information for phylogenetic analyses that will ultimately aid in the production of a natural classification.

Table III

Proposed revision of the classification of the Grimmiaceae. Incertae sedis is used for taxa of uncertain relationship within the family. The use of quotes indicates informal category status for monophyletic taxa. *Scouleria* is removed from the Grimmiaceae and placed in its own family, the Scouleriaceae. *Glyphomitrium* is excluded from the Grimmiaceae

Family	Subfamily	Tribe	Genus
Family Grimmiaceae			
	Subfamily Grimmioideae		
		Tribe "Guembelieae"	
			Genus "*Guembelia*"
			Genus "*Rhabdogrimmia*"
		Tribe "Grimmieae"	
			Genus *Grimmia*
			Genus *Schistidium*
	Subfamily Grimmioideae incertae sedis		
			Genus *Hydrogrimmia*
	Subfamily Coscinodontoideae		
			Genus *Coscinodon*
			Genus *Jaffueliobryum*
			Genus *Indusiella*[a]
			Genus *Aligrimmia*
	Subfamily Ptychomitrioideae		
			Genus *Racomitrium*[b]
			Genus *Campylostelium*
			Genus *Ptychomitrium*[c]

[a] Includes *Coscinodontella*.
[b] Includes *Bucklandiella* & *Dryptodon*.
[c] Includes *Ptychomitriopsis*.

The classification of the Grimmiaceae presented here is based on the phylogenetic analysis given in the previous section. The method used for classifying the genera follows the suggestions of Nelson (1973) and Wiley (1979). This method is based on a sequencing of monophyletic taxa that is directly transferred from the phylogenetic tree. The first branch point at the base of the tree is listed at the top. As one continues up the branching tree, each of the branch points are "read off" (e.g. compare Fig. 6 with Table III; note that those taxa placed incertae sedis are deleted from the phylogenetic tree). It is in this manner that the phylogenetic hypothesis, based on character analysis, allows us to erect a natural classification (Hennig, 1966, 1975).

The present classification recognizes three subfamilies for the Grimmiaceae (Table II). Subfamily Grimmioideae consists of four natural genera: "*Guembelia*" and "*Rhabdogrimmia*" under the tribe "Guembelieae," and *Grimmia* and *Schistidium* under the tribe "Grimmieae." The genus *Hydrogrimmia* is also placed within the Grimmioideae as incertae sedis until proper relationship can be established although, as stated, the genus is possibly related to "*Guembelia*." Next in the sequence is the subfamily Coscinodontoideae which consists of *Coscinodon, Jaffueliobryum, Indusiella* (including *Coscinodontella*), and *Aligrimmia*. The last sequenced subfamily is Ptychomitrioideae and includes *Racomitrium* (including *Bucklandiella* and *Dryptodon*), *Campylostelium* and *Ptychomitrium*. Finally one genus, *Pseudohyophila,* is placed incertae sedis within the family Grimmiaceae until its affinities can be properly resolved. The use of incertae sedis is employed in the traditional sense to mean of unknown or uncertain status (see

Wiley, 1979, for the use of this and other terms in phylogenetic classifications). *Scouleria*, subfamily Scoulerioideae of Brotherus (1924, see Table II), is removed from the Grimmiaceae and established in its own family—the Scouleriaceae. Finally, *Glyphomitrium* is also excluded from the Grimmiaceae, but its proper placement remains uncertain.

Synopsis

In the following synopsis of the genera of the Grimmiaceae, I have presented a brief diagnosis for each of the eleven accepted genera, as well as *Hydrogrimmia*, and given the number of species for each genus. Since *Schistidium* is the only genus for which there is a world-wide monograph a rough estimate is given for the other genera using Index Muscorum (Wijk et al., 1959–1969) and other sources. In the smaller genera, all species are listed along with their distribution. For the larger genera, only a generalized distribution is stated. A discussion of phylogenetic transformations is given for some of the genera and is only meant to suggest the relative plesiomorphic-apomorphic character relationships among the species and the character transformations observed. Further phylogenetic comments of inter- and intrageneric relationships are presented in the section on phylogenetic analysis.

No formal status is given to either "*Guembelia*" or "*Rhabdogrimmia*" until further work at the alpha-taxonomic level can be carried out on these species. It is maintained here that both taxa are monophyletic and should be given a generic level category once investigations can better define many of the species. Likewise, no formal status is given to the two tribes, "Guembelieae" and "Grimmieae," in the subfamily Grimmioideae.

In addition to personal observations and research, I have made use of several general references in addition to those cited with each genus. These studies are Brotherus, 1924, 1925; Deguchi, 1979; Hirohama, 1978; Jones, 1933; Lawton, 1971; Smith, 1978.

GRIMMIACEAE Arnott, Disp. Metn. Mousses 19. 1825.

A. Subfamily Grimmioideae

Peristome whole or partly divided in upper half (Fig. 7); outer peristomal layer thicker than inner layer; teeth lightly papillose; horizontal bars distinct. Calyptra mitrate, small, conic, covering only the operculum or just beyond, if cucullate then extending to the middle of the urn.

1. "*Guembelia*." The name *Guembelia* Hampe in C. Müll., Bot. Zeitung. 4: 124. 1846, appears to be available for the group; no nomenclatural status is intended at the present time.
   LT.: *Guembelia elliptica* (Funck) Hampe (*Grimmia elliptica* Funck) (fide Pfeiffer, Nom. 1: 1511. 1874)

Leaves stiff, lanceolate from an oval or elliptical base, concave, keeled or grooved; margins plane or incurved, rarely recurved; apex acuminate, usually with a hyaline awn. Dioicous or autoicous. Seta straight, short or long exserted. Calyptra mitrate or cucullate.

The number of species in this group is uncertain at present. I would tentatively include the following species now placed in *Grimmia* sensu lato: *Grimmia alpestris* (Web. & Mohr) Schleich. ex Hornsch in Somm., *G. affinis* Hornsch., *G. arizonae* Ren. & Card., *G. atrata* Mielichh. in Hoppe & Hornsch., *G. brevirostris* Williams, *G. brittoniae* Williams, *G. coloradensis* Aust., *G. donniana* Sm., *G. elongata* Kaulf. in Sturm, *G. laevigata* (Brid.), *G. montana* B.S.G., *G. olympica* Britt. ex Frye (see comments by Deguchi, 1978; certainly this taxon, if a member, is derived within "*Guembelia*"), *G. ovalis* (Hedw.) Lindb., *G. percarinta* (Dix. et Sak.) Nog. ex Deguchi, *G. plifera* P. Beauv., *G. subsulcata* Limpr., *G. tenerrima* Ren. & Card., and *G. unicolor* Hook. ex Grev.

This particular group is widespread in both hemispheres. Our understanding of this taxon, as for most groups of mosses, is better for the Northern Hemisphere. In the Southern Hemisphere much more work is needed at the alpha-taxonomic level before serious progress can be made in understanding phyletic transformations within this genus.

2. "*Rhabdogrimmia*." No generic name is presently available for this group; this sectional name within *Grimmia* sensu Brotherus (1924) should be used if available.

Leaves rigid, lanceolate from an elongate base, keeled upwardly; margins recurved on one or both sides; costa often dorsally projecting. Propagula often present, attached to leaves. Dioicous, occasionally autoicous. Seta long, arcuate or cygneous (but often erect when dry); capsule striate-ribbed, rarely smooth. Calyptra mitrate, rarely cucullate.

The number of species to be included in the "*Rhabdogrimmia*" group is yet to be fully determined. I would tentatively include the following *Grimmia* sensu lato taxa: *Grimmia anomala* Hampe ex Schimp., *G. decipiens* (Schultz) Lindb. in Hartm., *G. elatior* Bruch ex Bals. & De Not., *G. funalis* (Schwaegr.) B.S.G., *G. hamulosa* Lesq., *G. hartmanii* Schimp., *G. hermannii* Crum, *G. incurva* Schwaegr., *G. muehlenbeckii* Schimp., *G. olneyi* Sull., *G. orbicularis* Bruch ex Wils., *G. pulvinata* (Hedw.) Sm., *G. retracta* Stirt., *G. torquata* Hornsch. in Grev., and *G. trichophylla* Grev.

Like "*Guembelia*," "*Rhabdogrimmia*" is widespread, better understood in the Northern Hemisphere, and less well known and understood in the Southern Hemisphere. Most of the taxa assigned to *Grimmia* s.l. belong to either "*Guembelia*" or "*Rhabdogrimmia*." The many *Grimmiae* named in the last century and the early part of this century present the greatest obstacle, for many of the several hundred names will likely prove to be synonyms, and most of these are from the Southern Hemisphere.

3. *Grimmia* Hedw., Sp. Musc. 75. 1801.
LT.: *Grimmia plagiopodia* Hedw. (fide B.S.G.)

Leaves oblanceolate to broadly lanceolate, often with a broad hyaline awn; unistratose or bistratose; margins plane. Autoicous (dioicous in one species). Seta curved or sigmoid, shorter than the capsule; capsule immersed among the leaves, subglobose, capsule mouth wide, base gibbous; stomata large, at the base of urn; operculum mamillate. Calyptra mitrate, small, just covering the operculum.

The number of species is uncertain, but there are at least five: *Grimmia americana* Bartr., *G. anodon* B.S.G., *G. campylopoda* (Luz.) Saito, *G. plagiopodia* Hedw., and *G. poecilostoma* Card. & Sed. ex Sed. Brotherus (1924) mentioned three or four additional species that may belong to *Grimmia*.

Reference: Saito (1973).

4. *Schistidium* Brid., Mant. Musc. 20. 1819. Emend. B. & S. in B.S.G., Bryol. Eur. 3: 97. 1845.
LT.: *Schistidium pulvinatum* (Hedw.) Brid. (*Gymnostomum pulvinatum* Hedw.) (fide B. & S. in B.S.G.)

Stems with central strand present or absent. Leaves lanceolate to ovate; upper lamina flat or keeled; margin flat or recurved; costa differentiated or similar to lamina in cross-section; upper lamina unistratose or irregularly bistratose, mostly along margin. Autoicous or dioicous. Seta shorter than or equal to capsule length; capsule immersed among leaves, cyathiform to cylindrical; air space lacking between capsule wall and spore sac; stomata present or absent; columella remaining attached to operculum; annulus absent.

Thirteen species. *Schistidium agassizii* Sull. & Lesq., *S. angustifolium* (Mitt.) Herz., *S. apocarpum* (Hedw.) B. & S. in B.S.G., *S. cinclidodonteum* (C. Müll. in Roell.) B. Bremer, *S. falcatum* (Hook. f. & Wils.) B. Bremer, *S. holmenianum* Steere & Brassard, *S. maritimum* (Turn.) B. & S. in B.S.G., *S. occidentalis* (Lawton) Churchill, comb. nov., *S. pulvinatum* (Hedw.) Brid., *S. rivulare* (Brid.) Podp., *S. streptophyllum* (Sull.) Herz., *S. tenerum* (Zett.) Nyh., and *S. trichodon* (Brid.) Poelt.

This genus is widespread with the majority of taxa found in the Northern Hemisphere. The above treatment and accepted taxa are adapted from the recent studies by Bremer.

References: Bremer (1980a, 1980b, 1981) and Lawton (1967).

Subfamily Grimmioideae incertae sedis: *Hydrogrimmia*

Hydrogrimmia Loeske, Stud. Morph. Syst. Laubm. 108. 1910.
T.: *Hydrogrimmia mollis* (B.S.G.) Loeske (*Grimmia mollis* B.S.G.)

Central strand well developed, epidermis and cortex not distinctly differentiated. Leaves oblong to ovate-lanceolate, obtuse-rounded, ±cucullate; leaf margin plane to slightly incurved; unistratose to irregularly bistratose; costa subprecurrent; lamina cells thin walled, median cells rounded-quadrate, basal cells short rectangular to quadrate. Dioicous. Seta erect; capsule ovoid to cylindrical, smooth. Calyptra cucullate.

One species. *Hydrogrimmia mollis* (B.S.G.) Loeske, mountainous regions of the Northern Hemisphere.

Reference: Abramova (1969).

B. **Subfamily Coscinodontoideae** Churchill, subfam. nov.

Leaves either lanceolate and plicate or ovate and not plicate, margins recurved, flat or incurved to enrolled. Peristome perforated, densely papillose; inner and outer peristomal layers equal in thickness. Calyptra plicate campanulate, erose or lobate.

5. *Coscinodon* Spreng., Einl. Stud. Krypt. Gew. 281. 1804.
   T.: *Coscinodon pulvinatus* Spreng., nom. illeg. (=*C. cribrosus* (Hedw.) Spruce)

Leaves lanceolate, with a hyaline awn; plicate, with or without differentiated elongate cells; lamina bistratose in plicate area and at margin. Dioicous. Annulus present or absent. Calyptra at base erose or irregularly lobate.

Four species. *Coscinodon arctolimnia* (Steere) Steere, Northwest Territories of Canada; *C. calyptratus* (Hook.) C. Jens. in Kindb., western North America; *C. cribrosus* (Hedw.) Spruce (*C. australis* Dix. et Sainsb., Trans. Roy. Soc. New Zealand 75: 175. 1945. syn. nov.), widely distributed in the Northern Hemisphere, and from Africa west to New Zealand; *C. trinervis* (Williams) Broth. (*C. bolivianus* Broth. in Herz., Biblioth. Bot. 87: 52. 1916. syn. nov.; also includes the combination *Grimmia boliviana* (Williams) F. J. Hermann, Bryologist 79: 137. 1976), northern Andes of South America (Bolivia and Peru).

References: Churchill (manuscript in prep.) and Steere (1974).

6. *Jaffueliobryum* Thér., Rev. Bryol. n. ser. 1: 191. 1928.
   LT.: *Jaffueliobryum wrightii* (Sull. in Gray) Thér. (*Coscinodon wrightii* Sull. in Gray) (chosen here)

Branches mostly julaceous; leaves ovate to obovate with a hyaline awn as long as lamina or longer. Autoicous. Seta short or long, erect; operculum rostrate; calyptra base erose or in one case lobate.

Four species. *Jaffueliobryum arsenei* (Thér.) Thér., central Mexico; *J. latifolium* Thér., south central Siberia and northern Mongolia; *J. raui* (Aust.) Thér., western North America; and *J. wrightii* (Sull. in Gray) Thér., western North America with a disjunction in Bolivia.

References: Churchill (1982) and Thériot (1928).

7. *Indusiella* Broth. et C. Müll., Bot. Centralbl. 75: 321. 1898. emend. Churchill.
   T.: *Indusiella thian-schanica* Broth. et C. Müll.
   Coscinodontella Williams, Field Mus. Nat. Hist. Bot. 4: 129. 1927. syn. nov.
   T.: *Coscinodontella bryani* Williams

Stem in cross-section with central strand very indistinct or lacking. Leaves with upper and lower lamina differentiated, upper lamina incurved to enrolled, lower lamina very broad and sheath-like; bistratose only above. Peristome set below capsule mouth; calyptra plicate and deeply lobed.

Three species. *Indusiella andersonii* Delgadillo, northern Africa (Thchad); *I. bryanii* (Williams) Churchill, comb. nov., Peruvian Andes of South America; *I. thian-schanica* Broth. et C. Müll., eastern Europe and central Asia.

The reasons for the inclusion of Williams' *Coscinodontella* in the genus *Indusiella* were briefly discussed in the phylogenetic analysis section. *Indusiella bryanii* is considered the least derived taxon, and both *I. andersonii* and *I. thian-schanica* derived with the synapomorphies of a broader sheath and upper lamina margin that is strongly enrolled.

References: Brotherus (1898), Delgadillo M. (1976), and Williams (1927).

8. *Aligrimmia* Williams, Bull. New York Bot. Gard. 3: 123. 1903.
T.: *Aligrimmia peruviana* Williams

Stems with large central strand, ca. one third of stem width. Leaves closely imbricate, broadly ovate-lanceolate, obtuse, with margins incurved distally; 8–10 lamellae on adaxial surface. Annulus large and well developed; calyptra smooth above, plicate below, base lobate.

One species. *Aligrimmia peruviana* Williams, Peru, South America.

In establishing this taxon as a new genus, Williams pointed out the close resemblance to *Indusiella* which is in agreement with the present treatment.

Reference: Williams (1903).

C. Subfamily Ptychomitrioideae Brotherus emend. Jones in Grout, Moss Fl. N. Am. 2: 46. 1933.

Leaves lanceolate; costa with modified anatomy from the C-type (Kawai, 1968). Peristome divided nearly to the base, outer and inner peristomal layer equally thick, teeth densely papillose; basal membrane present.

9. *Racomitrium* Brid., Mant. Musc. 78. 1819.
LT.: *Racomitrium canescens* (Hedw.) Brid. (*Trichostomum canescens* Hedw.) (fide Pfeiffer, Nom. 2: 907. 1874)
Bucklandiella Roiv., Ann. Bot. Fenn. 9: 116. 1972.
Bucklandia Roiv., Arch. Soc. Zool. Bot. Fenn. "Vanamo" 9: 98. 1955. nom. illeg. (fide Robinson, 1974)
Dryptodon Brid., Bryol. Univ. 1: 191. 1826.

Plants cladocarpous. Stems erect to prostrate, often with abbreviated branches; lacking a central strand. Leaves often with an awn that is serrate or dentate; margins generally recurved; cells, both basal and median, sinuose, occasionally papillose. Dioicous. Seta twisted, usually to the right, smooth or occasionally scabrous; calyptra smooth mitrate, not plicate.

About 40 species. The 80 names listed in Index Muscorum (Wijk et al., 1959–1969) are likely an overestimate of the actual number. Probably not more than half this number are distinct. Phenotypically, *Racomitrium* is a notoriously variable group, equal possibly only to *Schistidium* (see Bremer, 1981). The genus is widely distributed throughout the world in mountainous or rocky places.

References: Deguchi (1977b), Noguchi (1974), Robinson (1974), Roivainen (1955a, 1955b, 1972), Lawton (1972) and Sloover (1977).

10. *Campylostelium* B. & S. in B.S.G., Bryol. Eur. 2: 25. 1846.
T.: *Campylostelium saxicola* (Web. & Mohr) B.S.G. (*Dicranum saxicola* Web. & Mohr) (fide B. & S. in B.S.G.)

Plants small, gregarious. Leaves linear-lanceolate from a slightly expanded base; entire; margins bistratose; basal cells pellucid. Cryptoicous. Seta long, cygneous; peristome inserted below capsule mouth. Calyptra mitrate, extending little beyond the operculum; smooth and lobate at base.

Two species. *Campylostelium saxicola* (Web. & Mohr) B.S.G., Europe and North America; and *C. strictum* Solms, Spain and Portugal.

Reference: Buck (1979).

11. *Ptychomitrium* Fuernr., Flora 12 Erg. 2: 19. 1829.
T.: *Ptychomitrium polyphyllum* (Sw.) B. & S. in B.S.G. (*Dicranum polyphyllum* Sw.)
Ptychomitriopsis Dix., Jour. Bot. 69: 284. 1931. Syn. nov.
T.: *Ptychomitriopsis africana* Dix.

Leaves crisped or occasionally appressed when dry; lanceolate or lanceolate-ligulate from an expanded base; apex obtuse or acute; upper margins serrate or entire; leaves rounded, not keeled; costa anatomy E-type (Kawai, 1968). Cryptoicous one or two sporophytes from a single perichaetium, perichaetial leaves similar to stem leaves in size or smaller. Peristome inserted below the capsule mouth; spores often granulate or papillose; calyptra campanulate, base deeply lobed.

Forty to fifty species. Widely distributed throughout the world with the greatest diversity found in the Gondwanaland areas. I believe there are two main monophyletic groups within the genus *Ptychomitrium.* The group of earliest origin includes those taxa with the plesiomorphic features of an entire margin and a usually single sporophyte from a perichaetium (the small species are possibly a derived group within the larger monophyletic group which also includes the derived taxon that Dixon named *Ptychomitriopsis*). The more derived group consists of those taxa with the apomorphic features of an upper serrate margin and two (rarely one or three) sporophytes from a single perichaetium.

References: Deguchi, 1977a; Dixon, 1931; Noguchi, 1954; and Sloover, 1976.

Summary of Nomenclatural Changes

Family

Scouleriaceae Churchill, fam. nov.
T.: **Scouleria** Hook. in Drumm.

Muscus rupestris aquaticus; caulis filo centrali destitutus; folia ligulata obtusa vel acuta, apice dentata vel serrata. Dioicus. Capsula magna primo sphaerica, peracta dehiscentia deflata ac depressa, columella costata tunc exserta, operculo columellae semper affixo; peristomium e dentibus 32 constans, cum formula Edwardsiana (8″: ′4–8″:′)4:′8c conveniens; sporae magnae 30–55 μm diam, papillis gemmatis.

Rupestral, aquatic moss; stem lacking a central strand; leaves ligulate, obtuse to acute, dentate or serrate at apex. Dioicous. Capsule large and spherical, when dehiscent, becoming deflated and depressed with columella exserted, columella ribbed, operculum remaining attached to columella; peristome of 32 teeth, peristomal formula of (8″:′4–8″:′)4″:′8c sensu Edwards (1979); spores large, 30–55 μm in diameter, with gemmate papillae.

Subfamily

Coscinodontoideae Churchill, subfam. nov.
T.: **Coscinodon** Spreng.

Folia plerumque e forma typica lanceolata transmutata, praesertim ovata vel lamina e basi lato angustata. Peristomium e dentibus perforatis 16 constans, stratis interiori et exteriori subaequicrassis, dense papillosum; calyptra campanulata plicata, basi erosa vel profunde lobata.

Leaves mostly modified from typical lanceolate shape, generally ovate or with a broad base and narrow upper lamina. Peristome of 16 perforated teeth, inner and outer layer equally or subequally thick, densely papillose; calyptra campanulate, plicate, erose to deeply lobed at base.

Species

Indusiella bryanii (Williams) Churchill, comb. nov.
Coscinodontella bryani Williams, Field Mus. Nat. Hist. Bot. 4: 129. 1927.
Ptychomitrium africanum (Dix.) Churchill, comb. nov.
Ptychomitriopsis africana Dix., Jour. Bot. 69: 284. 1931.
Schistidium occidentale (Lawton) Churchill, comb. nov.
Grimmia occidentalis Lawton, Bull. Torrey Bot. Club 94: 461. 1967.

Acknowledgements

I wish to thank William Buck (New York Botanical Garden) and Marshall Crosby (Missouri Botanical Garden) for their valuable comments and review of the manuscript. Rupert Barneby (New York Botanical Garden) was kind enough to translate the descriptions of new taxa into Latin. Birgittia Bremer (University of Stockholm) kindly provided published and unpublished data on *Schistidium.* Dale Vitt (University of Alberta) furnished unpublished data on the classification of moss families. A Grant-in-Aid from Sigma Xi provided funds for the SEM studies; help provided by Lorraine Hammer (University of Kansas) with the SEM was appreciated. Loans of specimens were obtained from MO, NY and SMU for study. Finally, I am very much indebted to Ed Wiley (University of Kansas) for comments and discussions on various stages of the manuscript; numerous discussions on the theory and practice of systematics have been invaluable to my understanding of the subject.

Biogeography and Cladistics

Fresh-Water Stingrays (Potamotrygonidae) and Their Helminth Parasites: Testing Hypotheses of Evolution and Coevolution

Daniel R. Brooks, Thomas B. Thorson and Monte A. Mayes

Table of Contents

Introduction

Although stingrays have been reported in fresh waters of all continents except Antarctica, most of them are not restricted to fresh water but are euryhaline forms capable of tolerating the full range of environmental salinity. Adjustments in body fluid chemistry made in response to lower ambient salinity involve primarily (*1*) a drop in the high urea content characteristic of marine elasmobranchs, and (*2*) cessation of salt excretion by the rectal gland. Serum may vary with ambient salinity, from about 500 mmol/liter urea content and 1000 mmol/liter total osmotic concentration when the rays are in full strength sea water, to approximately 150 mmol/liter urea and 600 mmol/liter total osmotic concentration when

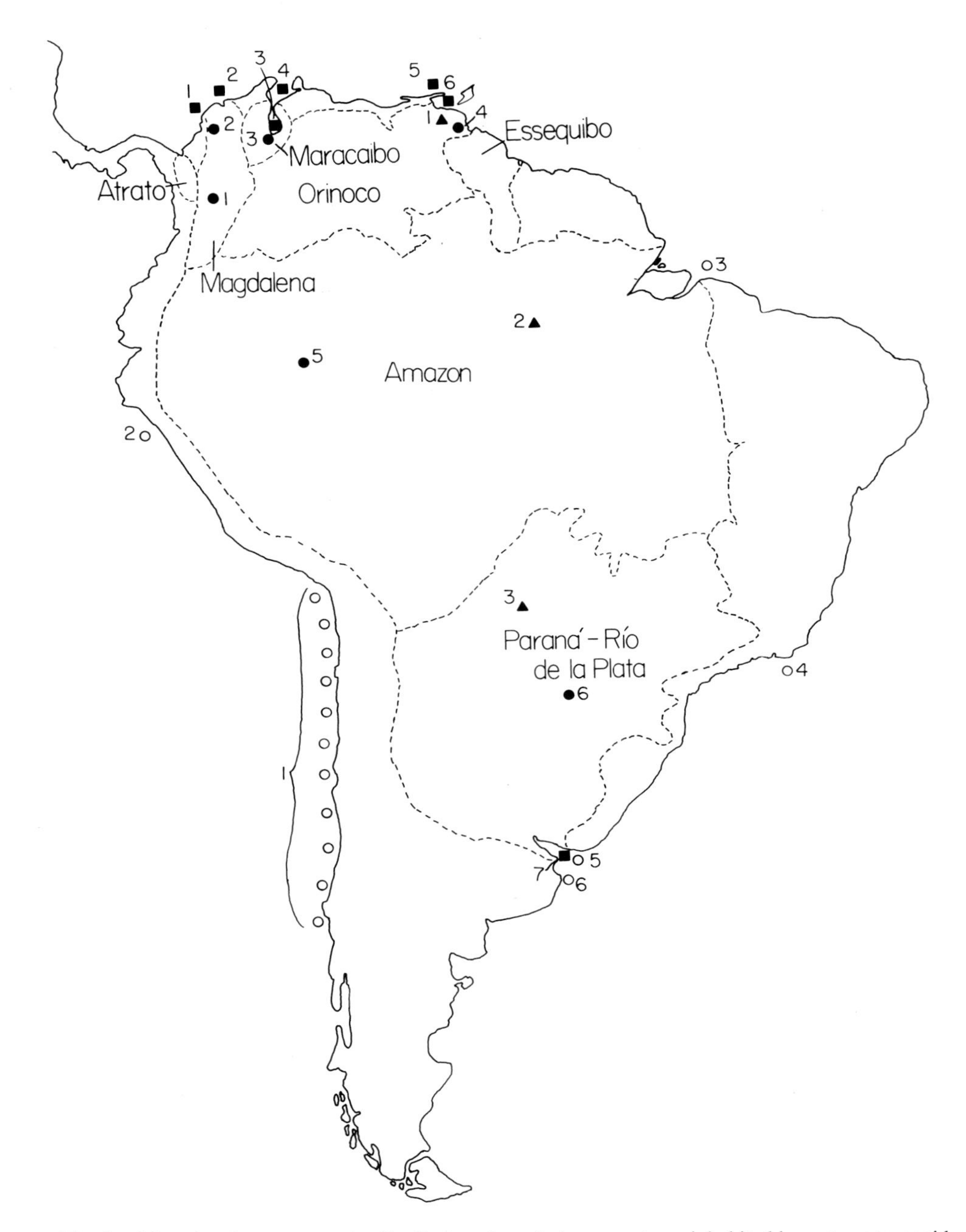

Fig. 1. Map showing seven major South American drainage systems inhabited by potamotrygonid stingrays and the localities from which helminth parasites infecting Neotropical elasmobranchs have been reported. *Closed circles* = reports of helminths infecting potamotrygonids made by authors: (1) LaDorada, Colombia; (2) San Cristóbal, Colombia; (3) several localities around Lake Maracaibo, Venezuela; (4) several localities in the Orinoco Delta, Venezuela; (5) Itacuai River, near its confluence with Javari River, Brazil; (6) Paraná River, Hohenau, Paraguay. *Closed squares* = reports of helminths infecting coastal batoids made by authors: (1) Cartagena, Colombia; (2) Ciénaga Grande de Santa Marta, Colombia; (3) several locations in Lake Maracaibo, Venezuela; (4) Caimare Chico, Gulf

in fresh water. These responses are similar to those which occur in the bull shark, *Carcharhinus leucas,* the sawfish, *Pristis perotteti* (Thorson, 1967; Thorson et al., 1973; Urist, 1962), and in various elasmobranchs reported by Homer Smith (1931, 1936).

Unique among elasmobranchs that occur in fresh water are members of the family Potamotrygonidae, the river rays of South America. These occur in all of the major river systems of that continent, excepting streams that empty into the Pacific, and into the Atlantic farther south than the Río de la Plata (Fig. 1). They are also absent from the Río Sao Francisco system in coastal Brazil.

Their uniqueness lies in the fact that they have adapted completely to life in fresh water. They not only live exclusively in fresh water throughout their life cycle, but they have completely abandoned the marine osmoregulatory features employed by all other elasmobranchs. They do not concentrate urea (Junqueira et al., 1968; Thorson et al., 1967), although urea is produced and excreted (Gerst and Thorson, 1977; Goldstein and Forester, 1971). Furthermore, the rectal gland has atrophied, although it is still present in reduced form (Thorson et al., 1978). If transferred to saline water, they are incapable of calling on their ancestral responses to the marine environment: they cannot retain urea, nor can they secrete salt via the rectal gland. They function essentially as fresh-water teleosts, and in an environment of more than 30–40% sea water their salt regulation breaks down and they die (Gerst and Thorson, 1977; Thorson, 1970; Griffith et al., 1973). In nature they are not known to occur in more than three parts per thousand salt (Thorson, Brooks and Mayes, In prep.).

The taxonomy and nomenclature of the potamotrygonids is in need of complete re-study. From whatever source they arose, they have been present in South America long enough for extensive speciation to have occurred. More than 20 species of the largest genus, *Potamotrygon,* have been named. However, some descriptions were based upon old and extremely limited accounts and/or illustrations; others upon very limited material, and in some cases only one specimen; and in one case (*P. hystrix*) the specimens on which the species was erected apparently represent as many as three species (Bailey, pers. comm.). Intraspecific variability in color patterns, probable hybridization, and the existence of large areas, including whole river systems, where no collecting has occurred, all add to the uncertainty and incompleteness of our knowledge about the group, frequently making identification extremely difficult.

The confusion extends also to higher taxa. In addition to *Potamotrygon,* two genera have been recognized, each including only one species: *Elipesurus spinicauda* and *Disceus thayeri.* A debate is before the International Commission on Zoological Nomenclature in which Castex (1968, 1969) recommends suppression of *Elipesurus* and retention of *Disceus,* and Bailey (1969) recommends retention of *Elipesurus* and relegation of *Disceus* to synonymy with it. On the familial level, the river rays are considered by some to represent a subfamily (Potamotrygoninae) of the Dasyatidae, and by others a separate family (Potamotrygoni-

---

←

of Venezuela; (5) Isla de Margarita, Venezuela; (6) Gulf of Paria, Venezuela; (7) Río de la Plata, off Montevideo, Uruguay. *Closed triangles* = reports of helminths infecting potamotrygonids made by other authors: (1) Orinoco Delta, Venezuela; (2) Lower Amazon, Brazil; (3) Rio Salobra, Mato Grosso, Brazil. *Open circles* = reports of helminths infecting coastal elasmobranchs made by other authors: (1) coast of Chile; (2) coast of Perú; (3) mouth of Amazon River, Brazil; (4) Rio de Janeiro, Brazil; (5) Río de la Plata, Uruguay; (6) Mar de la Plata, Argentina.

dae). In any case, their closest affinities have been considered to lie with the family Dasyatidae.

No fossil evidence provides clues to the history of the Potamotrygonidae. Dasyatid rays have been reported from fresh-water assemblages of the Tertiary, e.g., the Green River Formation of Wyoming (Cope, 1879; Marsh, 1877), but no fossils of potamotrygonids have been identified positively. Two references to fossil *Potamotrygon* have found their way into the literature, but both are highly questionable. A report of the genus from Tertiary deposits in the banks of the Río Paraná was based on a report by Larrazet (1886), who named two species of *Dynatobatis* from fossil dermal scales. Larrazet made no claims that they were of fresh-water origin and it is not known whether the specific deposits were of fresh-water or marine origin. However, Garman (1913), giving no evidence and without having seen Larrazet's material, referred to the latter's *Dynatobatis* as *Potamotrygon*. Garman apparently assumed that, because the fossils were found in a cut of a fresh-water river, where *Potamotrygon* are found today, they too were *Potamotrygon*. The weight of evidence does not support his conclusion [for a fuller discussion see Thorson and Watson (1975)]. The other reference was that by Arambourg (1947), who studied a fresh-water assemblage in Upper Tertiary deposits in southwestern Ethiopia including a species of stingray which he called *Potamotrygon africanus*. He assigned the fossil to *Potamotrygon*, not on the basis of the fossil spines which he pictured as syntypes of the new species, but on the basis of ecological and geographical parameters which he considered parallel to those of South American fresh-water rays. In view of the close correspondence between South American and African fresh-water fishes such as cichlids and characinids, he considered it possible that, at the end of the Tertiary, the Potamotrygonidae were also present in fresh-water habitats in Africa. But there is no direct support for this ray's assignment to *Potamotrygon*. Indeed, the wide occurrence of dasyatids in fresh water, both as fossil and extant forms, the restriction of living potamotrygonids to South America exclusively, and even Arambourg's (1947) illustrations of spines, strongly suggest that the ray should be assigned to the genus *Dasyatis* rather than *Potamotrygon* [for a fuller discussion see Thorson and Watson (1975)].

The lack of urea concentrating ability in the Potamotrygonidae has led to some speculation that these rays may have arisen from fresh-water ancestors that had never been marine. However, a marine ancestry is suggested by the presence of a rectal gland, even though it is considerably smaller than that of marine rays and has relatively few tubules scattered throughout a more extensive connective tissue matrix (Thorson et al., 1978). The elasmobranch rectal gland functions solely in salt excretion and would not be expected in forms lacking a marine ancestry. The atrophy of the rectal gland, the loss of urea-concentrating ability, and the substantial speciation that has occurred, all indicate a considerable length of time in fresh water for potamotrygonids. Nevertheless, the question of the fresh-water or marine ancestry bears further scrutiny.

Because all drainage systems inhabited by potamotrygonids empty on the Atlantic side of South America, the common assumption has been that the group arose from Atlantic ancestry. This view was strengthened with the description of a fresh-water stingray from the Benue River of Cameroon and Nigeria, called *Potamotrygon garouaensis* (Stauch and Blanc, 1962). This strongly suggested that the African species represented a relictual population surviving from the time Africa and South America started to move apart by the mid-Cretaceous, or that a common ancestor invaded both South American and African fresh-water habitats from the Atlantic during the same or a slightly later time period. However,

Castello (1973) and Thorson and Watson (1975) showed that the Benue ray was not a potamotrygonid, but was actually a member of the genus *Dasyatis*, both morphologically and physiologically. It has neither the median prepelvic process on the ischiopubic bar characteristic of all potamotrygonids, nor has it lost its urea-concentrating ability. It deals with fresh-water habitats as do all other euryhaline elasmobranchs (see above). The family Potamotrygonidae therefore is still known only from South America and there is no evidence available, including the geological history of the South American continent, that requires any specific geographical origin for the family.

Obviously, the evolutionary history of the family Potamotrygonidae is poorly known. There is no firm evidence concerning (*1*) whether they are monophyletic or were derived from two or more ancestral lineages, whether their geographical origin was (*2*) fresh-water or marine or (*3*) Atlantic, Pacific, or continental, (*4*) from what group of rays they were derived, or (*5*) whether they actually invaded fresh-water habitats or were trapped in them.

From 1975 to 1979 we conducted a field investigation of batoid elasmobranchs in major drainage systems of South America and the coastal areas near the outlets of those systems, in order to provide greater insight into various aspects of the evolution of fresh-water adaptations in elasmobranchs in general. Parts of that study will be published elsewhere (Thorson, Brooks and Mayes, In prep.). The portion of the study presented herein attempts to shed light on the five questions above, using evidence provided by the helminth parasite fauna of several potamotrygonids and various coastal ray species.

Four narrative hypotheses concerning the origins and evolution of the Potamotrygonidae are examined in terms of the helminth parasites found. Those hypotheses, differing from each other with respect to one or more of the five questions above, together with their critical assertions relevant to this study, are listed below.

Hypothesis 1. *Potamotrygonids represent a monophyletic or paraphyletic group which originated in fresh water, and are relicts of ancestral fresh-water stingrays.* Assertions: (*1*) monophyletic or paraphyletic parasite fauna; (*2*) parasites will be relatively most primitive members of their respective taxa whose other members inhabit other elasmobranchs or will represent a parasite fauna not occurring in marine elasmobranchs; (*3*) continental origin. Figure 2.

Hypothesis 2. *Potamotrygonids represent a polyphyletic grouping derived independently from several species or from populations of a single polytypic species of marine ancestor each of which invaded fresh water by penetrating a different river system.* Assertions: (*1*) polyphyletic parasite fauna; (*2*) closest relatives of potamotrygonid parasites found in one river system will be found in marine stingrays, presumably dasyatids, near the mouths of the rivers and not in potamotrygonids occurring in other river systems; (*3*) Atlantic origin; (*4*) sistergroups of all potamotrygonid taxa dasyatids; (*5*) geographic distribution consistent with a dispersal hypothesis. Figure 3.

Hypothesis 3. *Potamotrygonids represent a monophyletic group derived from a single species of marine dasyatid which invaded one river system (presumably the Amazon) and dispersed throughout South America.* Assertions: (*1*) monophyletic parasite fauna; (*2*) closest relatives of parasites occurring in one species of potamotrygonid occur in potamotrygonids in other river systems; (*3*) closest relatives of potamotrygonid parasites inhabit Atlantic dasyatids; (*4*) Atlantic origin; (*5*) geographic distribution consistent with a dispersal hypothesis. Figure 4.

Hypothesis 4. *Potamotrygonids represent a monophyletic group whose ancestor was a non-dasyatid marine stingray species trapped in South America by*

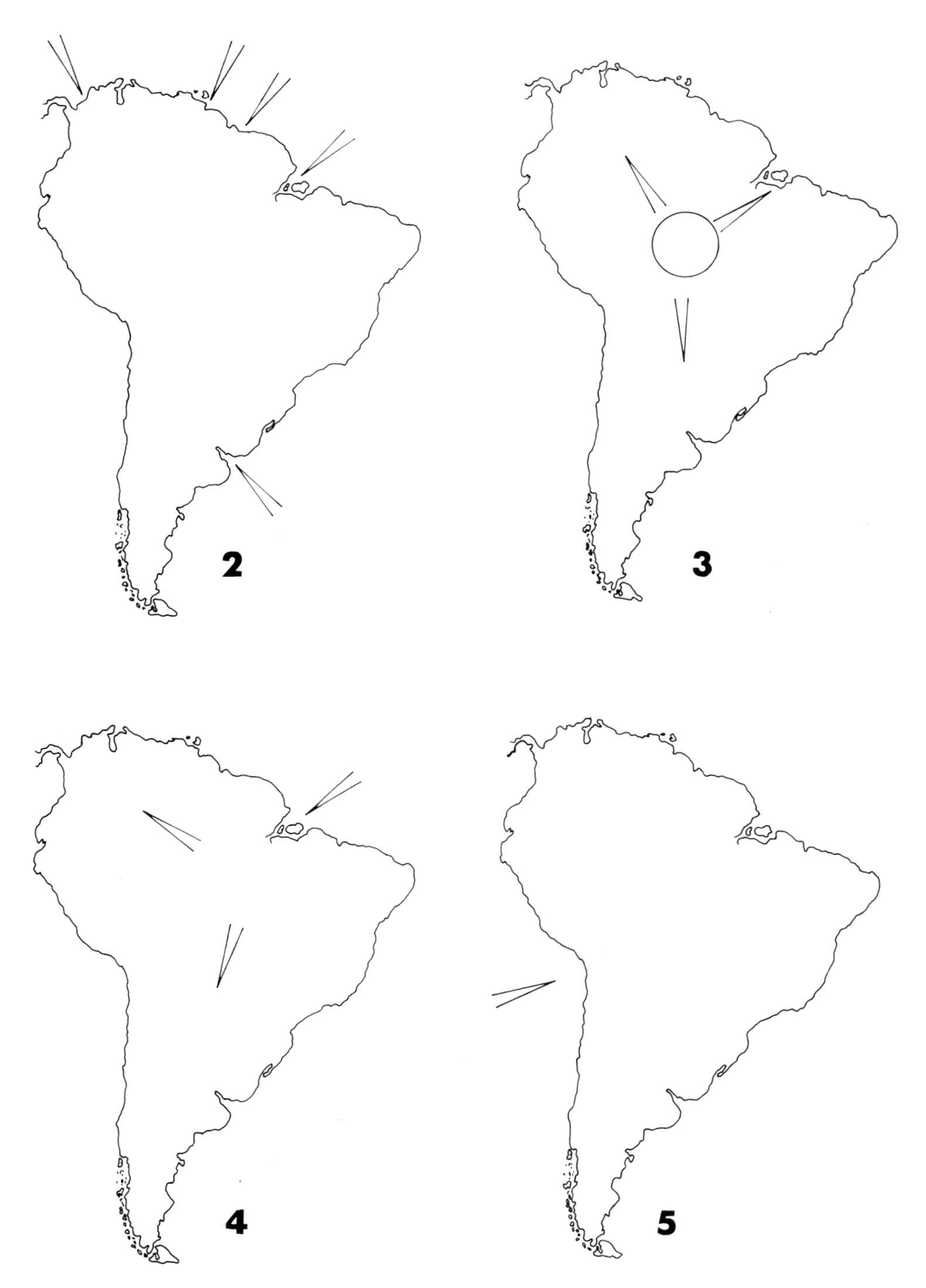

Figs. 2–5. Diagrams illustrating four different hypotheses concerning the origin and evolution of potamotrygonid stingrays. 2 = Hypothesis 1; 3 = Hypothesis 2; 4 = Hypothesis 3; 5 = Hypothesis 4. For descriptions of the hypotheses, refer to text.

Table I

Tabular representation of predictions made by each of four hypotheses concerning origin and evolution of fresh-water stingrays. FW = fresh-water; MA = marine; MO = monophyletic; PA = paraphyletic; PO = polyphyletic; CON = continental; ATL = Atlantic; PAC = Pacific; DIS = dispersed from fresh water; INV = invaded fresh water; ENT = entrapped in fresh water; AS = ancestral stingray; DA = dasyatid stingray; OS = other, non-dasyatid stingray

Hypothesis	Test points				
	1	2	3	4	5
1	FW	MO/PA	CON	DIS	AS
2	MA	PO	ATL	INV	DA
3	MA	MO	ATL	INV	DA
4	MA	MO	PAC	ENT	OS

Andean orogeny. Assertions: (*1*) monophyletic parasite fauna; (*2*) closest relatives of parasites occurring in potamotrygonids in one river system occur in potamotrygonids in other river systems; (*3*) closest relatives of potamotrygonid parasites inhabit Pacific non-dasyatid stingrays; (*4*) Pacific origin; (*5*) geographic distribution consistent with a vicariance hypothesis. Figure 5.

Although this study tests whole narrative evolutionary hypotheses, those hypotheses are presented in determinate form (cf. Brady, 1979) comprising five different elements of predictions. Table I presents the predictions of each hypothesis. We intend to demonstrate that parasitological data corroborate Hypothesis 4 and refute Hypotheses 1–3. Other hypotheses are possible based on various combinations of the five test categories. The four we present comprise three which embody the previous assumptions concerning the evolution of potamotrygonids (Hypotheses 1–3) and the only one which represents the uniformly corroborated combination of test elements.

Materials and Methods

Helminth parasites were taken from the alimentary tracts and gills of eight species of fresh-water stingrays taken at sites in five major drainage systems of South America (Magdalena, Maracaibo, Orinoco, Amazon, Paraná-Río de la Plata), and from eleven marine or euryhaline species of rays from coastal areas near the outlets of those systems (except the Amazon, Fig. 1). Additional data concerning helminths infecting South American elasmobranchs were obtained through a literature search. Our findings pertinent to the latter are presented, along with our collections, in Appendix I. Parasites were processed using standard procedures. Their identities are being reported in a series of publications. Cladograms were produced using the Wagner–78 program developed by James S. Farris, State University of New York, Stony Brook, whose cooperation and aid are gratefully acknowledged. Criteria upon which the various cladograms were constructed have been published elsewhere (Brooks et al., 1981b; Deardorff et al., 1981).

Point 1: Monophyletic or Polyphyletic?

The four hypotheses predict three types of phyletic histories for the potamotrygonids. Hypothesis 1 predicts a monophyletic or paraphyletic helminth fauna comprising relatively primitive members of various helminth taxa. Hypothesis 2 differs from the others by suggesting that fresh-water stingrays are polyphyletic,

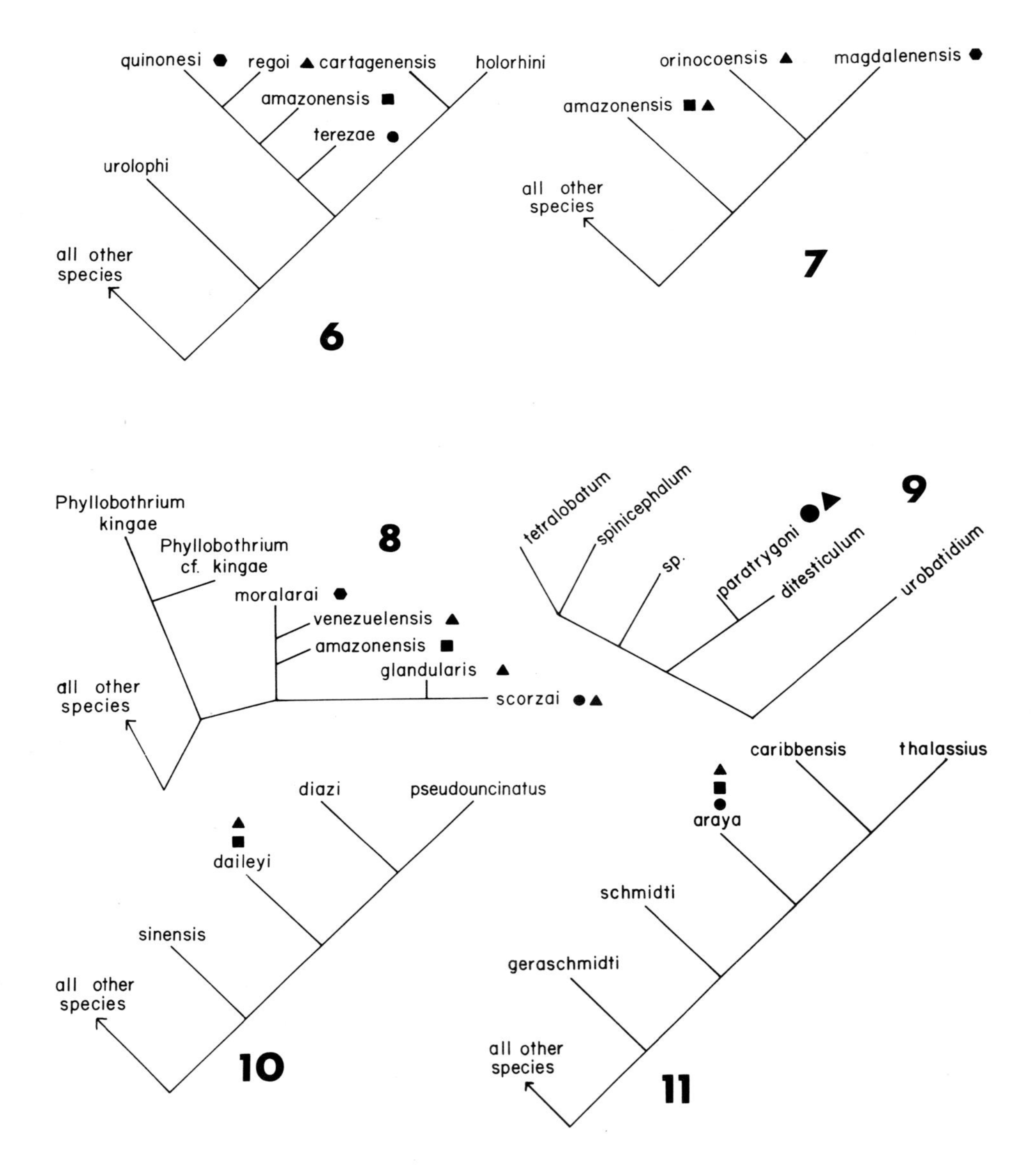

Figs. 6–11. Cladograms depicting phylogenetic relationships of helminth taxa including species inhabiting potamotrygonid stingrays. Species inhabiting potamotrygonids are indicated by geometric symbols above their names. Symbols refer to geographic areas in which the taxa occur. *Circles* = Paraná system; *Squares* = Amazon system; *Triangles* = Orinoco system; *Hexagons* = Magdalena system. 6. Cladogram of *Acanthobothrium* species-group. 7. Cladogram of *Potamotrygonocestus* species. 8. Cladogram of *Rhinebothroides* species and sister group. 9. Cladogram of *Rhinebothrium* species group. 10. Cladogram of *Eutetrarhynchus* species group. 11. Cladogram of *Echinocephalus* species group. Figures 6–9 represent tetraphyllidean cestode taxa, Figure 10 represents a trypanorhynchan cestode group, and Figure 11 represents a gnathostome nematode group.

and would thus assert that the helminth fauna would also be polyphyletic. Hypotheses 3 and 4 both predict a monophyletic helminth fauna including members of relatively advanced or derived taxa. Those predictions may be tested by constructing cladograms for each helminth taxon containing species infecting potamotrygonids, and then observing their relative phyletic positions.

The cladograms thus constructed (Figs. 6–11) clearly contradict the predictions of Hypotheses 1 and 2. Helminths infecting potamotrygonids comprise or are members of relatively highly derived monophyletic taxa. Those findings corroborate Hypotheses 3 and 4. Although Figures 6–11 depict only the monophyletic groups containing potamotrygonid parasites, complete analysis of the entire taxon was performed for four groups: *Acanthobothrium* (66 terminal taxa analyzed using 23 transformation series converted to 90 binary characters); *Rhinebothrium* (25 terminal taxa analyzed using 13 transformation series converted to 77 binary characters); *Eutetrarhynchus* (10 terminal taxa analyzed using 8 transformation series); and *Echinocephalus* (11 terminal taxa analyzed using 6 transformation series). Partial analysis was undertaken for the sister group of *Rhinebothroides*, the taxonomic catch-all *Phyllobothrium*, and a search is still underway to determine the sister group of *Potamotrygonocestus*.

Point 2: Fresh-Water or Marine?

In no case does a known helminth species inhabit both potamotrygonid species and any marine or euryhaline stingrays. Parasites infecting potamotrygonids further belong to relatively apomorphic lineages of their respective taxa, and when two or more members of the same genus occur in fresh-water stingrays, they comprise monophyletic groups within their inclusive taxa whose other members infect marine stingrays. There are two exceptions: *Potamotrygon magdalenae* hosts *Paravitellotrema overstreeti*, a hemiurid digenean representing a taxon whose other members infect fresh-water teleosts; *Potamotrygon hystrix* hosts *Megapriapus ungriai*, an echinorhynchid acanthocephalan related most closely to species occurring in fresh-water teleosts. However, no potamotrygonid hosted any specimens of proteocephalidean cestodes despite their obvious suitability as hosts for cestodes. Proteocephalideans infect primarily ostariophysans, and exhibit their greatest diversity in South America (Brooks, 1978). Thus, the two exceptions listed above do not seem to represent a general pattern of parasites infecting potamotrygonids which are not most closely related to parasites occurring in other elasmobranchs.

Biogeographic analysis discussed next shows that the relative positions of all species infecting potamotrygonids in their particular inclusive monophyletic group are equivalent. Potamotrygonids host a monophyletic and highly-derived helminth fauna. That helminth fauna comprises species whose closest relatives occur in marine elasmobranchs. The last two observations clearly refute Hypotheses 1 and 2. However, both Hypothesis 3 and Hypothesis 4 are corroborated by those observations. The next three predictions represent critical tests of Hypotheses 3 and 4 but also provide additional arguments against Hypotheses 1 and 2.

There are three main differences between Hypothesis 3 and Hypothesis 4. The former predicts a single invasion by a dasyatid ancestor from the Atlantic which could have occurred only after the Atlantic rifting had progressed to a point that marine or euryhaline stingrays occurred along the eastern shoreline of South America. The latter hypothesis predicts that ancestral non-dasyatids were entrapped by Andean orogeny along the Pacific coastline of South America, and

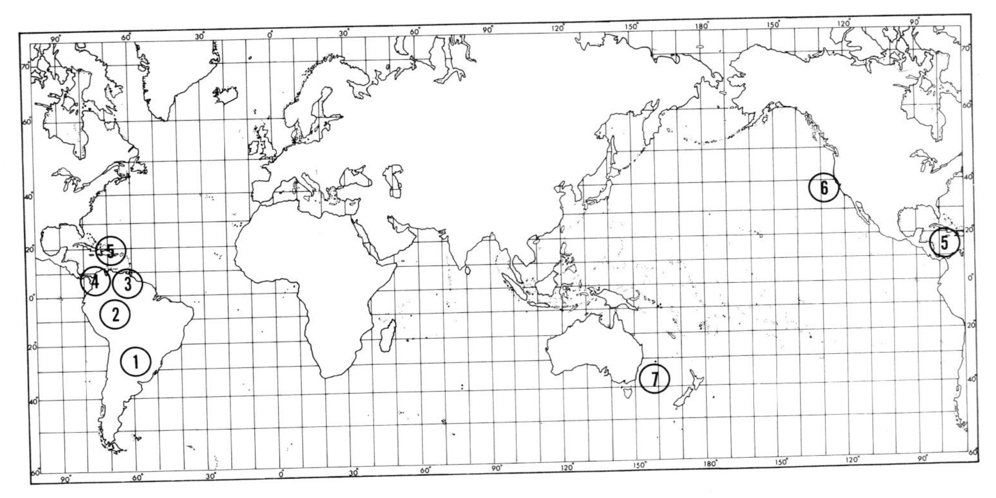

Fig. 12. Map depicting geographic distribution of members of *Acanthobothrium* species group containing those inhabiting potamotrygonid stingrays. 1 = *A. terezae*; 2 = *A. amazonensis*; 3 = *A. regoi*; 4 = *A. quinonesi*; 5 = *A. cartagenensis*; 6 = *A. holorhini*; 7 = *A. urolophi*.

that the ancestral distribution must have reached an appropriate extent for vicariance before Andean orogeny began. There appears to be no a priori reason to prefer one scenario over another. Only a bias toward vicariance and allopatric speciation would cause one to prefer entrapment over invasion as causal explanations for the origin and evolution of potamotrygonids. We next test each of the three areas of disagreement between the two remaining viable hypotheses.

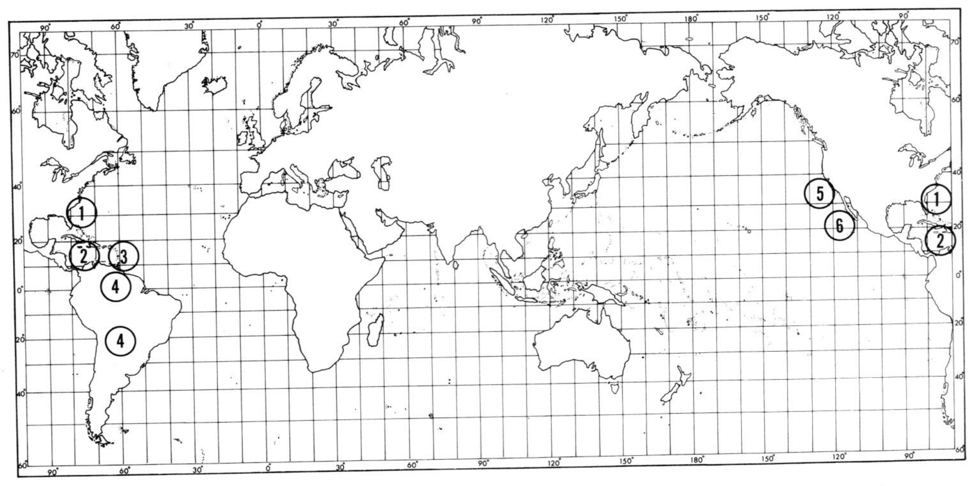

Fig. 13. Map depicting geographic distribution of members of *Rhinebothrium* species group containing *R. paratrygoni*, inhabiting potamotrygonid stingrays. 1 = *R. spinicephalum*; 2 = *R. tetralobatum*; 3 = *R. margaritensis*; 4 = *R. paratrygoni*; 5 = *R. ditesticulum*; 6 = *R. urobatidium*.

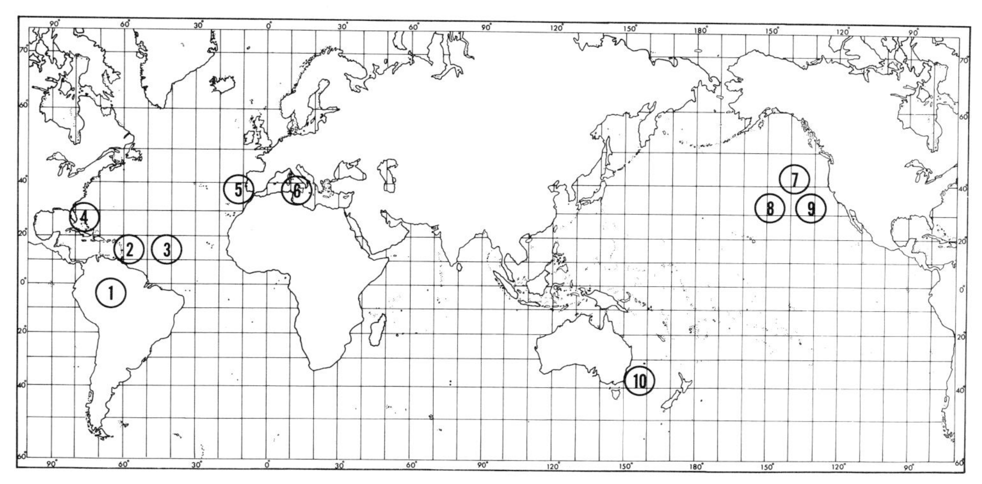

Fig. 14. Map depicting geographic distribution of members of *Eutetrarhynchus*. 1 = *E. araya*; 2 = *E. thalassius*; 3 = *E. caribbensis*; 4 = *E. lineatus*; 5 = *E. ruficollis*; 6 = *E. carayoni*; 7 = *E. macrotrachelus*; 8 = *E. litocephalus*; 9 = *E. schmidti*; 10 = *E. geraschmidti*.

Point 3: Area of Derivation

The general geographic distributions of the closest genealogical relatives of parasite taxa whose members infect potamotrygonids illustrate a redundant distribution pattern (Figs. 12–15). That redundant pattern is clearly circum-Pacific, with the exception of single species in the Caribbean or of sister taxa in the Caribbean and northern West Atlantic. This suggested to us that the geographic affinities of helminths infecting potamotrygonids lay in the Pacific and not the

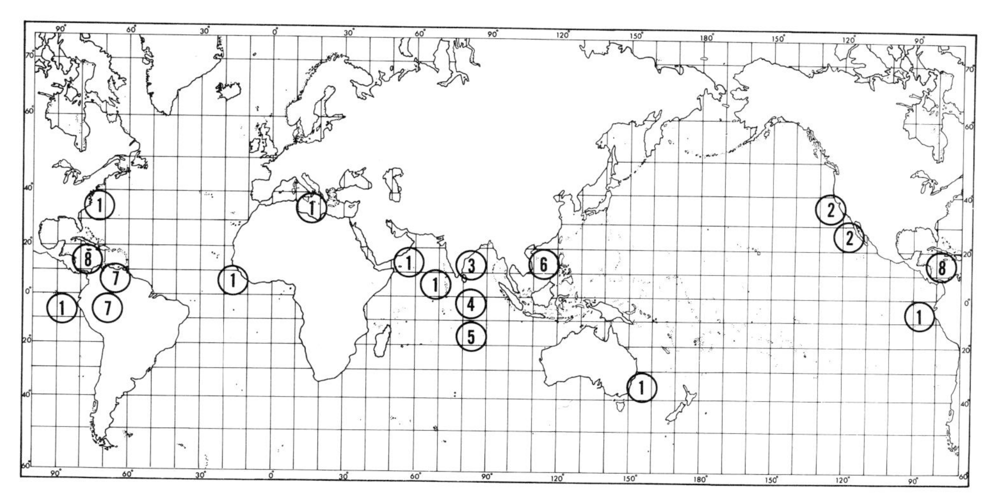

Fig. 15. Map depicting geographic distribution of members of *Echinocephalus*. 1 = *E. uncinatus*; 2 = *E. pseudouncinatus*; 3 = *E. mobulae*; 4 = *E. multidentatus*; 5 = *E. southwelli*; 6 = *E. sinensis*; 7 = *E. daileyi*, 8 = *E. diazi*.

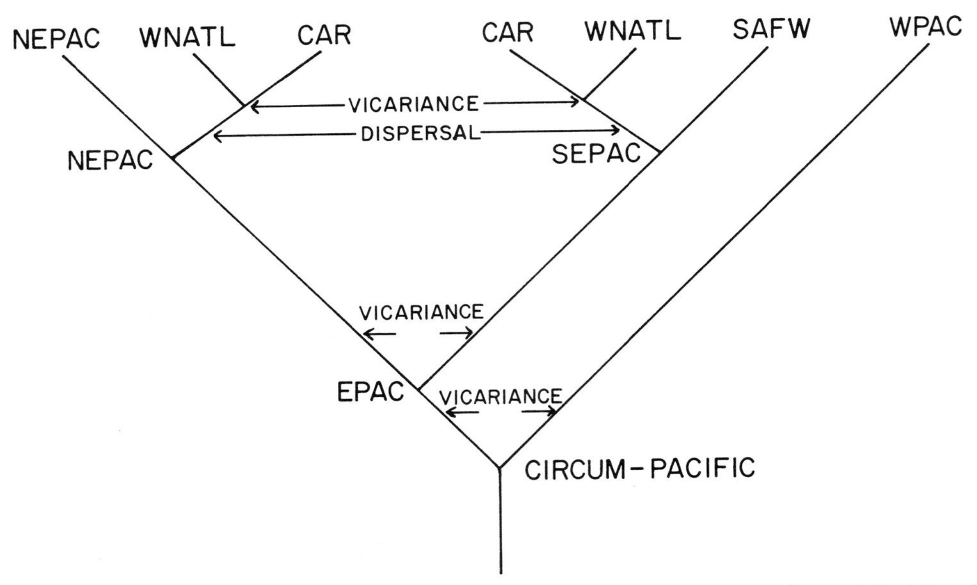

Fig. 16. Cladogram depicting historical relationships of various circum-Pacific and Caribbean/ Atlantic areas based on geographic distributions of taxa depicted in Figures 6–11 and 12–15. NEPAC = Northeastern Pacific; WNATL = Western North Atlantic; CAR = Caribbean; SAFW = South American fresh water; WPAC = Western Pacific; SEPAC = Southeastern Pacific; EPAC = Eastern Pacific.

Atlantic, corroborating Hypothesis 4 and refuting Hypothesis 3. A more critical look at the distribution patterns confirmed that suggestion.

The geographic distributions of helminths infecting potamotrygonids, and their closest relatives, involve six general regions: the western Pacific (WPAC), the northeastern Pacific (NEPAC), the southeastern Pacific (SEPAC), South American fresh-water habitats (SAFW), the Caribbean (CAR), and the northern West Atlantic (NWATL). The patterns of genealogical relationships depicted in Figures 6–11 relate the six areas in the manner shown in Figure 16. Viewed in a more specific, genealogical manner the Atlantic-Caribbean species are relatively more highly-derived than the species infecting potamotrygonids. Their occurrence in the Caribbean and Atlantic apparently results from dispersal of NEPAC and SEPAC elements into the Caribbean-Atlantic basin with subsequent isolation and divergence from their Pacific relatives. The NEPAC and SEPAC elements themselves represent apparent vicariant elements of an earlier, more widespread eastern Pacific fauna from which the helminth parasites endemic to potamotrygonids were derived. Thus, the history of the Atlantic-Caribbean species could not pertain directly to events producing the species of helminths inhabiting potamotrygonids. These data corroborate Hypothesis 4 and refute Hypothesis 3. Further, because the biogeographic status of each helminth species infecting potamotrygonids relative to its closest phylogenetic relatives is equivalent, the monophyletic nature of the helminth fauna of potamotrygonids is further underscored and Hypothesis 2 further refuted.

Table II

Summary of host genera for closest genealogical relatives of helminth parasites inhabiting fresh-water stingrays

Helminth taxon infecting potamotrygonids	Closest genealogical relative	Host genus (-era) for closest relatives
Rhinebothroides spp.	*Phyllobothrium kingae*	*Urolophus*
	Phyllobothrium cf. *kingae*	*Urolophus*, *Dasyatis*
Eutetrarhynchus araya	*Eutetrarhynchus thalassius*	*Urolophus*
	Eutetrarhynchus caribbensis	*Urolophus*
	Eutetrarhynchus schmidti	*Urolophus*
	Eutetrarhynchus geraschmidti	*Urolophus*
Echinocephalus daileyi	*Echinocephalus pseudouncinatus*	*Urolophus*, *Myliobatis*
	Echinocephalus diazi	*Himantura*
	Echinocephalus sinensis	*Aetobatus*
Acanthobothrium spp.	*Acanthobothrium cartagenensis*	*Urolophus*
	Acanthobothrium holorhini	*Myliobatis*
	Acanthobothrium urolophi	*Urolophus*
Rhinebothrium paratrygoni	*Rhinebothrium urobatidium*	*Urolophus*
	Rhinebothrium ditesticulum	*Urolophus*
	Rhinebothrium tetralobatum	*Himantura*
	Rhinebothrium spinicephalum	*Dasyatis*
	Rhinebothrium margaritensis	*Dasyatis*

Point 4: What is the Sister Group?

Hypothesis 3 stipulates that potamotrygonids are derived from dasyatid ancestors whereas Hypothesis 4 stipulates non-dasyatid ancestors. We tabulated the stingrays known to host helminth species closely-related to those occurring in potamotrygonids (Table II). In only three cases does a species of *Dasyatis* appear, and in only five cases does a dasyatid serve as host. For one of those five cases, a member of *Urolophus* is also a known host. By contrast, *Aetobatus* and *Myliobatis* (1 record and 2 records, respectively) of the Myliobatidae and *Urolophus* (11 records) of the Urolophidae represent the majority of hosts. An additional observation, from Point 3, also pertains. Species related to potamotrygonids' helminths and infecting dasyatids occur in relatively more derived positions on their respective cladograms than do those infecting potamotrygonids; they represent the Atlantic-Caribbean dispersal elements depicted in Figure 16. Those helminths most closely related to species occurring in potamotrygonids as plesiomorphic sister groups *all* inhabit members of *Urolophus*.

Such an observation is suggestive that at least some urolophids and all potamotrygonids harbor parasites derived from a common ancestral helminth fauna. Because the general pattern of host-parasite relationships, from a historical point of view, indicates concomitant allopatric speciation by both groups (Brooks, 1979), that common ancestral parasite fauna may have occurred in a common ancestral host, suggesting close phylogenetic relationships between urolophids and potamotrygonids. The alternative explanation requires that some dasyatids represent the sister group of potamotrygonids phylogenetically but that potamotrygonids lost all their parasites during divergence from dasyatids, then re-

Table III

Distribution of various batoid genera occurring in the Pacific region, with emphasis on their occurrence or absence along the west coast of South America

Genus	Occurs on west coast of South America	Circum-Pacific distribution	Trans-Pacific distribution
Pristis	yes (1 sp.)	yes	no
Rhinobatos	yes (1 sp.)	yes	no
Platyrhinoidis	yes (1 sp.)	yes	no
Torpedo	no	yes	no
Narcine	no	yes	no
Raja	yes (1 sp.)	yes	no
Psammobatis	yes (2 spp.)	yes	no
Dasyatis	no	yes	no
Gymnura	yes (1 sp.)	yes	no
Urolophus	yes (2 spp.)	yes	no
Urotrygon	yes (1 sp.)	no	yes (?)
Myliobatis	yes (1 sp.)	no	yes
Aetobatus	no	yes	no
Rhinoptera	no	no	yes
Mobula	no	no	yes
Manta	no	no	yes

acquired a parasite fauna from urolophids, and then became isolated in freshwater habitats. We prefer the more parsimonious explanation, namely, that potamotrygonids and urolophids share a unique common ancestor and that the close phylogenetic relationship between them is illustrated by their hosting closely related helminth parasites. This conclusion supports Hypothesis 4 and refutes Hypothesis 3. Further, we emend Hypothesis 4 to be more restrictive and hypothesize that potamotrygonids and at least some urolophids share a unique common ancestor (are sister groups) rather than simply stating that potamotrygonids are derived from non-dasyatids.

Figure 17 depicts the geographic distribution of urolophid stingrays. That distribution pattern coincides with the distribution patterns exhibited in Figures 12–15 and summarized in Figure 16. This observation further strengthens our contention that at least some urolophids and potamotrygonids are sister groups. Note also the restricted distribution of *Urolophus* along the western coastline of South America. Table III indicates that a number of batoid elasmobranchs exhibit similar distribution patterns, suggesting that the limited occurrence of *Urolophus* along that coast corresponds more to a general phenomenon than to an instance of random extinction. The general phenomenon, in our estimation, was the Andean orogeny, produced when eastward-moving tectonic plates were subducted by the westward-moving Americas plate. This resulted not only in orogenic activity but also in the formation of a trench along the subduction zone, extending along present-day South America. As a result of the trench formation, sediments were subducted along the western side of the Andes and organisms such as stingrays, which fed on sediment-dwelling organisms, were bound to find the area progressively less productive for food-gathering. However, any stingrays occurring east of the Andes during this initial orogeny would not have faced the problem of reduction in nutrient-rich sediments although they would experience progressive

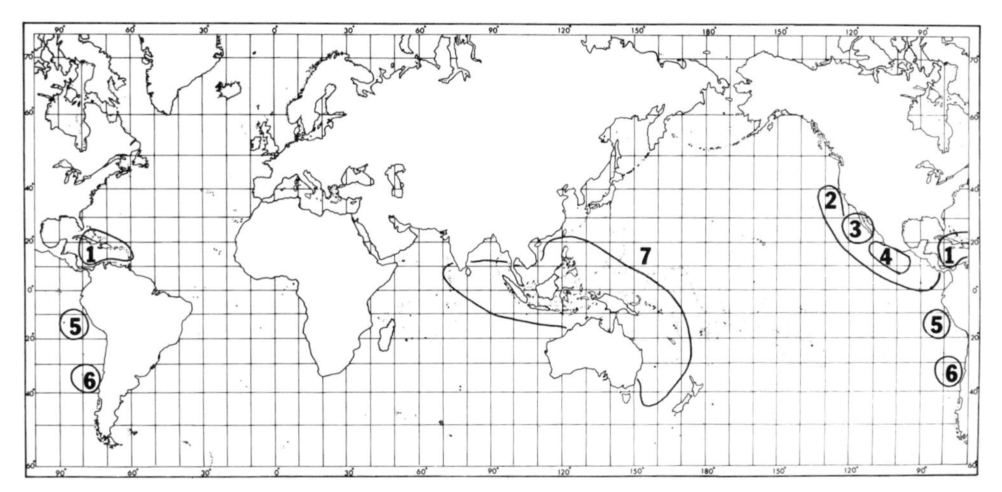

Fig. 17. Geographic distribution of stingrays comprising the genus *Urolophus*. 1 = *U. jamaicensis*; 2 = *U. halleri*; 3 = *U. maculatus*; 4 = *U. concentricus*; 5 = *U. tumbesensis*; 6 = *U. marmoratus*; 7 = western Pacific members of the genus, comprising 15 nominal species.

desalination of their habitats. Hypothesis 4 thus suggests that potamotrygonids represent the descendants of *Urolophis*-like ancestors which occurred along the Pacific coast of South America at the time of Andean orogeny. They represent part of a Pacific coastal elasmobranch fauna reduced and altered by Andean orogeny. They would, accordingly, be the product of entrapment and isolation, with subsequent evolution in place, rather than the result of invasion or dispersal. Thus, the next test is the most critical of the last three.

Point 5: Vicariance or Dispersal?

The above findings are suggestive but may be viewed as incomplete in some aspects. If Hypothesis 4 is correct, patterns of geographic distribution of the various parasite groups infecting potamotrygonids should be concordant and should be consistent with the geographic history of South America since the Andean orogeny. Following the method proposed by Platnick and Nelson (1978) and Rosen (1978), we first identified centers of endemism for parasites infecting potamotrygonids (Tables IV and V). We then constructed a cladogram, based on geological history studies, representing the areal relationships of the centers of endemism (Figs. 18–22). Finally, we converted the taxon cladograms into area cladograms by substituting geographic distributions for taxon names as terminal taxa (symbols above taxon names in Figs. 6–11), and compared the area cladogram with the taxon cladograms looking for concordance in branching patterns.

We identified four areas of endemism among known localities for helminths infecting fresh-water stingrays: (*1*) the Magdalena River, (*2*) the delta of the Orinoco River, (*3*) the Amazon River, and (*4*) the Paraná River system. Next, we formulated the following simplified geographic history of South America pertinent to the areas of endemism based primarily on information presented by Kummel (1962), Harrington (1965), and Putzer (1969).

1. *Beginning of the Cretaceous:* widest marine (Pacific) ingression, west coast of South America wrapped around Brazilian shield. Figure 18.

Table IV

Geographic distribution of helminth parasites inhabiting freshwater stingrays. 1 = Magdalena River system; 2 = Maracaibo region; 3 = Orinoco River system; 4 = Amazon River system; 5 = Paraná River system. + = collected and described by authors; [1] = collected by authors but previously described. X = present; – = absent from area

Helminth species	Areas of occurrence					
	1	2	3	4	5	Endemic
Rhinebothrium paratrygoni[1]	–	–	X	–	X	
Eutetrarhynchus araya[1]	–	–	X	X	X	
Potamotrygonocestus magdalenensis+	X	–	–	–	–	1
Potamotrygonocestus orinocoensis+	–	–	X	–	–	3
Potamotrygonocestus amazonensis+	–	X	X	X	–	
Rhinebothroides moralarai+	X	–	–	–	–	1
Rhinebothroides venezuelensis+	–	X	X	–	–	
Rhinebothroides amazonensis+	–	–	–	X	–	4
Rhinebothroides scorzai[1]	–	–	X	–	X	
Rhinebothroides freitasi	–	–	–	X	–	4
Rhinebothroides glandularis+	–	–	X	–	–	3
Acanthobothrium quinonesi+	X	X	–	–	–	
Acanthobothrium regoi+	–	–	X	–	–	3
Acanthobothrium amazonensis+	–	–	–	X	–	4
Acanthobothrium terezae	–	–	–	–	X	5
Paravitellotrema overstreeti+	X	–	–	–	–	1
Echinocephalus daileyi+	–	–	X	X	–	
Terranova edcaballeroi[1]	–	–	X	–	–	3
Brevimulticaecum sp.	–	–	–	–	X	5
Megapriapus ungriai	–	–	X	–	–	3
Leiperia gracile	–	–	–	–	X	5
Heteronchocotyle tsalickisi+	–	–	–	X	–	4
Potamotrygonocotyle amazonensis+	–	–	–	X	–	4

2. *Mid-Cretaceous:* Chilean cordillera uplifted, Brazilian "sea" cut off from ocean except at northern end. Figure 19.

3. *Late Cretaceous:* Brazilian, Bolivian, and Uruguayan interior continental and fresh water, and remain so for the rest of history.

4. *Late Cretaceous-Paleocene:* Paraná system isolated, Colombia and Venezuela under marine waters, Venezuela active tectonically. Figure 20.

5. *Paleocene:* Maracaibo area marine, Colombia and Venezuela mostly continental for the first time.

Table V

Levels of endemism for areas in which fresh-water stingray helminths have been collected, based on numbers and percentages of endemic helminth species

Locality	# Species	# Endemics	% Endemics
Magdalena River, Colombia	4	3	75
Maracaibo region, Venezuela	3	0	0
Delta of Orinoco River, Venezuela	10	5	50
Amazon River system, Brazil	8	5	62.5
Paraná River system, Brazil and Paraguay	6	3	50

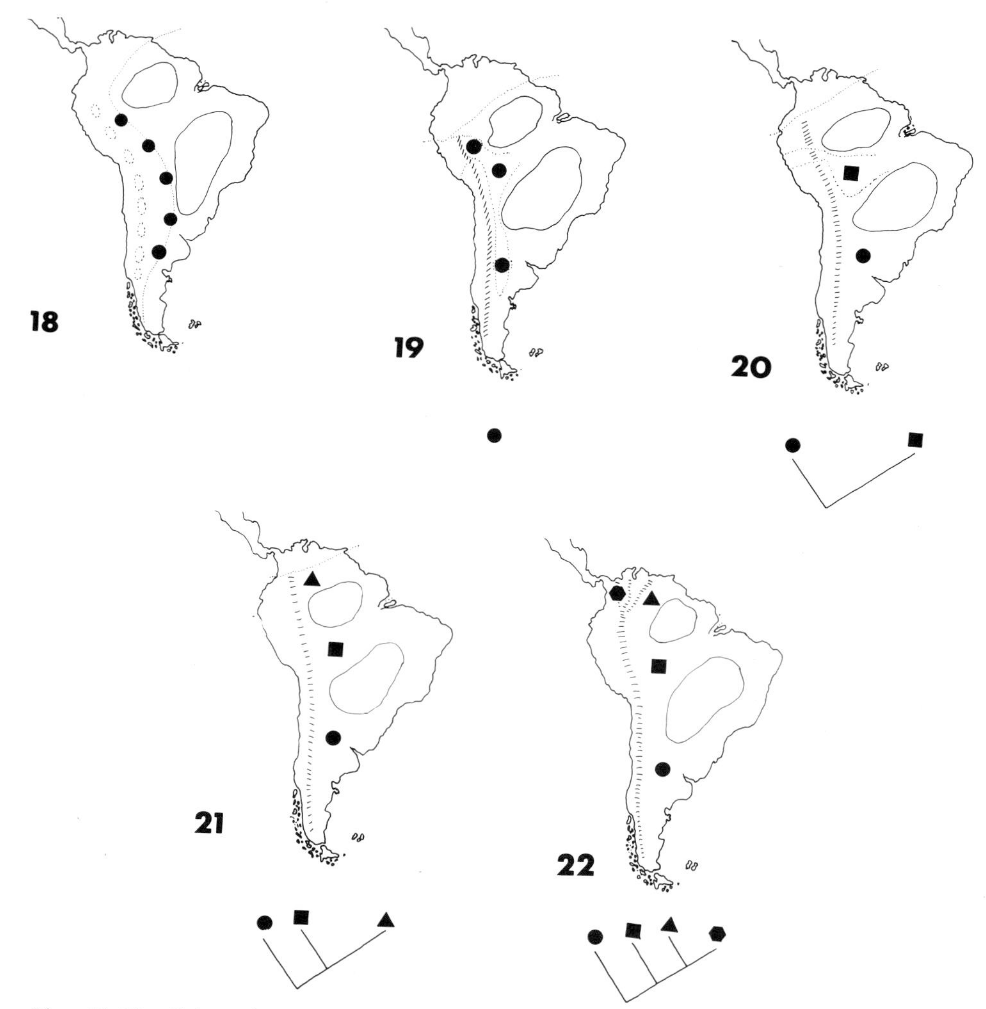

Figs. 18–22. Schematic diagrams depicting historical events producing areas of endemism for helminths inhabiting potamotrygonid stingrays. 18 = beginning of Cretaceous; 19 = mid-Cretaceous; 20 = late Cretaceous–Paleocene; 21 = Miocene–Pliocene; 22 = Pliocene–Pleistocene. Circles in Figures 18 and 19 represent distribution of ancestral stingrays. For Figures 20–22, circles represent vicariant elements in the Paraná system, squares represent vicariant elements in the Amazon system, triangles represent vicariant elements in the Orinoco system, and hexagons represent vicariant elements in the Magdalena system.

6. *Eocene:* Maracaibo area fresh water.

7. *Oligocene:* Maracaibo area marine again.

8. *Miocene-Pliocene:* Maracaibo area along with Colombia and Venezuela initially marine but become fresh water for rest of history at end; terminal uplifting of Andean cordillera closes off Pacific opening of Amazon; Amazon basin becomes distinct entity. Figure 21.

9. *Pliocene-Pleistocene:* finishing of cordilleran uplift isolates Magdalena and Orinoco systems, cuts off Maracaibo area from other river drainages. Figure 22.

Figure 23 contains the areal cladogram predicted by the above geological reconstructions. There is a high degree of concordance among the geographic distribution patterns of potamotrygonids' helminth parasites (Figs. 6–11) and Figure 22, suggesting a common evolutionary history for those helminths influenced by geographic changes since the beginning of the Cretaceous. That implies that the common ancestor for each parasite taxon and for the host taxon were already distributed along areas which could be cut off by Andean orogeny and associated geographic changes before such events began. A hypothesis of Mesozoic ancestral distributions over a wide area is not only consistent with predictions made by Hypothesis 4 for Point 5 but is also consistent with the distribution patterns of parasites related to those occurring in potamotrygonids but which themselves infect marine stingrays (Fig. 16). For a vicariance explanation it is necessary, of course, for the geological events affecting the fresh-water stingray ancestors to have occurred during the same time as those events affecting their marine relatives. Thus, a further prediction of the hypotheses, again corroborated only by Hypothesis 4, is that Point 3 and Point 5 refer to interdependent phenomena. Variations in concordance of branching patterns among the various cladograms reflect vagaries of collecting, or instances of apparent isolation without speciation or post-speciation dispersion (see Anderson and Evenson, 1979) (*Rhinebothrium paratrygoni, Eutetrarhynchus araya, Rhinebothroides venezuelensis, R. scorzai, Acanthobothrium quinonesi, Potamotrygonocestus amazonensis*). For three of those species, the pattern of endemism is broken by their occurrence in the Lake Maracaibo area, where no endemic parasites occur. We will address Maracaibo again later.

On the basis of the above discussion, Hypothesis 4 states that since the Cretaceous, potamotrygonids and their parasites have evolved in a manner congruent with the evolution of South American fresh-water habitats. The discovery that a single distribution pattern conforms to a generalized pattern constitutes a critical test of any such vicariance hypothesis. One could argue that the six helminth taxa comprise a generalized pattern, one including also their obligate intermediate and definitive hosts. One of us has argued such a case for studies of parasite biogeography (Brooks, 1979). However, if one wanted to consider a unified parasite fauna as a non-independent element, analysis of other taxa might be necessary to corroborate a vicariance hypothesis. Such a critical analysis is beyond the scope of this study, but two previously published works are suggestive of a generalized pattern. First, Gery (1969) discussed the zoogeographical affinities of South American fishes. He divided South America into eight regions, four of which correspond to the areas of endemism discovered for potamotrygonids' parasites (Fig. 23). In discussing those areas, Gery made the following observations: (*1*) the Magdalenean ichthyofauna was probably derived from a portion of the Amazonian fauna, (*2*) the Maracaibo area exhibits strong affinities with both the Orinocoan and Magdalenean ichthyofaunas, (*3*) the Orinocoan fauna is apparently derived from a portion of the Amazonian fauna, and (*4*) there are many pairs of apparent sister species represented in the Amazonian-Paranean faunas. All four of those points are consistent with the geological scenario presented herein and also consistent with Hypothesis 4. And secondly, Croizat (1964) presented a map detailing his views of the essential points of South American biogeography based primarily on terrestrial organisms. His map (redrawn in Fig. 24) indirectly depicted a number of lowlands-aquatic areas of biogeographic importance, four of which represent the four areas of endemism which we discovered for potamotrygonids' parasites.

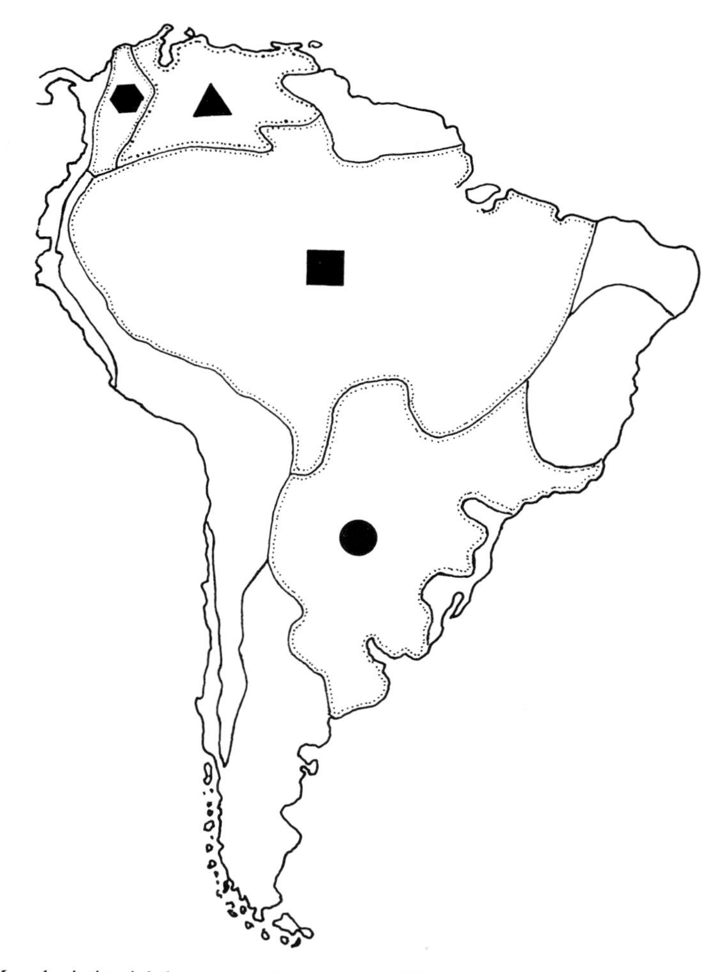

Fig. 23. Map depicting ichthyogeographic regions of South America as proposed by Gery (1969). Four of the regions correspond to the areas of endemism for helminths inhabiting potamotrygonid stingrays (symbols in regions with stippled outlines).

Our analysis of the relationships of endemic areas also produced evidence that the helminth fauna of the non-endemic area, the Maracaibo region, further corroborates Hypothesis 4. Gery's points 2 and 3 pertain directly to the helminth fauna we discovered in the Maracaibo area. We collected specimens of *Rhinebothroides venezuelensis,* known from the delta of the Orinoco River, specimens of *Acanthobothrium quinonesi,* known from the Magdalena River, and specimens of *Potamotrygonocestus amazonensis,* known from the Upper Amazon River system. That collection represents exactly the faunal mix predicted by Gery's study.

We have examined all four hypotheses in terms of five categories of predictions. Only for Hypothesis 4 were the predictions for all five categories confirmed by our observations (Table VI). For Hypothesis 1 no predictions were confirmed,

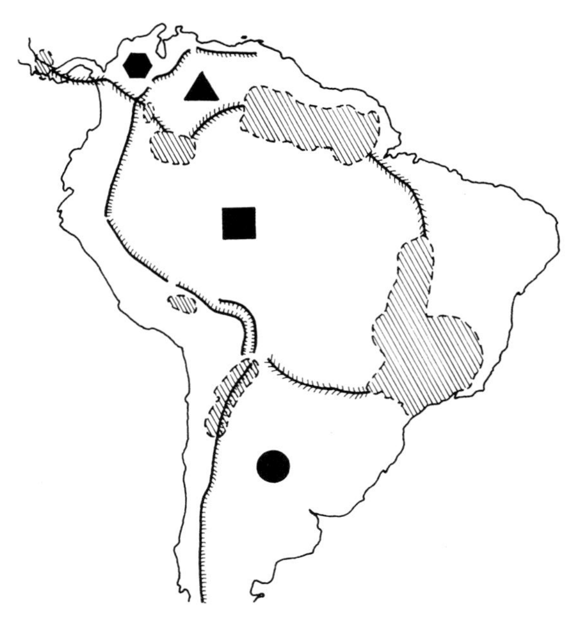

Fig. 24. Map depicting Croizat's (1964) concept of the essential points of South American biogeography. Symbols represent the four areas of endemism for helminths inhabiting potamotrygonid stingrays.

for Hypothesis 2 one prediction was confirmed, and for Hypothesis 3 two predictions were confirmed. However, the predictions confirmed by Hypotheses 2 and 3 were general predictions which also satisfied at least one other hypothesis. Each of those hypotheses was refuted by at least three sets of observations.

If the distribution patterns indicated by the genealogical relationships of helminths infecting potamotrygonids and their close relatives actually falsify an hypothesis of invasion or dispersal from the Atlantic Ocean, the distribution patterns and parasite faunas of organisms which actually did invade from the Atlantic ought to contrast with those displayed by potamotrygonid helminths. Otherwise, the severity with which our protocol actually tests the various hypotheses would have to be questioned. It seemed reasonable that if potamotrygonids are derived from ancestors which were widespread by the end of the Mesozoic, Neotropical fresh-water cetaceans (Platanistidae: *Inia* and Delphinidae: *Sotalia*) ought to exhibit such alternative distribution patterns because Romer (1966) listed the earliest fossil cetaceans of any kind as Lower Eocene. Thus, if cetaceans are considerably younger than potamotrygonids, the only way in which their fresh-water representatives could have gotten from marine to fresh-water habitats was via invasion.

Sotalia species occur throughout the Atlantic, including the Atlantic coast of South America. Additionally, two species occur as apparently fresh-water inhabitants in the Amazon and Orinoco systems. They are absent from the Paraná or Magdalena drainages. Zam et al. (1970) reviewed the helminth parasites of fresh-water *Sotalia* and discovered that they hosted some species also occurring in Atlantic marine delphinids. They concluded:

> While both of these hosts live exclusively in fresh water, it was somewhat surprising to find that both support, at least in part, a helminthofauna also found in marine cetaceans . . . . We

Table VI

Tabular summary of results of five tests of the four hypotheses concerning the origin and evolution of fresh-water stingrays. 0 = refuted by observations; + = confirmed by observations

Hypothesis	Test points 1	2	3	4	5
1	0	0	0	0	0
2	+	0	0	0	0
3	+	+	0	0	0
4	+	+	+	+	+

> can only assume . . . that the occurrence of common parasites from the ecologically divergent cetacean species (freshwater and marine) is a real phenomenon.

Neotropical fresh-water *Sotalia* exhibit the traits predicted by Hypothesis 3, differing markedly from Hypothesis 4 and the patterns demonstrated by potamotrygonids' helminths.

Pilleri and Gihr (1977) reviewed the taxonomic status of *Inia* and recognized three allopatric taxa: *I. boliviensis, I. g. geoffrensis,* and *I. g. humboldtiana.* Figure 25 presents the geographic distributions of *Inia* spp. with a cladogram of their relationships superimposed. *Inia boliviensis* is restricted to the Beni region of Bolivia, isolated from the rest of the Amazon by a series of rapids and waterfalls which apparently originated in the Pliocene or Pleistocene. Other *Inia* occur in the Amazon proper (*I. g. geoffrensis*) and in the Orinoco (*I. g. humboldtiana*). Other platanistids occur in the La Plata estuary, North America (Eocene deposits), India, and China. Some *Inia* parasites also occur in Atlantic marine cetaceans, whereas the rest exhibit closest relationships with parasites of Atlantic marine cetaceans. *Inia* species maintain a wider distribution in South America than do *Sotalia* species, the latter being absent from the Beni region. However, both groups exhibit geographic distribution patterns, genealogical relationships with other cetaceans, and helminth faunas consistent with Hypothesis 3 and differing from Hypothesis 4, which is most highly corroborated by potamotrygonids' helminths.

From the examination of fresh-water cetaceans, it appears that the technique used in this study is in fact not biased towards one type of conclusion. We must therefore conclude that if any of the parasitological data indicated an Atlantic, dasyatid, or dispersal influence in the evolution of potamotrygonids, we would have recognized it.

Testability of Hypothesis

Hypothesis 4, which we adopt as the best corroborated of those considered, may be tested in a variety of manners. First, additional collections of potamotrygonids and their parasites could be made and incorporated into the raw data base; new area and taxon cladograms would then be constructed and either corroborate or refute the original hypothesis. Secondly, a survey of potamotrygonids observing their urea retention patterns and state of atrophy of their rectal glands could provide data indicating the presence or absence of a gradation in salinity tolerance. Such a study has been undertaken as part of the overall project and the results will be published elsewhere. Thirdly, examination of parasites infecting urolophids would determine whether or not the closest relatives of helminths

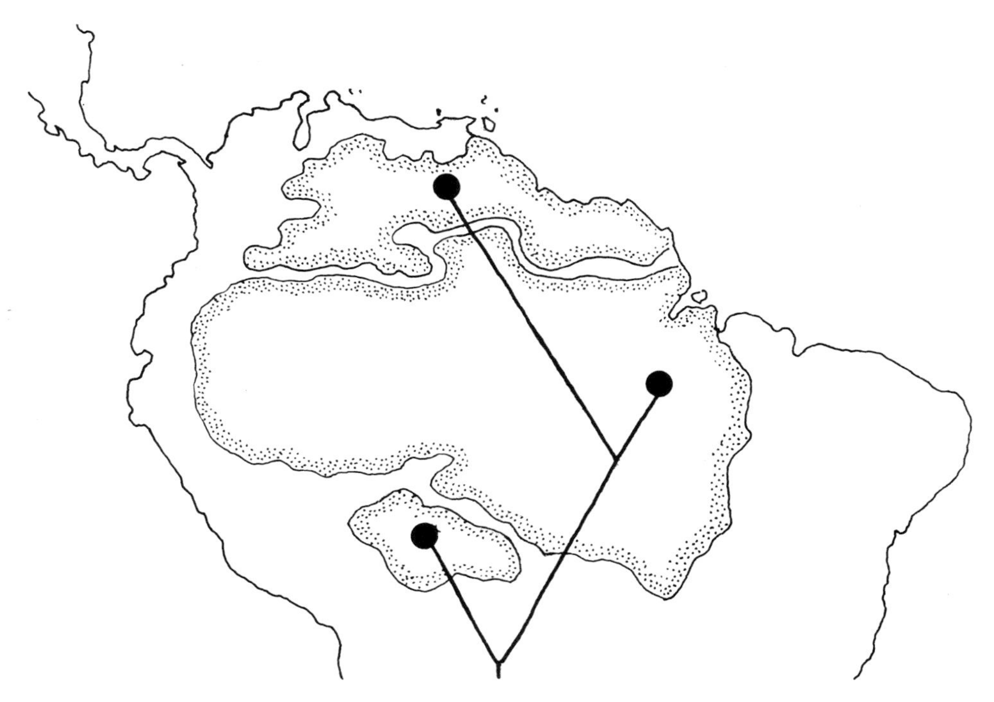

Fig. 25. Map depicting geographic distribution of member of *Inia* (Cetacea: Platanistidae). Superimposed cladogram represents relationships of three nominal taxa: *Inia boliviensis* as sister species of *Inia geoffrensis geoffrensis* and *I. g. humboldtiana*.

infecting potamotrygonids actually occur in urolophids. Fourthly, according to Hypothesis 4 the parasites infecting *Dasyatis garouaensis*, the African freshwater ray, should have no genealogical connection with those infecting potamotrygonids. Examination of the Benue River rays for parasites constitutes a critical test of the hypothesis. And lastly, if phylogenetic analysis of potamotrygonids, urolophids, and dasyatids produced synapomorphies shared by urolophids and potamotrygonids alone, our hypothesis would be corroborated. In fact, if our hypothesis is correct *Urolophus* can be considered a monophyletic group only if the potamotrygonids are included in it. This is because potamotrygonids' parasites exhibit phylogenetic relationships with parasites infecting members of *Urolophus* suggesting that *Urolophus* species as currently recognized constitute a paraphyletic group.

Evolutionary Scenario

The foregoing analysis of the phylogenetic characteristics of potamotrygonids' helminth parasites allows inference of the evolutionary scenario or set of processes and events by which potamotrygonids and their helminths attained their present existence. We present only the minimum parameters necessary to explain the patterns outlined in this study, i.e. the most parsimonious explanation.

By the beginning of the Cretaceous, *Urolophus*-like ancestral stingrays comprised part of a coastal stingray fauna occurring along the west coast of South America. By the mid to late Cretaceous, Andean orogeny had created a shallow

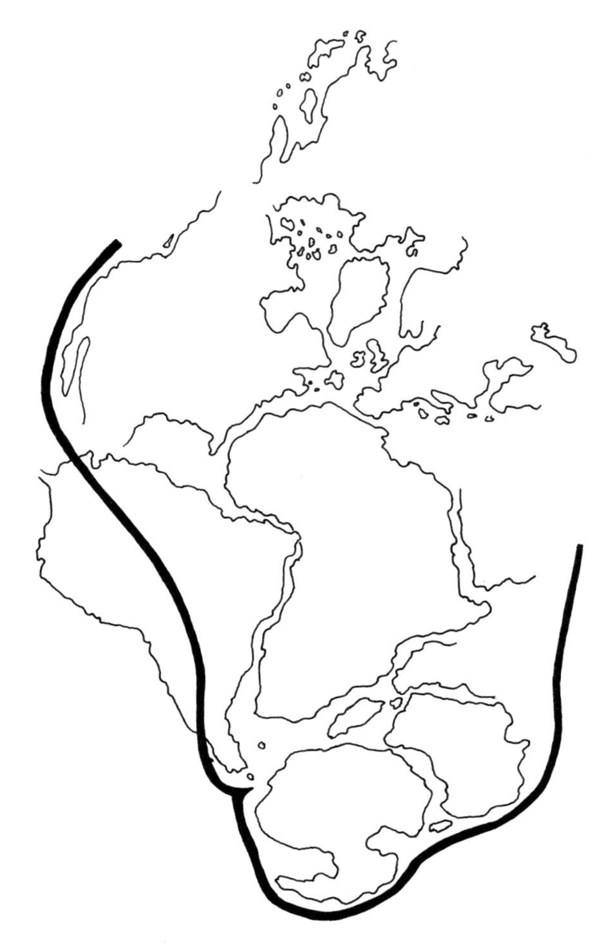

Fig. 26. Map depicting Mesozoic continental configurations with postulated distribution of ancestral urolophid-like stingrays superimposed.

saline inland body of water communicating with the Pacific in the north, which was to remain open until the Miocene. From later Cretaceous to Miocene, continuing orogeny and fresh-water runoff resulted in progressive desalination of the inland water and sediment subduction along the western edge of the emerging Andes. During this period a number of batoid elasmobranchs which occurred sympatrically with the urolophid ancestors disappeared from the area, leaving the incipient potamotrygonids as the only remnant of that portion of the ancestral elasmobranch fauna.

Paleocene to Miocene geographic changes in South America produced isolated fresh-water habitats beginning in the southern part of the continent (Paraná drainage) and proceeding northward. Potamotrygonids and their parasites were isolated by those same geographic changes; the Paraná-Paraguay system was probably isolated in the Paleocene. The Amazon basin became a separate eastward-flowing entity in the Miocene. Terminal Andean orogeny in the Pliocene-

Fig. 27. Map depicting Mesozoic continental configurations with contemporary distribution of urolophid stingrays and of potamotrygonid stingrays superimposed.

Pleistocene along with eustatic drops in sea level produced, as fresh-water habitats, the Magdalena River system, the Orinoco River system, and the Maracaibo area. Since the Paleocene, there has been apparent post-isolation dispersion by potamotrygonids and their parasites from fresh-water isolated areas of greater age to those of lesser age, such as from the Paraná to the Amazon or the Amazon to the Orinoco. Some cases of apparent dispersion may also be the result of isolation without speciation. In either case, the "noise" produced by plesiomorphic or convergent dispersion has not overshadowed the "signal" of evolution consistent with the geographical evolution of South America since the beginning of the Cretaceous. In fact, the entire scenario which we espouse may be illustrated in two maps, a "before" or Mesozoic distribution (Fig. 26) and an "after" or contemporary distribution (Fig. 27) both placed on a representation of Mesozoic continental configurations. The two maps clearly show that the offshoot surviving group of urolophids which we call potamotrygonids occupy roughly the same areas today that their ancestors did during the Mesozoic. Only the land has changed around them.

Further Implications of the Study

Although no a priori resolution of a center of origin was needed or attempted in the present study, we examined the question in retrospect using the parasitological data. Following the assumption that the most highly derived taxa occur at or near the center of origin (Darlington, 1957) one would postulate that the Magdalena River system of Colombia was the center of origin. On the other hand, one using Hennig's (1966) "biogeographic progression rule" would be forced to postulate that the center of origin was the Paraná River system because that is where the most plesiomorphic taxa occur. And alternatively, if one considers the area of greatest taxic diversity as the most likely candidate for the center of origin or original source area, one would accept the Delta of the Orinoco River as the center of origin.

The first resolution would provide difficulties for anyone attempting to find geological support for either a vicariance or a dispersalist explanation for the present-day distribution of potamotrygonid helminths. The third resolution is clearly an artifact of collecting in this case; at least two helminth species occurring in the Orinoco and in the Paraná also may occur in the Amazon and further new taxa remain to be discovered but have not yet been collected. The second resolution, following Hennig's technique, postulates that the center of origin is the area we recognize as the first part of an ancestral distribution isolated by geographic changes.

Thus, if our hypothesis is correct none of the above areas could correctly be called the center of origin for fresh-water stingrays' parasites. Hennig's formulation provides the best estimate of any, but it is still not the most highly corroborated. We view this as a clear practical indictment of methods of biogeographical analysis relying on a priori resolution of centers of origin, in agreement with Nelson (1974), Croizat (1964), Croizat et al. (1974), Platnick and Nelson (1978), and Rosen (1975, 1978).

A further clear implication of our study is that, if our hypothesis is correct, the classification of *Urolophus* and the potamotrygonids must be re-evaluated. The genus *Urolophus* excluding the potamotrygonids is a paraphyletic assemblage and therefore not a natural taxon. Thus, *Urolophus* should be subdivided or the potamotrygonids included in it. Alternatively, if one wished to recognize the distinctive characters of potamotrygonids by maintaining their familial status, at least two groups of *Urolophus* would have to be recognized at the same level. The above suggestions might appear rather sinister, promising to do great damage to existing and presumably stable classifications. We do not think such fears would be realized. For example, such a new classification would be consistent with biogeographic and parasitological data as well as the morphological traits upon which the new diagnoses would be founded. The new classification would actually be more stable, consistent, natural, and informative despite appearing somewhat strange to workers used to thinking of potamotrygonids in very remote terms. Farris (1979a) has shown empirically that classifications comprising paraphyletic groups must distort or suppress information available in classifications composed only of monophyletic groups. Thus, it is impossible for a new classification of these stingrays to be anything but more informative than the existing one.

Acknowledgements

We appreciate support for most of our work from National Geographic Society grants to T. B. Thorson. We are also grateful for the financial backing of the University of Nebraska–Lincoln Research Council and for permission to accompany the R/V Eastward on a collecting cruise in the Orinoco Delta financed by a National Science Foundation grant to John G. Lundberg and Jonathan N. Baskin.

For help in all stages and aspects of our investigation, we are indebted to far too many individuals and institutions to mention, but for major aid in obtaining stingrays and providing facilities, we must mention the following: In *Venezuela,* Dr. Francisco Mago Leccia, Antonio Ríos, Héctor Barrios, José Mosco M., Alfredo Goméz G., Donald G. Taphorn, Craig Lilyestrom, Eric Sutton, and the owner, captain, and crew of the trawler, Dantjole; In *Colombia,* Orlando Mora Lara, Augusto Samper M., Alvaro Boada Guarín, Guillermo Quiñones González, Adolfo Barón Porras, Alberto Villaneda, Dr. James M. Kapetsky, Dr. David A. Conroy, Hans Heinrich, Mike Tsalickis, William Mackay, and Dean Hendrickson; In *Paraguay,* Dr. Hernando Bertoni, Minister of Agriculture and Livestock, Antonio Torres, Juan Pío Rivaldi, Phil and Peg Myers, Ana Aurora Galli, Erik Raynears, Philippe Legris, Wilfredo Richter, and Bruno Ruggero Fornells; In *Uruguay,* Dr. Raúl Vaz-Ferreira, Dr. Hebert Nion, Alex Schwed Olin, and the owner, captain and crew of the trawler, Carla; In the *United States,* Drs. Jamie E. Thomerson, Phillip W. Myers, Reeve M. Bailey, Jeffrey M. Taylor, C. J. D. Brown, and Robert L. Martin.

Appendix I. Annotated list of helminth parasites infecting South American marine and euryhaline elasmobranchs. + = endemic to South America

Chile
- *Psammobatis scobina*
  - +*Acanthobothrium psammobati* Carvajal and Goldstein, 1969
  - +*Rhinebothrium scobinae* Euzet and Carvajal, 1973
- *Psammobatis lima*
  - +*Rhinebothrium leiblei* Euzet and Carvajal, 1973
  - +*Rhinebothrium chilensis* Euzet and Carvajal, 1973
  - +*Echinobothrium euzeti* Campbell and Carvajal, 1980
- *Raja chilensis*
  - +*Grillotia dollfusi* Carvajal, 1971
  - +*Acanthobothrium annapinkiensis* Carvajal and Goldstein, 1971
  - +*Echeneibothrium multiloculatum* Carvajal and Dailey, 1975
  - +*Echeneibothrium williamsi* Carvajal and Dailey, 1975
  - +*Echeneibothrium megalosoma* Carvajal and Dailey, 1975
- *Myliobatis chilensis*
  - +*Caulobothrium myliobatidis* Carvajal, 1977
  - +*Rhodobothrium mesodesmatum:* Campbell and Carvajal, 1979
- *Hexanchus griseus*
  - *Phyllobothrium dohrnii:* Carvajal, 1974
  - *Phyllobothrium sinuosiceps:* Carvajal, 1974
  - *Grillotia heptanchii:* Carvajal, 1971; Carvajal, 1974

Appendix I.
Continued

Mustelus mento
+*Prochristianella musteli* Carvajal, 1974
Phyllobothrium lactuca: Carvajal, 1974
Orygmatobothrium musteli: Carvajal, 1974
Crossobothrium triakis: Carvajal, 1974
Calliobothrium verticillatum: Carvajal, 1974
Alopias vulpinus
Crossobothrium angustum: Carvajal, 1974
Prionace glauca
Crossobothrium angustum: Carvajal, 1974
Platybothrium auriculatum: Carvajal, 1974
Hepatoxylon trichiuri: Carvajal, 1974
Triakis maculata
Lacistorhynchus tenuis: Carvajal, 1974
Centroscyllium granulosus
Gilquinia squali: Carvajal, 1974
Rhinobatos planiceps
+*Prochristianella heteroacantha* Dailey and Carvajal, 1976
+*Rhinebothrium rhinobati* Dailey and Carvajal, 1976
Proleptus acutus: Dailey and Carvajal, 1976
Prochristianella monomegacantha: Dailey and Carvajal, 1976
Acanthobothrium olseni: Dailey and Carvajal, 1976

Peru

Myliobatis chilensis
+*Rhodobothrium mesodesmatum:* Rego et al., 1968 (as *Anthobothrium peruanum* Rego et al., 1968)
+*Acanthobothrium chilensis* Rego et al., 1968

Colombia

Dasyatis americana
+*Phyllobothrium* cf. *kingae* Brooks and Mayes, 1980
Polypocephalus medusius: Brooks and Mayes, 1980
Lecanicephalum peltatum: Brooks and Mayes, 1980
Himantura schmardae
+*Acanthobothroides thorsoni* Brooks, 1977
+*Acanthobothrium tasajerasi* Brooks, 1977
+*Acanthobothrium himanturi* Brooks, 1977
+*Caulobothrium anacolum* Brooks, 1977
+*Rhinebothrium magniphallum* Brooks, 1977
+*Rhinebothrium tetralobatum* Brooks, 1977
+*Echinocephalus diazi:* Deardorff et al., 1981
Urolophus jamaicensis
+*Acanthobothrium cartagenensis* Brooks and Mayes, 1980
+*Phyllobothrium* cf. *kingae* Brooks and Mayes, 1980
+*Rhinebothrium magniphallum:* Brooks and Mayes, 1980
Urotrygon venezuelae
+*Acanthobothrium urotrygoni* Brooks and Mayes, 1980
+*Rhinebothrium magniphallum:* Brooks and Mayes, 1980
Aetobatus narinari
+*Acanthobothrium colombianum* Brooks and Mayes, 1980
Narcine brasiliensis
+*Acanthobothrium electricolum* Brooks and Mayes, 1978
Acanthobothrium lintoni: Brooks and Mayes, 1978

Venezuela

Dasyatis guttata
+*Acanthobothroides thorsoni:* Mayes and Brooks, 1981
+*Acanthobothrium urotrygoni:* Mayes and Brooks, 1981
+*Acanthobothrium tasajerasi:* Mayes and Brooks, 1981
+*Rhinebothrium margaritensis* Mayes and Brooks, 1981
+*Rhinebothrium magniphallum:* Mayes and Brooks, 1981
Rhodobothrium pulvinatum: Mayes and Brooks, 1981

Appendix I.
Continued

Dasyatis americana
  †*Rhinebothrium margaritensis* Mayes and Brooks, 1981
  Acanthobothrium americanum: Mayes and Brooks, 1981
  Rhinebothrium corymbum: Mayes and Brooks, 1981
  Rhodobothrium pulvinatum: Mayes and Brooks, 1981
  Phyllobothrium centrurum: Mayes and Brooks, 1981
Himantura schmardae
  †*Echinocephalus diazi* Troncy, 1969; Deardorff et al., 1981
  †*Rhinebothrium magniphallum:* Mayes and Brooks, 1981
  Parachristianella cf. *monomegacantha:* Mayes and Brooks, 1981
Gymnura micrura
  Acanthobothrium fogeli: Mayes and Brooks, 1981
Aetobatus narinari
  Acanthobothrium tortum: Mayes and Brooks, 1981
  †*Disculiceps* sp. Mayes and Brooks, 1981
Rhinoptera bonasus
  †*Dioecotaenia campbelli* Mayes and Brooks, 1981
  †*Rhodobothrium paucitesticularis* Mayes and Brooks, 1981
  †*Tylocephalum* sp. Mayes and Brooks, 1981
  Rhinoptericola megacantha: Mayes and Brooks, 1981
Narcine brasiliensis
  †*Acanthobothrium electricolum:* Mayes and Brooks, 1981

Jamaica

Urolophus jamaicensis
  †*Phyllobothrium kingae* Schmidt, 1978
  †*Eutetrarhynchus caribbensis* Kovacs and Schmidt, 1980
  †*Eutetrarhynchus thalassius* Kovacs and Schmidt, 1980
  †*Acanthobothrium cartagenensis:* Schmidt, pers. comm.
  †*Discobothrium* sp. n.: Schmidt, pers. comm.

Brazil

Carcharhinus longimanus
  Phyllobothrium filiforme: Rego, 1977
  Anthobothrium laciniatum: Rego, 1977
  Tentacularia coryphaenae: Rego, 1977
Unknown shark
  Phyllobothrium prionacis: Rego and Mayer, 1976
  Platybothrium auriculatum: Rego and Mayer, 1976
Unknown stingray
  †*Acanthobothrium* cf. *dasybatis* Rego, Dos Santos and Silva, 1974

Uruguay

Psammobatis microps (as *Raja m.*)
  †*Calicotyle microcotyle* Cordero, 1944
  †*Neoerpocotyle tudes* Cordero, 1944
Myliobatis goodei
  †*Acanthobothrium* sp. Brooks et al., 1981a
  †*Acanthobothrium* sp. Brooks et al., 1981a
  †*Caulobothrium uruguayensis* Brooks et al., 1981a
  †*Caulobothrium ostrowskiae* Brooks et al., 1981a
  †*Phyllobothrium* sp. Brooks et al., 1981a
  †*Phyllobothrium myliobatidis* Brooks et al., 1981a
  †*Discobothrium arhynchus* Brooks et al., 1981a

Argentina

Psammobatis microps
  †*Calicotyle microcotyle:* Ostrowski de Núñez, 1971
  Otodistomum pristiophori: Ostrowski de Núñez, 1971
  †*Nybelinia* sp. Ostrowski de Núñez, 1971
  †*Phyllobothrium* sp. Ostrowski de Núñez, 1971
Zapteryx brevirostris
  †*Acanthobothrium zapterycum* Ostrowski de Núñez, 1971
  †*Echinobothrium pigmentatum* Ostrowski de Núñez, 1971

Appendix I.
Continued

+*Acanthobothrium* sp. Ostrowski de Núñez, 1971
+*Phyllobothrium* sp. Ostrowski de Núñez, 1971
Squatina argentina
Otodistomum pristiophori: Ostrowski de Núñez, 1971
Mustelus schmitti
+*Calliobothrium verticillatum australis* Ostrowski de Núñez, 1973
Calliobothrium esrichtii: Ostrowski de Núñez, 1973
Calliobothrium lintoni: Ostrowski de Núñez, 1973
Orygmatobothrium velamentum: Ostrowski de Núñez, 1973
Mustelus asterias
+*Paracalicotyle verticillatum* Szidat, 1970

Biogeographical Methods and the Southern Beeches (Fagaceae: *Nothofagus*)

C. J. Humphries

Table of Contents

Introduction

"The key to the history of terrestrial life in the far south may be *Nothofagus*"

P. J. Darlington (1965)

"*Nothofagus* is uninformative on the interrelationships of southern hemisphere areas"

C. Patterson (1981)

In the introductory essay to his *Flora Novae Zelandiae* (1853) Sir Joseph Dalton Hooker attempted for the first time to explain the problem of why so many different groups of unrelated organisms should show similar, widely disjunct distribution patterns in the southern hemisphere areas of southern South America, Tasmania, Australia and New Zealand. In an effort to find a general explanation that could account for geographic disjunctions in one hundred or more plant genera, Hooker provided the novel theory that all of the different groups originated and subsequently vicariated in the southern hemisphere from a continuous tract of land. Although this was the simplest possible explanation for a major biological problem Hooker's theory was unacceptable to many of his contemporaries because fashionable geological theories of the day advocated that continents and intervening oceans are and always had been stable. Even Charles Darwin, who was to publish his theory of evolution only six years later, could not agree with Hooker on how different groups of plants moved over the globe. Darwin was completely influenced by stable globe theories and his explanations involved many origins in the northern hemisphere with individual long-distance dispersal events for repeated disjunctions in the austral biota.

Even though these extreme viewpoints are now some 130 years old and literally thousands of biogeographical papers have since been published, there are still many similar differences of opinion amongst contemporary theorists on the origins of austral plants and animals. It is impossible to review all of these papers comprehensively, but certain groups of organisms such as the marsupials and the southern beeches (*Nothofagus,* family Fagaceae) have frequently figured in them as important examples. This article, which is a more detailed paper of a short essay published in a student text (*Chance, Change and Challenge,* Humphries *in* P. Forey, 1981), attempts to review the theoretical history of phytogeography since the middle of the 19th century to the present day by examining the different explanations for disjunct distributions of *Nothofagus.*

This genus has been selected as a review subject for a variety of reasons. Firstly it has attracted the attention of a wide range of biogeographers working from different theoretical frameworks. The quotation given above from Darlington is one of many to be found in the writings of classical biogeographers who use existing theories on earth history, traditional classifications, fossils and dispersal hypotheses as the basis for explaining disjunct distribution patterns. By contrast, the quotation from Patterson represents the views of a much smaller group of biogeographers who test phylogenetic theories of relationship and models of distribution patterns in relation to models of earth history in an effort to find general biogeographical explanations for disjunctions. Secondly, the interrelationships of *Nothofagus* species and their relationships with other members of the beech family are, on the whole, quite well known and comparatively well documented. Thirdly, despite the similarity of the supporting evidence for each biogeographical explanation the hypotheses are so different from one another that methodological differences can be sharply contrasted.

Taxonomy and Biogeography

Historical biogeography, as defined by Rosen (1978), is the study of distribution patterns as connections between areas in terms of time and space. Taxonomy is the vehicle used to establish historical distribution and since biogeography can only be as good as the taxonomy it uses, the subject becomes a slave to the theoretical foundations of systematics. For this reason, biogeography must be regarded as an integral part of systematics and indeed changes in biogeography have followed the changes in taxonomic theory.

An analogy may be drawn between the development of taxonomic methods and biogeographic methods over the last 130 years. Ball (1976) and Patterson (1981) recognise three sequential phases, the empirical (or descriptive), the narrative and the analytical. Empirical or alpha taxonomy is the datum gathering enterprise of bread-and-butter systematics which seeks to distinguish and describe species. Narrative taxonomy, by contrast, attempts to explain the evolutionary relationships of different organisms by using what may be called traditional methods of analysis for assessing overall similarities and differences. The degree and rate of morphological divergence in contemporary organisms are estimated by comparison with fossils or by considering that some extant organisms are more primitive than others and therefore ancestral. Narrative taxonomy is common to many undergraduate textbooks where evolutionary relationships are presented as nice stories rather than scientific hypotheses. Analytical taxonomy describes the methods originally developed by such people as Camin and Sokal (1965), Hennig (1965, 1966) and Farris (1970) which have formalised the concepts and procedures for the estimation of cladistic relationships in order to present the results as explicit testable hypotheses of genealogy.

Narrative biogeography is the practice of using known historical (geological or climatical) events and various biological assumptions as a basis for explaining distribution patterns. Analytical biogeography, by contrast, compares cladistic patterns of different groups of endemic organisms occupying similar areas in attempts to find general biogeographical explanations from a combined geological and biological point of view.

In narrative approaches the historical distribution of contemporary species is expressed in terms of the taxonomy of modern species as compared with the distribution and taxonomy of fossils. Narrative historical biogeography is mostly concerned about providing explanations in terms of six variables: mobile or fixed continents, dispersal or sometimes vicariance, and climate or ecology. Usually the fashionable geological or climatological theory of the day determines the biogeographical explanation. In a world of mobile continents intercontinental disjunctions have come about either as a dispersal event prior to continental breakup or as a splitting event—which is the division of a formerly continuous community by continental rafting. In a world of fixed continents dispersal explanations interpret disjunctions in terms of one group of plants living in one area giving rise to the plants occupying another area. Disjunctions between one or more areas are thus caused by at least one sexually mature individual crossing from one area over a pre-existing barrier and establishing itself in a new second area. In other words the first area is considered to be the 'centre of origin,' a feature which can be determined by a number of different criteria (Table I). For example, in traditional Angiosperm classifications the species of the Magnoliaceae family are considered primitive because they have free petals, indeterminate numbers of stamens and free carpels. By contrast, flowers with united petals, a low, determinate number of stamens and fused carpels are considered derived and therefore only remotely related to the Magnoliaceae. Because there is a preponderance of 'primitive' *Magnolia*-like plants in South-east Asia and the western Pacific, this region has been considered the 'centre of origin' for all of the flowering plants (Smith, 1970; Taktajhan, 1969) (Table I). Sometimes the reverse has been thought to be true, when the 'centre of origin' is that area where the greatest variety of taxa or the most derived taxa are formed in 'a centre of diversity.' The ancestral or primitive groups are assumed to have migrated to peripheral areas (Table IAii) (see Matthew, 1915; Oliver, 1925). Evidence to support this idea comes from some of the relictual 'living fossil' floras found on oceanic islands on the periphery of mainland continental areas with relatively derived floras. The introduction of fossils gives a different picture since they are of cardinal importance to dispersal biogeographers. The mere presence of fossils in one area with modern species in two or more areas can determine the centre of origin (Table IBi). When there is a tolerably good fossil record, i.e. occurring in more than one area, those areas containing the oldest fossils are said to be at the centre of origin (Table IBii). In groups with restricted modern distributions, fossils play a very important role for determining dispersal routes on the basis of relative age—in other words the oldest taxa determine the centre of origin (Table IBiii, iv).

The alternative to dispersal for explaining disjunct distribution patterns is the phenomenon of vicariance. Wulff (1943) defined vicarious taxa as those which occupy mutually exclusive areas but share a relatively recent common ancestor.

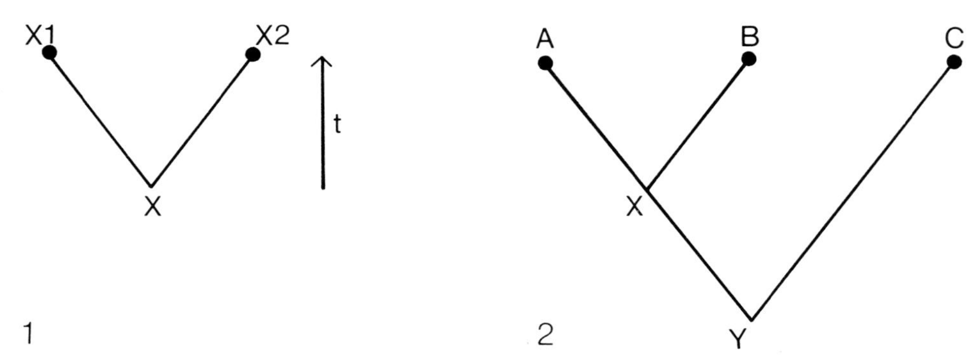

Figs. 1–2. **1.** A rooted tree diagram. With time (t) one species x will divide into two (or more) daughter species x_1 and x_2. **2.** The definition of a phylogenetic relationship. Species A and B share homologies at x which are not shared by C.

The genesis of vicariants or vicariads was seen by Croizat (1952) to be the historical separation of a formerly continuous population, biota or community by the development of a natural barrier. He went on to state that if a natural barrier such as a climatic change or continental drift separated a formerly continuous biota or community into two different groups, the organisms would show the same track of relationship across the barrier. In other words, correspondence between several tracks in both fossil and modern groups would give a "generalized track" linking the vicariant products of a formerly continuous biota or community. Following from the vicariance view of disjunction is the idea that historical biogeography is more about the biotic history of areas rather than particular groups or organisms.

Notable advances in historical biogeography have come through the two developments of explicit phylogeny reconstruction (cladistics; Hennig, 1965, 1966) and the combination of vicariance biogeography with the reasoning of cladistics (Platnick and Nelson, 1978; Rosen, 1978).

Explicit cladistic relationships are based on a concept which states that two species x_1 and x_2 are related by the possession of unique homologous characters (see x, Fig. 1). For any group there will be many species, some of which will share more inclusive homologies when compared to other species. This can be demonstrated with three species (Fig. 2). In the group of three species ABC, A and B are more closely related to one another because they share unique homologies (synapomorphies) at point x which are not shared by the related species C. AB form a nested set in a wider group ABC defined by the unique homologies

Table I

Reasons for dispersal in narrative biogeographic hypotheses (see text for explanation)

		Area 1	Dispersal route	Area 2
A	i	Primitive species	→	Advanced species
	ii	Advanced species	→	Primitive species
B	i	Modern species and fossils	→	Modern species
	ii	Modern species and old fossils	→	Modern species and young fossils
	iii	Fossils	→	Modern species
	iv	Modern species and old fossils	→	Young fossils

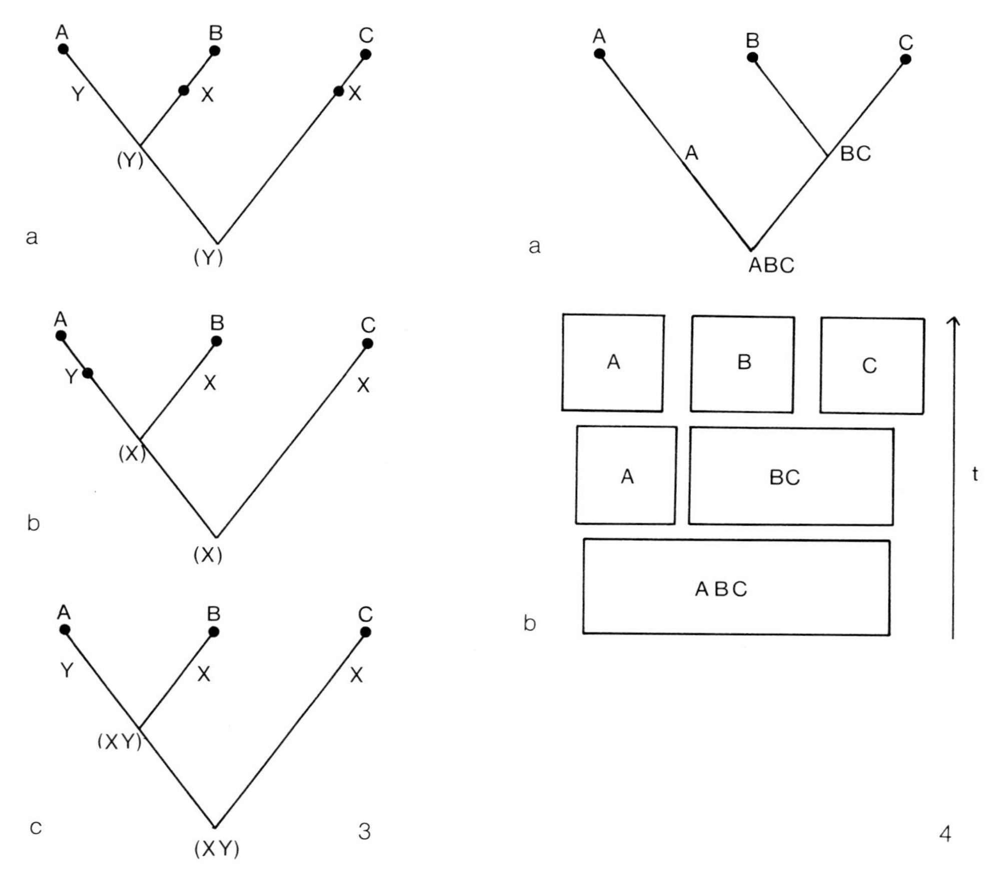

Figs. 3–4. **3.** Dispersal and vicariance concepts in cladistic models of relationship. **A.** For three modern species A, B and C occupying two areas X and Y, estimates of ancestral centres of origin are determined by combining descendant distributions when they are completely different (XY) and eliminating one or more of the unshared elements when they are not completely different. **B.** The second hypothesis is considered to be the simplest since it involves one dispersal event. ● = dispersal. **C.** For three modern species A, B and C occupying two areas x and y the simplest vicariance hypothesis would postulate one formerly continuous population ABC splitting into two, (A + B, C) and eventually into three (A, B, C). **4.** Geological cladograms. A formerly continuous land mass ABC which sequentially fragments into two areas A, BC and eventually into three areas A, B, C (a) can be formally represented in a cladogram (b).

at point Y. Such groups are called monophyletic groups. As an example from the angiosperms, we can make the hypothesis that apples (*Malus*) and pears (*Pyrus*) are more closely related to one another when compared to roses (*Rosa*) because they share unique characters of swollen mature fruits (pomes) and a few-seeded (multilocular) ovules which are not shared by the roses. At a higher level we can say that all of the angiosperms form a nested set within land plants because only they share the characters of double fertilisation and companion cells in the phloem which are not found in any other plant group (Bremer and Wanntorp, 1978; Hill and Crane, In press; Parenti, 1980). From such explicit cladistic hypotheses and knowledge of modern distribution patterns, ancestral distribution patterns can be inferred and the biogeographic history reconstructed.

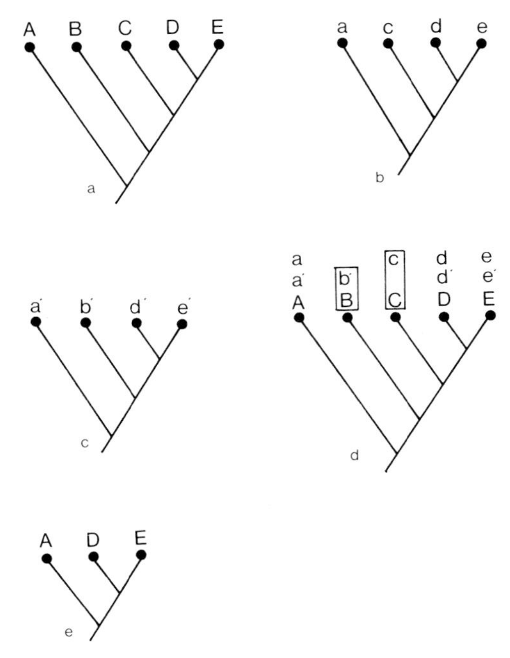

Fig. 5. The comparison of geological cladograms with taxa cladograms. **A.** A five area geological cladogram showing a gradual sequential breakup of a continuous land mass ABCDE into five areas, A, B, C, D, E. **B and C.** Two taxa cladograms of two unrelated taxonomic groups. **D.** Congruency for all three cladograms. **E.** A reduced geological area cladogram showing the branch points common to all three cladograms.

Initially, explicit phylogenetic hypotheses were used by Brundin (1966) and others to establish different 'centres of origin' for particular groups. In Figure 3 the method for determining ancestral distributions is shown. The descendant (contemporary) distributions are combined at each branch point and the ancestral 'centre of origin' established by eliminating the unshared or unique elements. The most parsimonious solutions, i.e. those hypotheses with the lowest number of steps, are considered to be the best biogeographical explanations. In the hypothetical biogeographical reconstruction of the three species ABC in Figure 3a, species A is considered to have originated from a common ancestor with species B. The diagrams indicate that species B and C occupy area X and species A area Y. If area Y is considered to be the centre of origin then there must have been two separate dispersal events for B and C to area X. If, on the other hand, the centre of origin is considered to be area X then a parsimonious solution would be a single dispersal event for A to Y (Fig. 3b). Should, however, the speciation of A and B have been due to the formation of a natural barrier then they would be considered vicariant taxa. The method for determining ancestral histories of vicariant taxa is described in Figure 3c. If the origin of the two areas X and Y was due to the continental drift of a formerly continuous land mass the biogeography can be established by adding the ancestral distributions together.

In realising that biogeography is about patterns of areas and dispersal is about unknowable processes that have taken place in the past, Platnick and Nelson (1978) proposed a method of historical biogeography which was put into practice by Rosen (1978). The method is one which combines the cladistic method with the vicariance approach exploiting the analogy between systematics and biogeography to the full (Patterson, 1981). The method applies cladistics to biogeography by producing cladograms of areas which are compared to cladograms of different groups of organisms occupying those areas. Figure 4b shows the sequential geological history of one area ABC which breaks up in two stages; first into two areas A + BC and eventually into three, A + B + C. Figure 4a shows that the area can be drawn as a cladogram in much the same way as those drawn for organisms. Areas B and C are held to be more closely related to one another because they have a longer lasting connection than A. In other words the biotas of areas B and C show a more recent common ancestry with each other than they do with A.

The basic difference between a dispersal hypothesis and a vicariance hypothesis lies in the relationship between the age of taxa and the age of the barrier limiting the area. In vicariance events the barriers and the taxa have the same age, whilst in dispersal events the barriers pre-date the taxa. Taxa cladograms congruent with the geological cladograms provide the most parsimonious explanation of the data. This is illustrated in Figure 5a, which shows a geological cladogram for five areas A, B, C, D and E; Figure 5b and c shows two different taxon cladograms showing different relationship patterns and the lower case letters indicate taxa which occupy the areas with corresponding upper case letters in Figure 5a; Figure 5d shows congruence between the two taxon cladograms and the geological cladograms; and Figure 5e shows a reduced area geological cladogram with the branching points common to all of the cladograms. If we state that the geological cladogram is the most parsimonious explanation for a pattern of earth history, congruency between the most parsimonious taxa cladograms and the geological cladograms is a test for a general vicariance hypothesis. However, incongruent cladograms do not necessarily falsify the general explanation but might instead indicate dispersal or vicariance events different from those under study. In this method the only useful taxa are those endemics in each of three or more areas whose cladistic (genealogical) relationships are established.

Theories of *Nothofagus* Biogeography

Narrative Biogeographical Theories and Explanations Involving Stable Continents and Dispersal

In contrast to Hooker's theory of southern origins and geological continuity during the Mesozoic, Darwin (1859) preferred to consider that wholly southern genera such as *Nothofagus* originated in the northern hemisphere and migrated by dispersal to the southern hemisphere. This general view was rapidly adopted by zoologists, and particularly well articulated by Wallace (1876):

> . . . The north and south division of the modern biota truly represents the fact that the great northern continents are the seat and birthplace of all the higher forms of life, while the southern continents have derived the greater part, if not the whole, of their vertebrate fauna from the north . . . .

This is an interesting comment with startling conclusions since, despite striking austral affinities in many different taxonomic groups, Wallace went on to say: ". . . but it implies the erroneous conclusion that the chief southern lands—Australia and South America—are more closely related to each other than to the northern continent . . . ." Even though there seemed to be little evidence for north-south migration routes, the views of Darwin and Wallace were quickly adopted by phytogeographers. For example, the 'monoboreal relict hypothesis'

Table II

The modern species of *Nothofagus* (based on Allan, 1961; Hanks and Fairbrothers, 1976; McQueen, 1976; Soepadmo, 1972; and van Steenis, 1953, 1971)

Name	Distribution	
N. alpina (Poepp. & Endl.) Oerst.	South America	(S Am)
N. procera (Poepp. & Endl.) Oerst.	South America	(S Am)
N. obliqua (Mirb.) Oerst.	South America	(S Am)
N. antarctica (Forst) Oerst.	South America	(S Am)
N. pumilio (Poepp. & Endl.) Krasser	South America	(S Am)
N. betuloides (Mirb.) Oerst.	South America	(S Am)
N. nitida (Phil.) Krasser	South America	(S Am)
N. dombeyi (Mirb.) Oerst.	South America	(S Am)
N. glauca (Phil.) Krasser	South America	(S Am)
N. alessandri Espinosa	South America	(S Am)
N. cunninghamii (Hook. f.) Oerst.	Australia	(Aus)
N. moorei (F. v. Muell) Krasser	Australia	(Aus)
N. gunnii (Hook. f.) Oerst.	Tasmania	(Tas)
N. cliffortioides (Hook. f.) Oerst.	New Zealand	(NZ)
N. solandri (Hook. f.) Oerst.	New Zealand	(NZ)
N. menziesii (Hook. f.) Oerst.	New Zealand	(NZ)
N. fusca (Hook. f.) Oerst.	New Zealand	(NZ)
N. truncata (Colenso) Cockayne	New Zealand	(NZ)
N. aequilateralis (Baum.-Bod.) Steen.	New Caledonia	(NC)
N. codonandra (Baill.) Steen.	New Caledonia	(NC)
N. balansae (Baill.) Steen.	New Caledonia	(NC)
N. baumanniae (Baum.-Bod.) Steen.	New Caledonia	(NC)
N. discoidea (Baum.-Bod.) Steen.	New Caledonia	(NC)
N. resinosa Steen.	New Guinea	(NG)
N. flaviramea Steen.	New Guinea	(NG)
N. carrii Steen.	New Guinea	(NG)
N. wormersleyi Steen.	New Guinea	(NG)
N. pullei Steen.	New Guinea	(NG)
N. pseudoresinosa Steen.	New Guinea	(NG)
N. crenata Steen.	New Guinea	(NG)
N. grandis Steen.	New Guinea	(NG)
N. perryi Steen.	New Guinea	(NG)
N. nuda Steen.	New Guinea	(NG)
N. rubra Steen.	New Guinea	(NG)
N. brassii Steen.	New Guinea	(NG)
N. starkenborghii Steen.	New Guinea	(NG)

of Schröter (1913) was an early attempt by a botanist to justify Darwin's original idea. In its simplest form this hypothesis stated that isolated austral groups were invariably primitive forms, driven southwards from the northern hemisphere through the development of new aggressive groups in the north. The evidence came from the distribution of such genera as the Gymnosperms, *Podocarpus* and *Araucaria,* with modern distributions in the southern hemisphere and Mesozoic fossil relatives in the northern hemisphere. For the Hooker supporters, like Engler (1882), groups with northern and southern complements could best be assumed to have simply been descendants of an earlier, worldwide group. However, in 1909 Thistleton-Dyer stated "that the extraordinary congestion in the peninsulas of the Old World points to the long continued action of a migration southwards." By strengthening the dictum with "the theory of southwards migration is the key to the interpretation of the geographical distribution of plants," dispersal explanations become central dogma to phytogeography. Indeed, Oliver (1925) provides a more detailed exposition of Schröter's view for *Nothofagus*:

> "It is probable that *Fagus* and *Nothofagus* originated in North America and spread thence east, south and west. The western moiety passed, *via* Japan, round the Pacific reaching Australia and New Zealand." He went on to say, "if the characters by which *Nothofagus* is separated from *Fagus* be considered primitive, then these two genera exemplify the principle enunciated by Matthew, which states that a group should be most advanced at its point of original dispersal, the most conservative stages being farthest from it" (see Table IAii).

The thirty-six species of the southern beech genus *Nothofagus* (Table II) are the only members of the family Fagaceae to occur in the southern hemisphere, south and east of New Guinea and New Britain. The closest relative is the northern beech genus, *Fagus*, widely distributed in Europe and North America; the remaining genera of the Fagaceae are important trees of the north temperate and south-east Asian to Papuasian forests. The species of *Nothofagus* are almost always tall trees dwarfed only at the latitudinal and altitudinal limits of tree growth when excessively severe weather conditions prevail. They are mainly evergreen but several are deciduous species. They have small seeds, which do not travel very far and do not survive in sea water. All species are poorly fitted for dispersal and for this reason alone have always raised intriguing questions regarding their wide distribution in the southern hemisphere. They form extensive forests in three principal areas of the southern cold-temperate zones—southern South America, Australia (including Tasmania) and New Zealand—and also occur in the cooler tropical highlands of New Guinea, New Britain and New Caledonia (Fig. 6) (see also van Steenis, 1953, 1971).

There are ten species in southern South America forming almost continuous forest on the tops and along the western slopes of the Andes, variously distributed from latitude 56°S to latitude 33°S. According to McQueen (1976) the southern limit is due to the lack of suitable land and the northern limit due to the aridity which accompanies the Mediterranean climate of central Chile. There are three species in Tasmania and Australia; the deciduous *N. gunnii* is endemic to the mountains of Tasmania. *N. moorei* occurs in New South Wales (the McPherson-Maclay overlap of Burbridge, 1960) and *N. cunninghamii* in Victoria and Tasmania. The five New Zealand species occur in rather isolated patches but from the lowlands to the timberline on both main islands, particularly on the western side. There are some 18 tropical evergreen species in the mountains of New Guinea, New Britain and New Caledonia. Here, they are important co-dominants of the rain forest forming extensive, but patchy stands, especially in the lower montane areas, between about 2300 and 2800 m.

Most phytogeographical discussion on *Nothofagus* revolves around establishing the 'centre of origin' and then deciding, on the basis of the taxonomic relationships, how the modern species arrived where they are today. Craw (1978) noted that there are generally two types of dispersal hypothesis depending on the acceptance of either stable or mobile continents. In stable continent theories disjunctions are seen to have originated by trans-oceanic long-distance dispersal. Acceptance of mobile continents, however, provides pre-drift overland dispersal routes, except when it is believed that continental drift occurred too early to account for the disjunction.

Croizat (1952), originally working within a stabilist framework, suggested that there were several centres for the origin of the Angiosperms. The Beech family (Fagaceae), as with several other 'primitive' families were said to have originated

Fig. 6. A map showing modern (black) and fossil distributions (▲) of *Nothofagus* in the southern hemisphere.

in New Caledonia from a Neo-Caledonian centre. Although no precise directions for dispersal were given, *Nothofagus* was considered to have moved along a circum-Antarctic 'track,' the other members of the family moving through south-east Asia to Europe and North America.

Since it is generally believed that fossils are amongst the best indicators of evolutionary direction the particularly good fossil pollen record of *Nothofagus* has been used extensively in phytogeographical discussion. There are three recognisable pollen morphs in *Nothofagus* (Fig. 7) named after three species, "*brassii,*" "*fusca*" and "*menziesii,*" considered typical of each type (see Hanks and Fairbrothers, 1976). All three pollen types of *Nothofagus* are very distinctive because they have unique structure within the Fagaceae. They have 5–8 small colpi as compared with the typical 3-colpate pattern of the nearest relative *Fagus,* a feature easily seen in fossil material. The "*brassii*" type correlates with the present-day tropical leathery-leaved species of New Guinea and New Caledonia whereas the "*menziesii*" and "*fusca*" types are found in both the present day

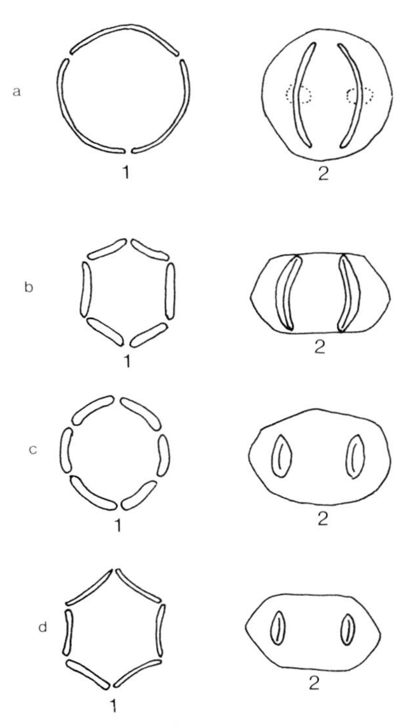

Fig. 7. Pollen morphology in the Fagoideae. **a.** "*Fagus*" type: *Fagus sylvatica*. (i) Polar view, (ii) equatorial view. **b.** "*brassii*" type: *Nothofagus grandis*. (i) polar view, (ii) equatorial view. **c.** "*fusca*" type: *Nothofagus alessandri*. (i) polar view, (ii) equatorial view. **d.** "*menziesii*" type: *Nothofagus menziesii*. (i) polar view, (ii) equatorial view (after Hanks and Fairbrothers, 1976).

evergreen and deciduous species of the cold-temperate zone. Despite a number of reports for *Nothofagus* pollen fossils in the northern hemisphere (see Cranwell, 1963) reliable records are to be found in areas of contemporary forest with appreciable extensions of range only in western Australia, Patagonia and Antarctica (Fig. 6; Table III). Acceptance or rejection of northern fossil records greatly influences narrative interpretations. Amongst the earliest fossil records (Table III) the "*brassii*" and "*fusca*" type occurs throughout the Upper Cretaceous of New Zealand. The earliest record for the "*menziesii*" type is the Eocene of Australia and South America. It appears, from fossil evidence, that *Nothofagus* was reasonably widespread in the Cretaceous becoming more widespread during the Tertiary. Today the genus is absent from Antarctica, the "*brassii*" type is restricted to New Guinea and New Caledonia, and only the "*fusca*" and "*menziesii*" types occur in the cooler temperate floras of Australia, New Zealand and South America.

Darlington's (1965) interpretation of the history of *Nothofagus* combined fossil pollen evidence with stable continent geology. Although he acknowledged continental drift as a means of causing disjunction, his dismissal of it rested on the fashionable belief of the day that it occurred too early for the angiosperms. Despite a reliable micro-fossil record present only in the austral continents, Darlington believed that *Nothofagus* could not have originated there because its nearest relative *Fagus* lives in the northern hemisphere. Since the 'centre of diversity' for the Fagaceae is in south-east Asia, Darlington (1965, p. 145) said:

Table III

Stratigraphic record for *Nothofagus* pollen (based on Hanks and Fairbrothers, 1976 and van Steenis, 1971). **b**—"*brassii*"-type pollen, **f**—"*fusca*"-type pollen, **m**—"*menziesii*"-type pollen

		Upper Creta-ceous	Palaeo-cene	Eocene	Oligo-cene	Lower Mio-cene	Upper Mio-cene	Pliocene	Recent and Pleisto-cene
Australia	b	+	+	+	+	+	+	+	
	f			+	+	+	+	+	+
	m			+	+	+	+	+	+
New Zealand	b	+	+	+	+	+	+	+	
	f	+	+	+	+	+	+	+	+
	m				+	+	+	+	+
Fuegia and Patagonia	b	+	+	+	+				
	f			+	+				+
	m			+	+				+
New Guinea	b						+	+	+
	f								
	m								
New Caledonia	b								+
	f								
	m								
Seymour Is.	b		+	+					
	f		+	+					
	m								
McMurdo Sound	b			+					
	f			+					
	m			+					

> *Nothofagus* may have originated in Asia primarily in sub-tropical parts of south-east Asia during the Cretaceous. It probably was never widespread or dominant in the northern hemisphere. It probably crossed the tropics once, by way of the Indo-Australian archipelago, in the Cretaceous. The *Nothofagus* of the '*brassii*' group now on New Guinea may be descendants of the original tropics crossers.

He went on to say that an origin in Asia followed by southwards dispersal across the Indo-Australian archipelago would bring *Nothofagus* into the southern hemisphere to Australia or New Zealand or both. By postulating the species of the "*brassii*" group as ancestral, Darlington then went on to postulate three separate late Cretaceous long-distance dispersal events around the austral land-masses to account for the modern distributions.

Other writers had similar views; Takhtajhan (1969), for example, agreed with Darlington that the "cradle of the flowering plants" was south-east Asia and suggested that *Nothofagus* "reached the southern hemisphere by way of Malaysia and Australasia."

Explanations Involving Stable Continents and Land Bridges

The "land-bridge" theory in botany as described by van Steenis (1962) represents an intermediate stage between the stable and mobile continent theorists. Although van Steenis believed in a steady state world, he simultaneously realized that the major plant disjunctions in both tropical and temperate floras could not

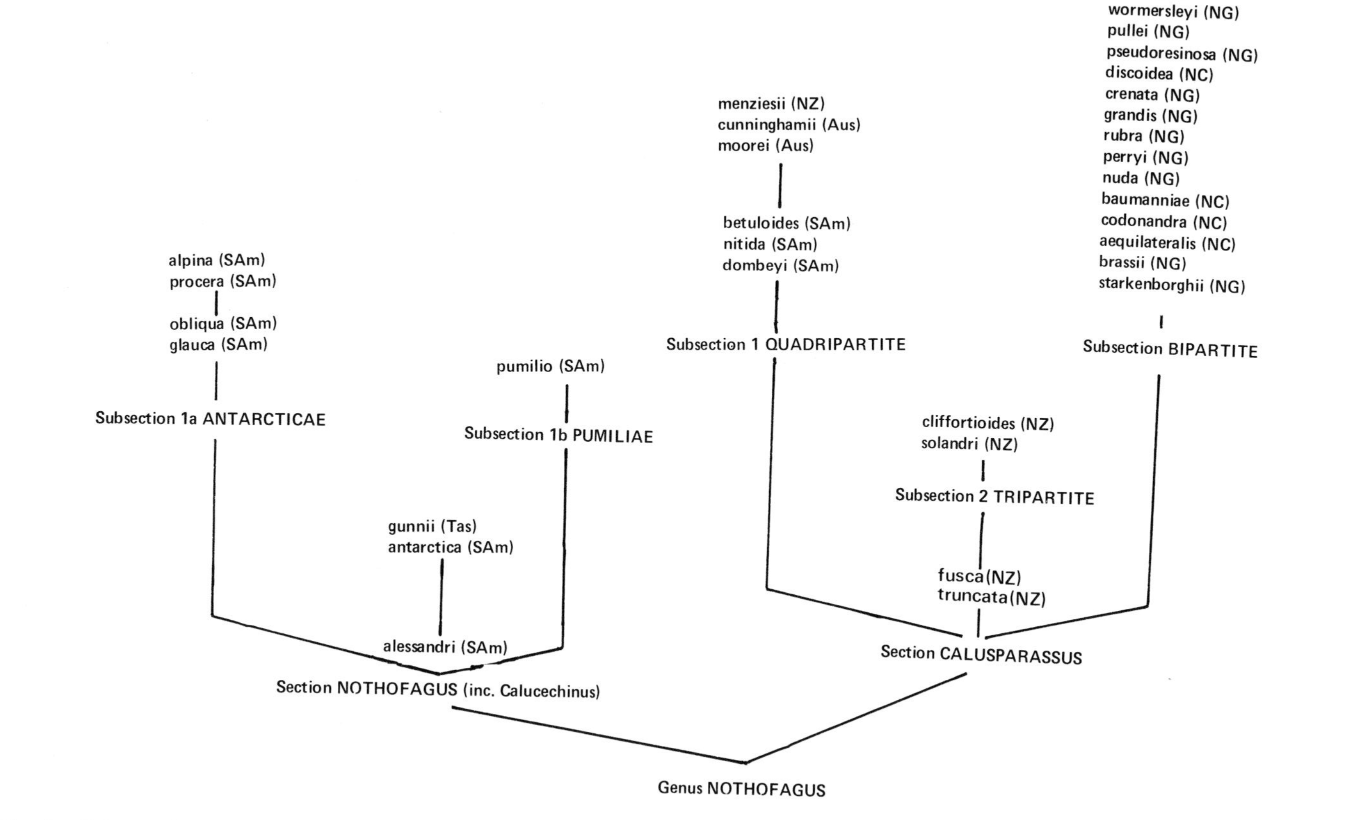

Fig. 8. Scheme of morphological interrelationships in the cupule structure of *Nothofagus*. The diagram is re-drawn from van Steenis (1953) and modified from van Steenis (1971) and Soepadmo (1972).

be accounted for by trans-oceanic dispersal. In rejecting long-distance dispersal and cumbersome polyphyletic origins for similar floras on different continents, he developed the land-bridge theory. To account for the relationships of the austral biota, land bridges were postulated connecting Australia, Tasmania, New Zealand, Antarctica and South America, and the presence of submerged continental shelves was used as evidence for a former Mesozoic highway. His classification of *Nothofagus* (van Steenis, 1953, as revised in 1971 and modified by Soepadmo, 1972), based on variations in the leaves, the female flower and the number of florets (Fig. 8), is that most generally accepted today. The eight deciduous species of South America and Tasmania were considered to be related to one another and grouped in a section *Calucechinus* (=*Nothofagus*). All of the evergreen species were placed in another section *Calusparassus* which was divided into three subsections on the basis of the number of lobes in the female fruit—the cupule. The first subsection is a grouping of eight South American, Australian and New Zealand species with four lobes in the mature cupule. The second subsection is a small group of two species, *N. cliffortioides* and *N. solandri,* from New Zealand with only three lobes in the female cupule, and the third subsection is a group of approximately 18 species occurring in New Guinea, New Britain and New Caledonia, characterized by having just two lobes in the female cupule. The "*brassii*" pollen type is restricted to the last subsection whilst the "*menziesii*" and "*fusca*" types are dispersed throughout the other groups.

According to van Steenis, all of the Fagaceae originated in Indo-Malesia in an area bounded roughly by Yunnan (China) and Queensland (Australia). As *Nothofagus* is the only austral genus of the Fagaceae, because the Papuasian species are a distinct group with "*brassii*" pollen, and because the "*brassii*" pollen is considered primitive, a land-bridge migration from south-east Asia into the temperate south was postulated to account for present-day distribution patterns. Also, since the fossil record (Table III) shows the "*brassii*" pollen type to have become extinct in the austral areas during the Pliocene, van Steenis considers the New Guinea and New Caledonia species as relict survivors of an older, more widespread group.

Explanations Invoking Mobile Continents

Other biologists were not wholly convinced that the world was in a steady state or that continental drift had occurred too early. In 1963, Cranwell commented that "Flotsam of sea and sky, including the gravid female . . . carried a heavy burden in pan-biogeographic argument." By initiating a fresh look at the pollen fossils of *Nothofagus,* she expanded the ideas of Florin (1940) and Couper (1960) who said that the areas with the oldest fossils are the best sites for the origin of particular groups. Therefore, for them, *Nothofagus* had a wholly southern origin and must have migrated to other continents by overland trails.

Much of the 1970's literature reflects a general acceptance of plate tectonic theory and the idea of mobile continents as a causal agent of disjunction. In terms of biogeographical theory however, this simply provides an alternative, albeit more plausible basis for dispersal, but this time with overland routes rather than over water. Whereas stable continents posed dire transport problems for such poor dispersers as *Nothofagus,* former overland dispersal events become more plausible. Most of the major movements of continents deemed to have affected the distribution of modern biota occurred, at least to some geologists, some 120 million years ago. The main effect of continental rafting has been to shift the emphasis of timing to earlier dispersal events when land masses were closer together. Nevertheless, the new interpretations remain narrative explanations with the same subjective elements as one found in stabilist theories. Hence, there is considerable variety in each explanation and all are basically untestable since they are utterly dependent on the initial assumptions.

For *Nothofagus* there are basically two types of narrative explanation—those which accept northern origins and those which favour southern origins. Raven and Axelrod (1972, 1974), on the evidence that the relatives of *Nothofagus* occur in the northern hemisphere coupled with the fact that various present-day austral gymnosperm groups have Mesozoic and Cenozoic fossil relatives in Europe, suggested that *Nothofagus* "probably passed between the northern and southern hemispheres by way of Africa and Europe since land connections were absent in Middle America" (Raven and Axelrod, 1972, p. 1382). In the southern hemisphere, *Nothofagus* migrated from Africa into South America and eventually to Australasia. In other words, this is a band-wagon theory fitting Schröter's monoboreal relict hypothesis into a pre-continental drift narrative to fit in with present geological theory.

In a more detailed essay, Schuster (1976) similarly accepts a northern origin for *Nothofagus* but gives an alternative migration route because there were land connections between North and South America until the Tertiary. Working on the principle that 'centres of diversity' are equivalent to 'centres of origin' Schuster adopted a North American origin for our well-travelled genus since the Fagaceae are Eurasian except for *Nothofagus*. He assures us that the ancestor of *Nothofagus* arrived in Gondwanaland by overland transport from North to South America, along the so-called Marsupial route, possibly by mid-Cretaceous times. This far-fetched explanation continues, completely without evidence, to say that: "By Upper Cretaceous times, *Nothofagus* had diffused not only to Antarctica but, more than eighty million years ago, across to Australia and New Zealand." Its entry into New Caledonia and New Zealand "must have occurred well before the end of the Cretaceous because of their separation at this time but did not reach New Guinea until well into the Tertiary" (Schuster, 1976, p. 120). Both the Raven and Axelrod and Schuster hypotheses accept the validity of northern hemisphere fossils despite their general rejection (for instance, Cranwell, 1963).

The main alternative is a southern origin and Moore (1971) completes the narrative picture by giving us the post-tectonic version of Hooker's original observations outlined in the introduction. Moore sets a starting date for the origin of *Nothofagus* by saying that "at least some of the Palaeo-austral distributions date from the Cretaceous" (Moore, 1971, p. 131). Since the oldest fossils occur in New Zealand and Antarctica and because of the poor dispersal qualities of *Nothofagus* seed "one is forced to the conclusion that *Nothofagus* achieved migration between New Zealand and Antarctica before their late Cretaceous separation, at a time when its major groups were already differentiated, crossing to Australia and South America while connections were still available during the late Cretaceous-early Tertiary and subsequently moving northwards in the east and west" (Moore, 1971, p. 131). In an independent review of pollen data, Hanks and Fairbrothers (1976), who incidentally accept northern pollen fossil evidence, strengthen Moore's view and provide a new twist by saying that *Nothofagus* "could have developed in a region between New Zealand, Antarctica, and Australia then migrated through Antarctica into South America; and north through Tasmania, Australia and New Zealand and eventually arrive in Europe" (Hanks and Fairbrothers, 1976, p. 7); presumably it then became extinct in the northern hemisphere.

To sum up the narrative approach, there are many possible centres of origin for *Nothofagus* involving almost every major continent on the globe, the main places being North America, Europe, South-east Asia, New Caledonia, somewhere between Yunnan and Queensland, Antarctica and New Zealand. In all cases the present-day distribution of *Nothofagus* has occurred by dispersal in a variety of trans-oceanic or overland routes ranging from South America to Japan depending on the acceptance or rejection of continental drift. In each case, there is no attempt to use a refined taxonomy showing relationship at the species level and only vague notions are given as to what are 'centres of origin.' Furthermore, there is little appreciation of the idea that biogeography is more about areas of the globe than one particular group of organisms.

Analytical Theories of *Nothofagus* Biogeography

In 1966, Brundin, in his classical investigation of the trans-antarctic relationships of chironomid midges, said that until such time as there are explicit phylogenetic hypotheses for groups such as *Nothofagus* it will be impossible to even start to reconstruct precise distribution patterns. In other words, until phylogenetic schemes can be hypothesised which account for *all* of the variation in modern species and fossils, it is neither possible to provide historical details for particular genera nor to make comparisons, hence biogeographical statements, between distantly related groups which occupy similar areas.

Melville (1973) was the first botanist to attempt to reconstruct a fully resolved model of species relationships in *Nothofagus*. His hypothesis, which is an extension of van Steenis's classification (1953) is shown in Figure 9. The relationships are based on three evolutionary trends of deciduous to evergreen leaves, a gradual reduction in the parts of the female flower and an elaboration of the scales on the lobes of the female cupule. The reconstruction is based on Forman's (1966) thesis that *Trigonobalanus,* an unusual tree genus from Celebes, Malaya, Sarawak, Sabah, Thailand and recently discovered in the Neotropics of Colombia (Lozano-C. et al., 1980), has the most primitive female infructescence structure in the Fagaceae. The mature infructescence of *Trigonobalanus* has seven nutlets, provided with a more-or-less continuous cupule derived from nine lobes. The most common condition in *Nothofagus* is a cupule of four lobes, enclosing an infructescence of two lateral triangular nuts and a median, lenticular, or flattened one. The most primitive condition is found in the South American species *N. allessandri* which has four lobes in the cupule but seven nutlets in the mature infructescence. Melville considers that reduction from four to three to two cupule lobes is a continuous transformation which puts van Steenis' subsections *Tripartite* and *Bipartite* amongst the most derived taxa (Fig. 9). Other changes include the development of three lenticular nutlets, nutlet loss and gradual reduction in the cupule scales. The ultimate conditions are achieved by the New Guinean *N. grandis* which has one lenticular nutlet and two cupule scales, and *N. cornuta,* in which the scales are minute and the solitary nutlet virtually naked. Characters used for arranging the cool temperate species include elaborations of the female cupule, such as branched lamellae and resin glands, and modifications to leaf venation.

Cracraft (1975) used Melville's model to reconstruct the historical biogeography of the modern species of *Nothofagus*. In Figure 9 Melville's cladogram is given Cracraft's selected species names for the four trans-antarctic groups; the "*brassii*" group includes the evergreen *N. dombeyi* from South America and its sister group in New Zealand, New Caledonia and New Guinea; the "*menziesii*" group comprising *N. betuloides* in South America and its sister species in New

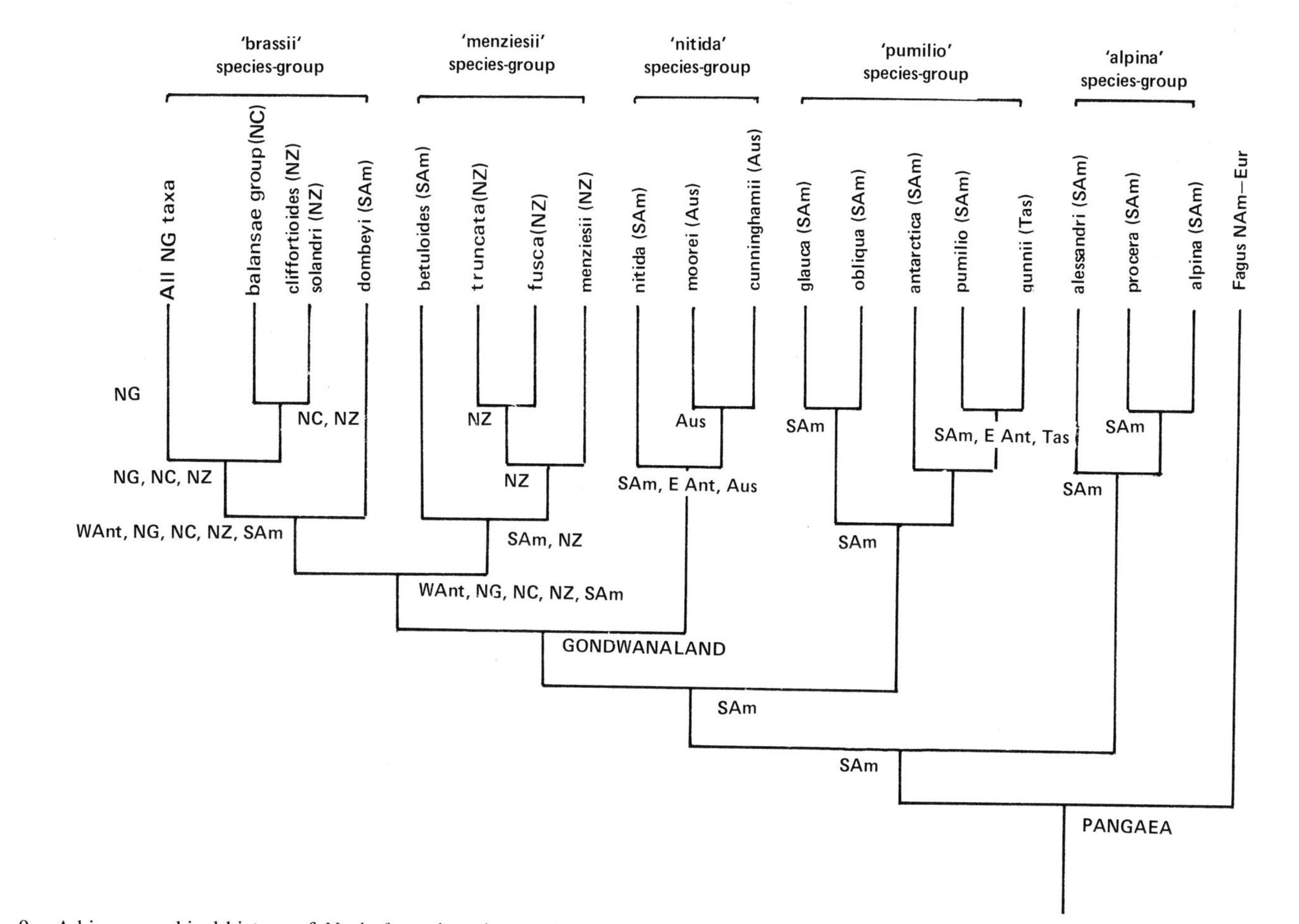

Fig. 9. A biogeographical history of *Nothofagus* based on Melville's cladogram as redrawn from Cracraft (1975) and corrected for modern species from van Steenis (1971) and Soepadmo (1972).

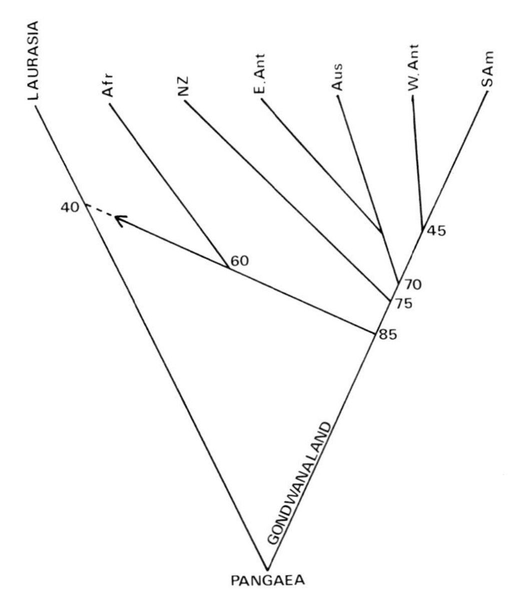

Fig. 10. A cladistic representation of continental breakup redrawn from Rosen (1978) based on the maps of Dietz and Sproll (1970), Rich (1975) and Ballance (1976). Figures refer to millions of years.

Table IV

Characters used for the cladogram in Figure 15

		Primitive	Advanced
1	Tracheids	present	absent
2	Wood fibres	×1.48 mm	×0.7–1.16 mm
3	Leaves	evergreen	deciduous
4	Terminal flowers	present	aborted
5	Male dichasia	many-flowered	few-flowered
6	Male perianth	6-lobed	campanulate fused, tubular
7	Male inflorescence	long spikes	hanging clusters
8	Pollen aperture number	3	5–8
9	Pollen apertures	colpate	colporate
10	Pollen apertures	long	medium short
11	Ovule integuments	2	1
12	Cupules	4-lobed	3-lobed, 2-lobed
13	Female flowers per cupule	2–7	1
14	Cupules	reaching nutlet tip	shorter than nutlet
15	Cupules	along rachis	in leaf axils
16	Cupules	not branched	branched
17	Cupule lamellae	not modified	gland tipped, recurved
18	Cupule lamellae	entire	simple branches, elaborated branches
19	Infructescence	7 fruits	3 fruits
20	Central fruits	trimerous	dimerous (or absent)
21	Lateral fruits	trimerous	dimerous
22	Styles	flattened with long stigmatic surface	cylindrical with short stigmatic surface

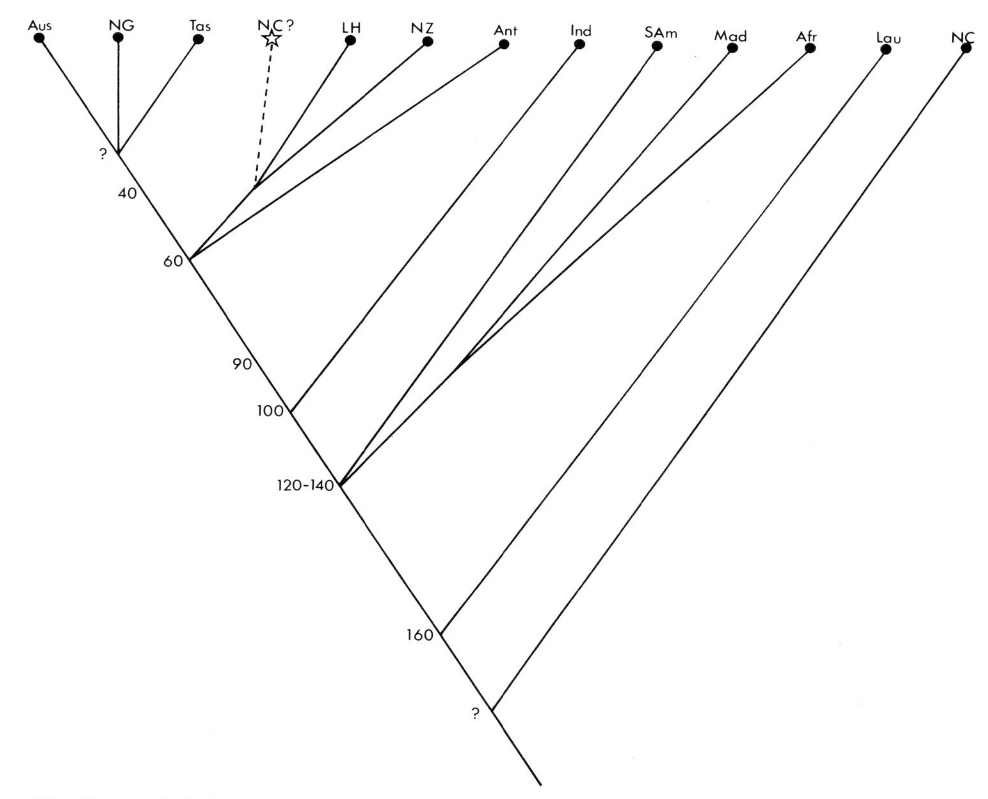

Fig. 11. A cladistic representation of continental breakup based on the maps of Smith et al. (1973). Figures refer to millions of years.

Zealand; the "*nitida*" group comprising *N. nitida* in South America and its sister group of two species in Australia; and finally the "*pumilio*" group occurring in South America. The method of reconstruction was an attempt to determine the centre of origin by the Hennig-dispersal method, but the result actually shows a combination of vicariance and dispersal events. According to this reconstruction the ancestors of the "*brassii*" group must have been distributed throughout a land-mass comprising South America, western Antarctica, New Zealand, New Caledonia and New Guinea.

Continental drift is used as an explanation for the present disjunct pattern. Separation of the New Zealand and South American continental blocks could account for the isolation of *N. dombeyi* from its sister group. A different, later event isolated the ancestor of *N. flaviramea* and *N. brassii* in New Guinea and the ancestor of *N. codonandra* and *N. solandri* in New Caledonia and New Zealand. The ancestor of the "*menziesii*" group was spread across South America, western Antarctica and New Zealand. Continental break-up could also account for vicariance speciation isolating *N. betuloides* in South America and leaving the ancestor of *N. menziesii* and its relatives in New Zealand. A similar pattern can be reconstructed for the "*nitida*" group but over eastern rather than western Antarctica. However, because with the "*pumilio*" group the stem-group species *N. glauca, N. obliqua* and *N. antarctica* all occur in South America and

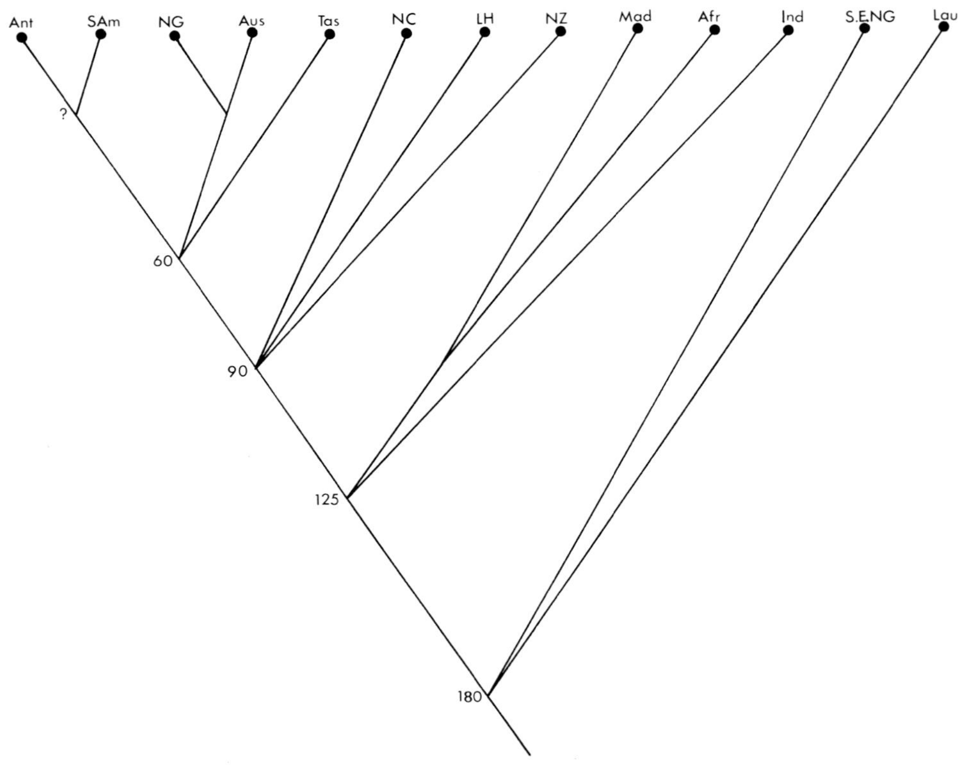

Fig. 12. A cladistic representation of continental breakup based on the maps of Owen (1976). Figures refer to millions of years.

the crown-group species *N. pumilio* and *N. gunnii* in South America and Tasmania, Cracraft suggests that the most parsimonious explanation for the crown-group is a dispersal event from South America to Tasmania. The "*alpina*" group has a wholly South American origin. Because the taxa in New Zealand, New Guinea, New Caledonia and Australia figure as unique events, the obvious deduction for the origin of the whole group is pre-drift west Gondwanaland with subsequent dispersals into other areas. The presence of at least four species groups in pre-drift Gondwanaland means that dispersal to east and west Antarctica and into Australia, New Guinea and New Zealand occurred by at least the Turonian, in the upper Cretaceous some 90 million years ago (Figs. 10–12). According to this model there are no sister group relationships between New Zealand and Australia but across Antarctica to South America, a pattern commensurate with several extant groups occurring at the time of continental break-up.

Although Cracraft's analysis is perhaps the most explicit hypothesis to date it is difficult to assess its validity because it is based on dubious taxonomy and incorporates untestable dispersal explanations for some of the ancestral distribution patterns. Since the cladogram is based on general interpretations of advancement in some morphological characters rather than on synapomorphies for establishing monophyletic groups there are several cumbersome and unlikely interpretations in the scheme. For example, the South American species *N. dombeyi* is, on the basis of cupule and floret morphology, one of the most primitive

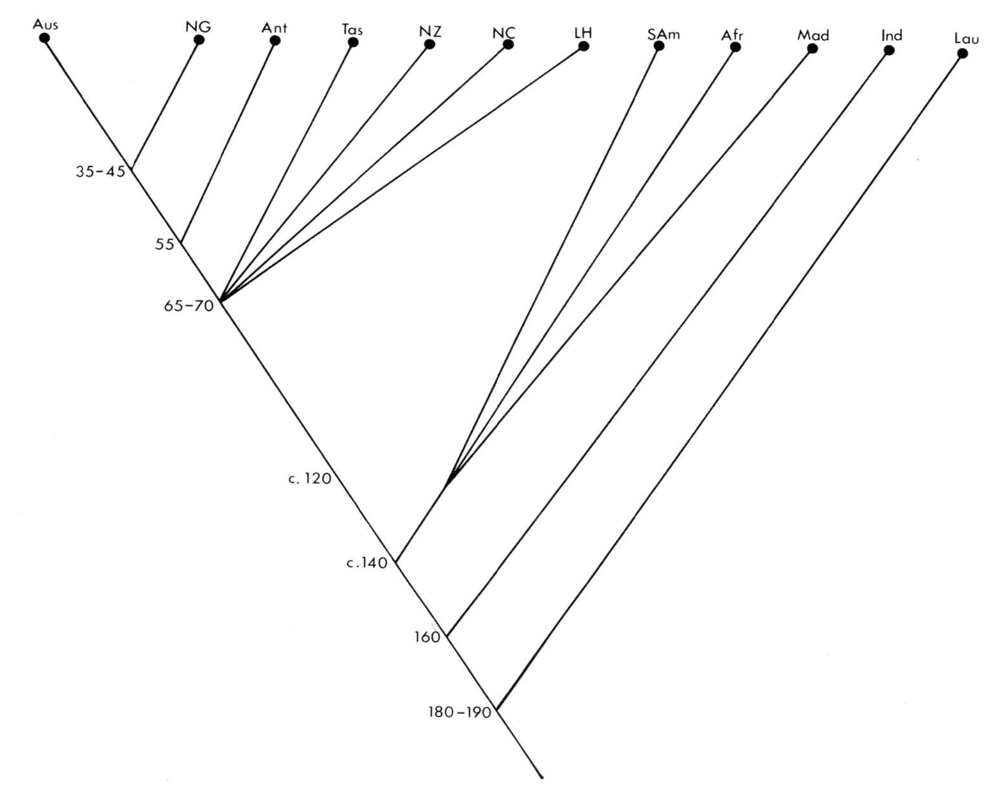

Fig. 13. A cladistic representation of continental breakup based on the paper of Shields (1979). Figures refer to millions of years.

species in the genus and yet is placed as the sister species to some, but not all, of the taxa with derived cupules. Furthermore, there are a number of dichotomies in the diagram which have no evidence at all for their inclusion. For these two reasons and because various other sources of valuable data on wood anatomy and the male inflorescence have been disregarded, another cladogram produced on transformed cladistic lines is given in Figure 15 and Table IV which will be used in subsequent discussions.

The scheme of interrelationships favoured here differs from that proposed by Melville in a number of points. All of the cool temperate species show a 'specialised' wood anatomy and are more closely related to one another than they are to the New Caledonian and New Guinean species. This means that the tripartite cupule in the two New Zealand species involves an independent origin from the bipartite cupule seen in the tropical taxa. Within the two temperate groups there is insufficient evidence to establish the relative order of some main branching points and so the divisions are given as trichotomies rather than resolved dichotomies. The net result from the rearrangement is that there are only two trans-antarctic relationships at the species level rather than four as predicted in the Melville cladogram, only one of which (*gunnii-pumilio*) is identical.

To test whether the dichotomies in the new *Nothofagus* phylogeny can be accounted for by known events in earth history, the vicariance-cladistic method

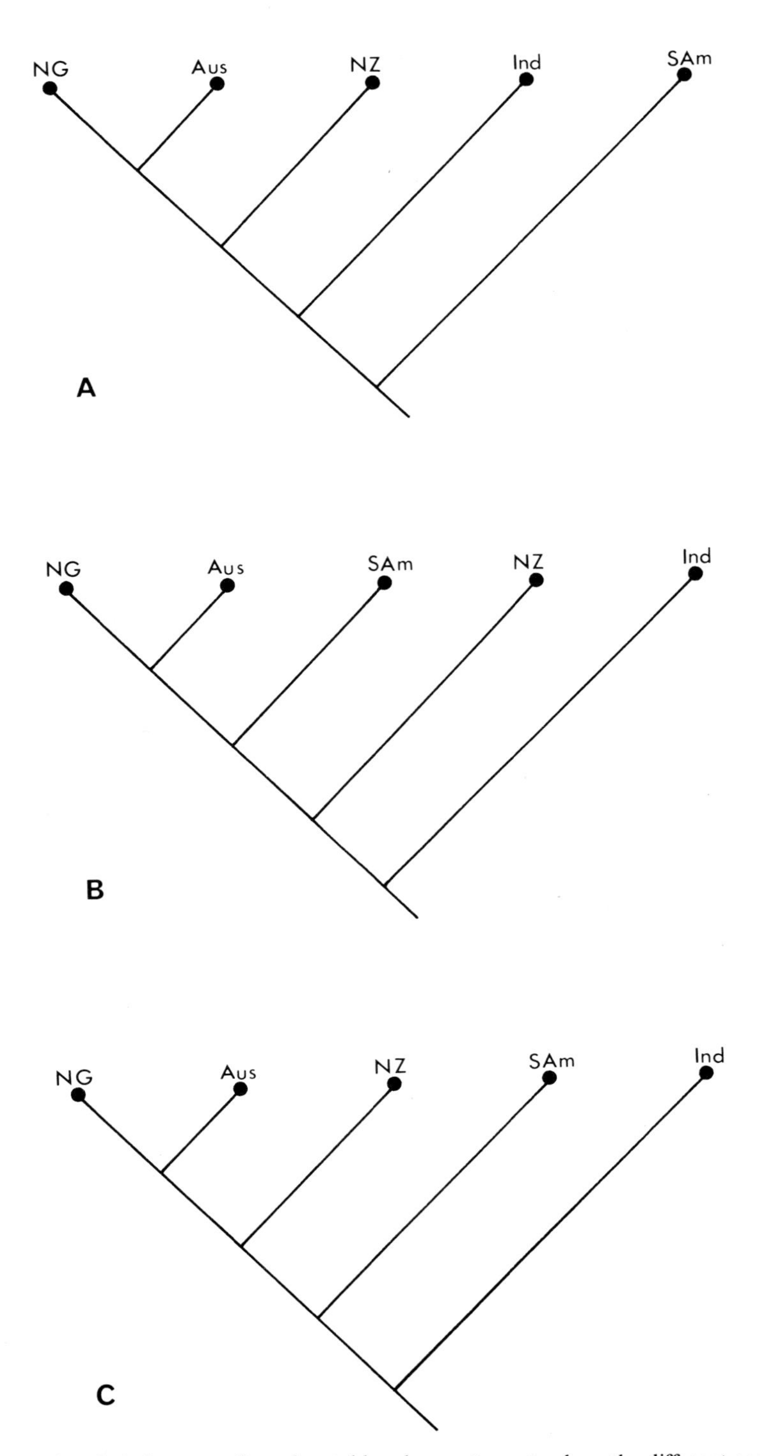

Fig. 14. Reduced cladograms of continental breakup patterns to show the different sequences for different mapping schemes. **A.** Based on Smith et al. (1973), see Figure 11. **B.** Based on Rosen (1978) and Owen (1976), see Figures 10 and 12. **C.** Based on Shields (1979), see Figure 13.

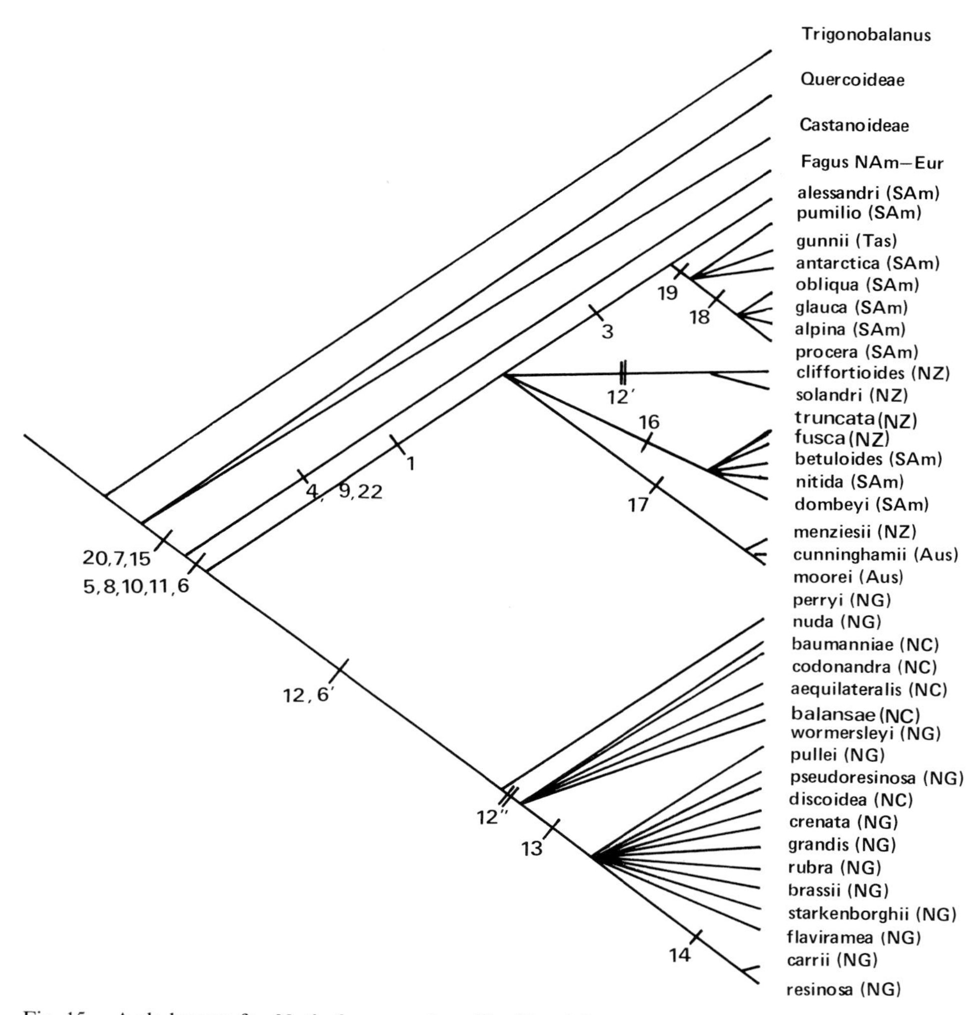

Fig. 15. A cladogram for *Nothofagus* produced by Hennig's method (1966) whereby monophyletic groups are identified by synapomorphies (see Table IV for key to characters; autapomorphies are left out).

can be applied. The additive area sequence for the new *Nothofagus* cladogram is given in Figure 16a. From such models produced in this way, Nelson (1974) suggests that within a vicariance paradigm, the unshared elements estimate not ancestral distributions but the exact positions of the barriers which caused divergence.

By combining the northern and the southern beeches together and by substituting taxa for areas in a reduced area cladogram, *Fagus* and *Nothofagus* generally form an eight-area pattern of distribution (Fig. 16b). Two of the areas represent North America and Europe in Laurasia and the other six the principal land masses of Gondwanaland. This cladogram can now be compared to the geological area cladograms.

Recent renewed interest in continental drift has led to a number of temporal mapping schemes. Four different schemes are given in Figures 10–13 as geological

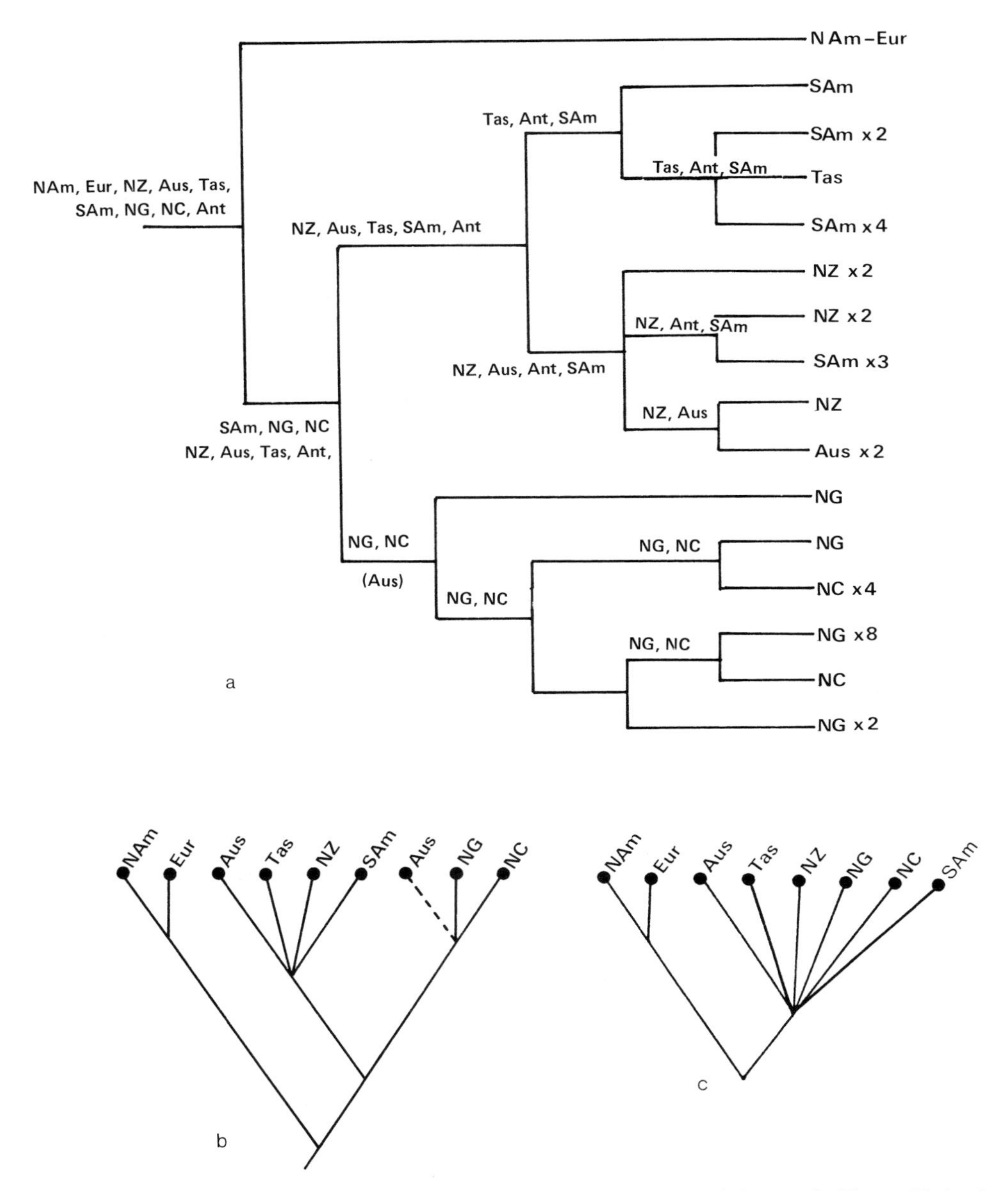

Fig. 16. **a.** Vicariance biogeography for *Nothofagus* based on the cladogram in Figure 15. **b.** A primary reduced area cladogram based on Figure 16A. **c.** A primary reduced area cladogram based on reconstruction of Melville's cladogram in Figure 9.

cladograms to demonstrate the break-up sequences based on two fixed earth models (Dietz and Sproll, 1970; Smith et al., 1973) and two expanding earth models (Owen, 1976; Shields, 1979). Figure 10 was originally published in Rosen (1978) in a greatly simplified form. It is interesting to compare the different schemes since New Zealand, India and South America give three different sequences on the cladograms (Fig. 14). As Rosen (1978) points out, geological data arranged in nested cladograms should reflect the history of the areas under study. Within a historical context related geological areas are two or more areas that arose by the disruption or fragmentation of an ancestral one. If, therefore, the geological and biological cladograms are constructed on special similarities which have arisen for the same historical reasons the cladograms should be congruent. By deleting the unique components of the cladograms to be compared means that they should be sequentially congruent for the areas. Thus, in addition to the possibility of testing congruence between different endemic groups occurring in similar areas congruence tests between biological and geological cladograms should determine the most plausible break-up sequence for Gondwanaland.

Since there are three groups in Figure 16A showing different disjunction patterns for South America-Australia, South America-Tasmania and New Zealand-Australia, they have to be treated in Figure 16B as branching from one point because such an incongruent pattern could not have occurred through sequential continental breakup. It is possible to resolve the next branching point because all of the species of the cool temperate austral areas are sister to the New Guinea-New Caledonia group. Although this model gives only low resolution for *Nothofagus,* a similar area diagram for the Melville/Cracraft model (Fig. 16C) gives even less resolution because the South America-New Guinea/New Caledonia disjunction in the "*brassii*" species group represents yet another pattern.

The only identifiable microfossils which can be affiliated with particular modern monophyletic groups are the pollen grains of the "*brassii*" type. Adding these fossils into the area diagram does not extend the areas occupied by *Nothofagus* but does extend the tropical "*brassii*" group into South America, Australia and New Zealand. The net effect is to produce a primary reduced area cladogram identical to the Melville/Cracraft model (Fig. 16c). Nevertheless, fossils do give the minimum age for some of the area relationships. The Australia-South-America-New Zealand pattern is at least as old as the late Cretaceous and, if the new cladogram in Figure 16a is correct, then it must be predated by the New Guinea/New Caledonia-Australia/South America/New Zealand dichotomy. The northern hemisphere-Gondwanaland dichotomy for the *Fagus-Nothofagus* split does not have a fossil record at all, but it must predate the Australia/South America/New Zealand split. In other words, the beeches relate northern and southern areas and within the vicariance framework probably diverged at the breakup of Pangaea. This being the case means that Patterson's comment is nearer to the truth; because of the age of the endemic taxa of contemporary *Nothofagus* the genus, as a whole, is uninformative about the history of the breakup of Gondwanaland.

By using reduced area cladograms selected at random from other taxonomic groups (Fig. 17) with endemic representatives in at least three of the areas occupied by *Nothofagus,* it should be possible to see better resolved general area patterns of relationship. The cladograms in Figure 17l, m, n give four 5-area tests, Figure 17a, e, g, h, i, j, k, o, and p give twelve 4-area tests, Figure 17b, d give four 3-area tests of the *Fagus/Nothofagus* pattern. Figure 17c, f although resolved for four areas are uninformative for the *Fagus/Nothofagus* pattern since they occur only in two areas occupied by the group. From these cladograms it does seem certain that the area relationships are generally similar in all but a few situations regarding the areas of interest. Exceptions include, for example, the last dichotomy in the genus *Leptocarpus* (Restionaceae) (Fig. 17c). Since this genus is the only one to occur in Malaysia, a parsimonious conclusion demands that, because all the other groups show austral connections, the crossing of Wal-

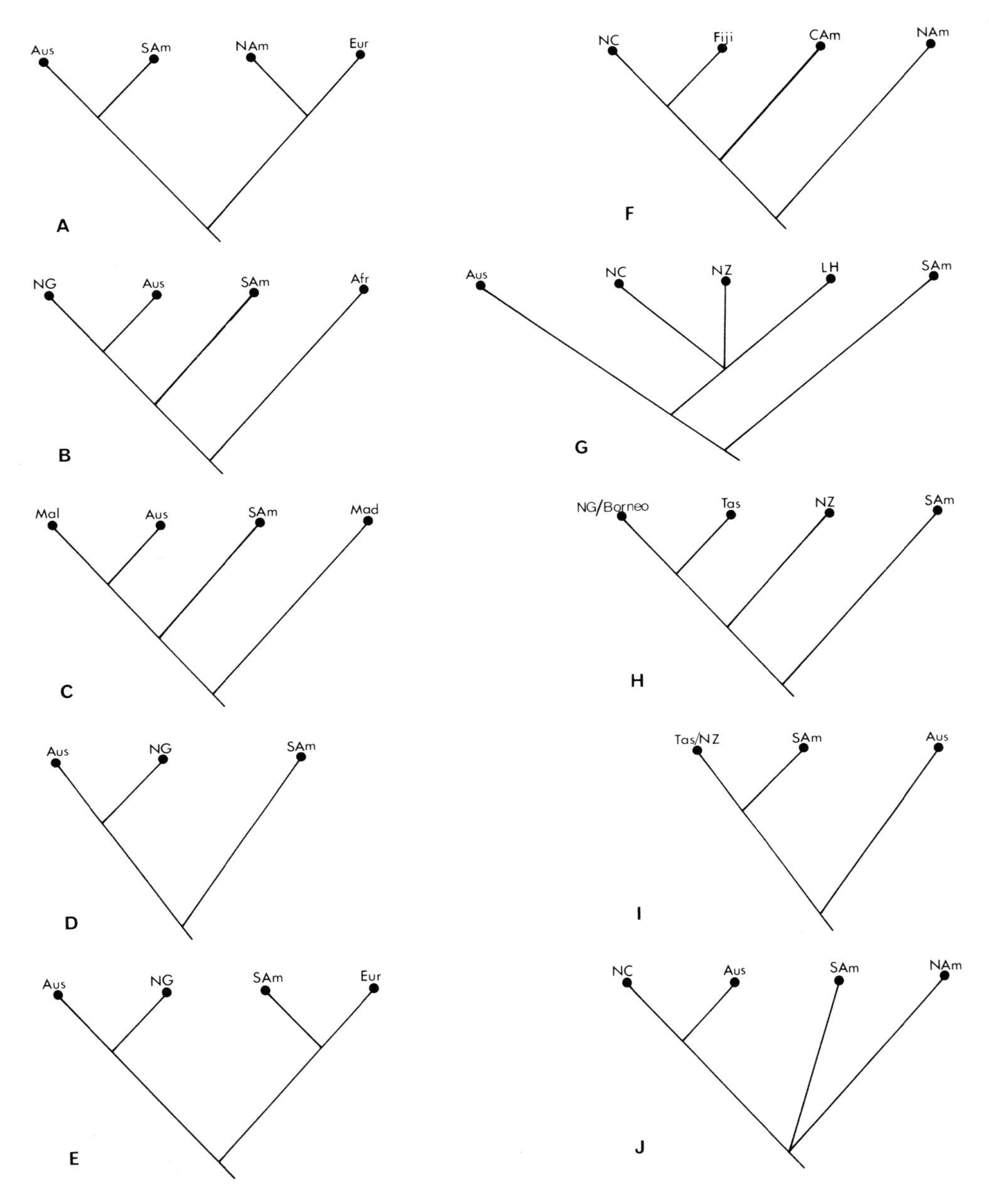

Fig. 17. Primary reduced area-cladograms for randomly selected groups. **A.** Chironomid midges (based on Brundin, 1966). **B.** **(i)** *Drimys* (Winteraceae), **(ii)** Proteaceae tribe Gevuineae, **(iii)** Proteaceae tribe Macadamiae (ii and iii based on Johnson and Briggs, 1975 and redrawn from Patterson, 1981). **C.** *Leptocarpus* in Malaysia, Australia and South America and *Calopsis* in Madagascar (after Cutler, 1972). **D.** Osteoglossine fishes (*Osteoglossum* and *Scleropagus*) omitting S.E. Asia, the sister area of Australia and New Guinea (after Gaffney, 1977 and redrawn from Patterson, 1981). **E.** Higher ratite birds omitting Africa as the sister area of Europe (from Cracraft, 1972, 1973 and redrawn from Patterson, 1981). **F.** *Lindenia* (Rubiaceae) (after Darwin, 1976 and Seeman, 1862). **G.** A combined

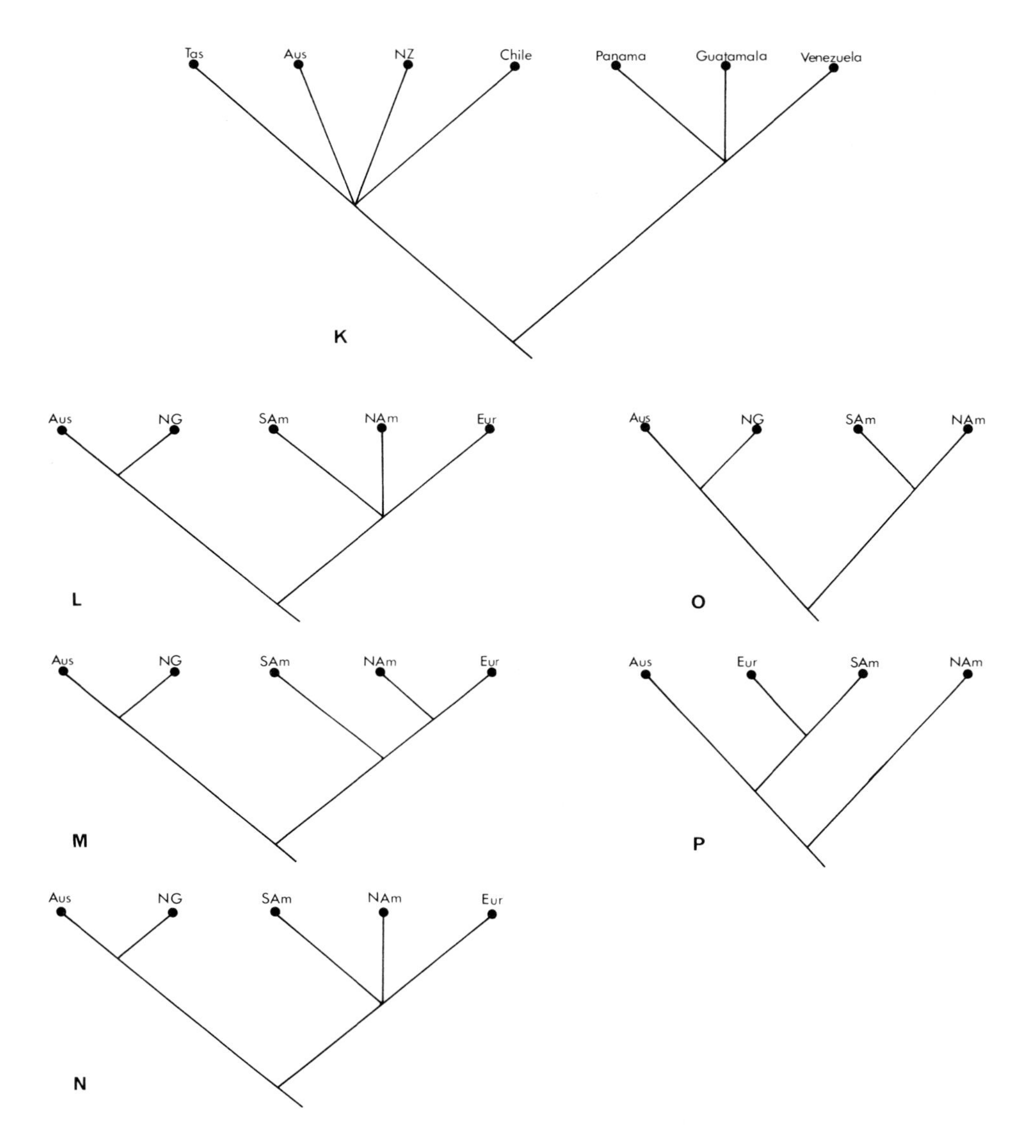

cladogram for *Negria, Depanthus, Rhabobthamnus, Corathera* and *Fielda* (Gesneriaceae). **H.** *Drapets* (Thymeleaceae) (based on data of Moore, 1968 and Ding Hou, 1960). **I. (i)** *Phyllacne,* **(ii)** *Donatia,* (Stylidiaceae) (based on Van Balgooy, 1975 and Allen, 1961). **J.** *Nicotiana* (Solanaceae) (based on the monograph of Goodspeed, 1954). **K.** Combined data set for *Lagenifera* and *Solenogyne* (Compositae) (after Drury, 1974). **L.** Galliform birds (from Cracraft, 1973 as redrawn from Patterson, 1981). **M. (i)** Hylid frogs (after Maxson and Wilson, 1975 and redrawn from Patterson, 1981), **(ii)** *Chaleosyrphus* group of xylotine syrphid flies (Syrphidae) (after Hippa, 1978 and redrawn from Patterson, 1981). **N.** Recent and fossil groups of marsupials (from Patterson, 1981). **O. (i)** Recent marsupials (from Patterson, 1981), **(ii)** Hydrobiosine caddisflies (after Ross, 1956 and redrawn from Patterson, 1981), **(iii)** *Anisotarsus* (Carabidae) (after Noonan, 1973 and redrawn from Patterson, 1981).**P.** Sciadocerid flies (Sciadoceridae) (after McAlpine & Martin, 1966 and redrawn from Patterson, 1981).

lace's line must be a unique, perhaps stochastic, phenomenon such as a dispersal event rather than a case of extinction in Malaysia for *Nothofagus* and the twenty other groups. A somewhat similar situation is found in Figure 17h indicating a reduced area cladogram for *Drapetes* in the Southern Hemisphere. The genus consists of four species, with *D. ericoides* restricted to New Guinea and Borneo. It is more reasonable to deduce a dispersal event from New Guinea to Borneo as a parsimonious solution rather than extinction of all the other groups.

One thing most of the cladograms have in common is the Australia-New Guinea connection. Although most of New Guinea is of very recent origin, the southern part of the island has had two historical connections with Australia during its geological history, one in the Pleistocene and the other in the Jurassic (Fig. 13). From the *Nothofagus* cladogram in Figure 16a and fossil evidence, it would appear that the austral groups are at least as old as the Cretaceous and that the New Guinea connection belongs to the first connection. On the other hand, since the Australia-New Guinea dichotomies are the most derived in the cladograms of Figure 17, it is likely, in these groups, to represent the more recent connection.

The only cladograms to show a similar pattern with *Fagus/Nothofagus* division are those of the chironomid midges and the Ratite birds (Fig. 17a, e). The four four-area cladograms for *Drimys,* the two tribes of the Proteaceae and *Leptocarpus/Calopsis* (Fig. 17b, c) do show the same general sequence but with southern African or Madagascan rather than northern hemisphere connections for the first dichotomy, a general area pattern which agrees more with the breakup sequence of Gondwanaland.

It becomes more obvious from looking at the geological cladograms that there are in fact a number of different patterns for area relationships particularly with regard to the position of South America in the galliform birds, hylid frogs, xylotine flies and the Marsupials (Fig. 17l, m, n, o), and in North America for sciadocerid flies (Fig. 17p). In fact, three general patterns can be observed by collapsing all of the cladograms of Figure 17 into one another as shown in Figure 18. The result is that each shows serious incongruences with one another. A question one might ask is can the incongruencies be accounted for by historical events? With regard to making comparisons between the biological cladograms and the biogeographic maps none of the cladograms contain Indian distributions thus negating any prospect of considering India's position in historical sequences. In the general picture, New Zealand and South America in relation to Australia and New Guinea, the geological cladograms based on Smith et al. (1973) and Shields (1979) show congruence with the biological cladograms having resolved sequences for these areas (see Fig. 14a, b, c and compare with Fig. 18a). The patterns for *Nothofagus,* chironomid midges, *Drimys,* both tribes of the Proteaceae, the osteoglossine fishes, and *Lagenifera* (Fig. 16b, c; 17a, b, d, k) agree with all four geological cladograms since they include only one of the test areas. The worst incongruences occur in five of the cladograms (Fig. 17e, l, m, n, o) which all give more recent common ancestries for South America with North America or Europe. When both New Zealand and South America are included in the cladogram (e.g., Fig. 17h, i) they give a relatively earlier date for the detachment of South America rather than New Zealand from Gondwanaland.

At this point it is worth asking the question what does incongruency between the cladograms and biogeographic maps mean? There are various answers. Unique distribution patterns can mean dispersal after the continental breakup patterns as in the cases of *Leptocarpus* (Fig. 17c) and *Drapetes* (Fig. 17h). In other cases poor resolution or incongruence can mean dispersal or vicariance events occurring prior to the specific historical events such as the breakup of

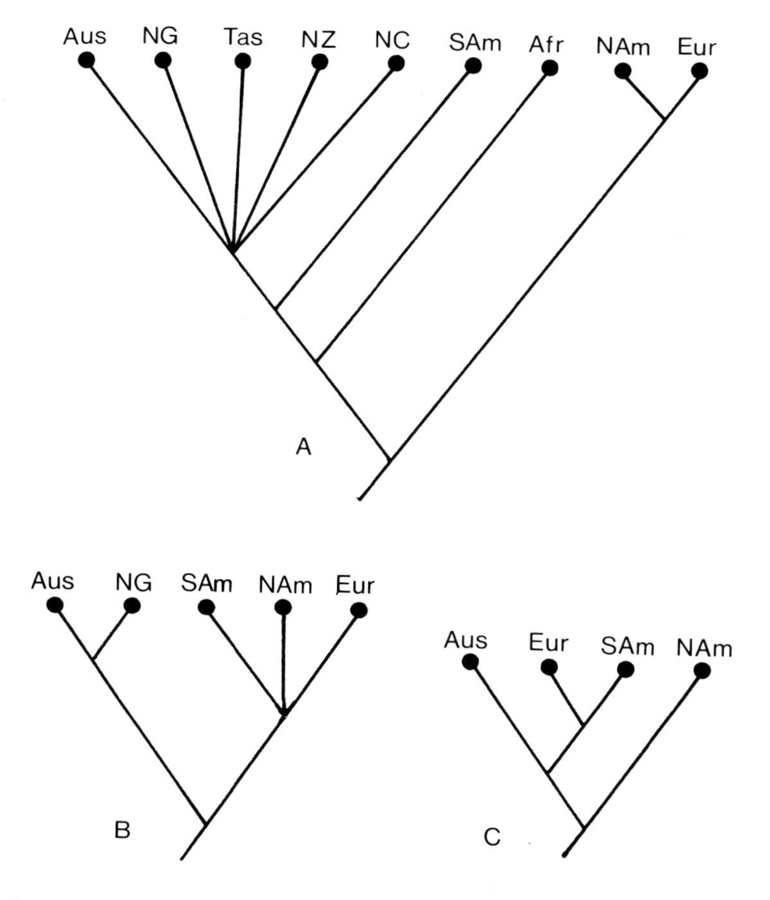

Fig. 18. General patterns of area relationship for the cladograms of Figure 17. **A.** General reduced area cladogram for the taxa included in Figure 17A, B, C, D, F, G, H, I, K. **B.** General reduced area cladogram for the taxa included in Figure 17E, J, L, M, N, O. **C.** Unique area cladogram for *Anisotarsus* in Figure 17P.

Gondwanaland examined in this paper. There is always the possibility too that in some groups one is dealing with poor taxonomy, extinction and faults in the mapping schemes.

By adopting an alternative principle of considering only the resolved patterns, a more resolved cladogram of the *Nothofagus* areas can be drawn (Fig. 19). By comparison with the geological cladograms (Figs. 10–13), the most congruent patterns are found in the sequences from the maps of Smith, Briden and Drewry and Shields, both incongruent only for New Caledonia. Rosen's cladogram is incongruent for South America, but deals with only five of the eight areas. The sequence from Owen's maps is incongruent for two areas, South America and New Caledonia.

The two positions for South America (shown as two dotted lines in Fig. 19) are probably due to the fact that it is a huge composite area and shouldn't be treated as a single area of endemism.

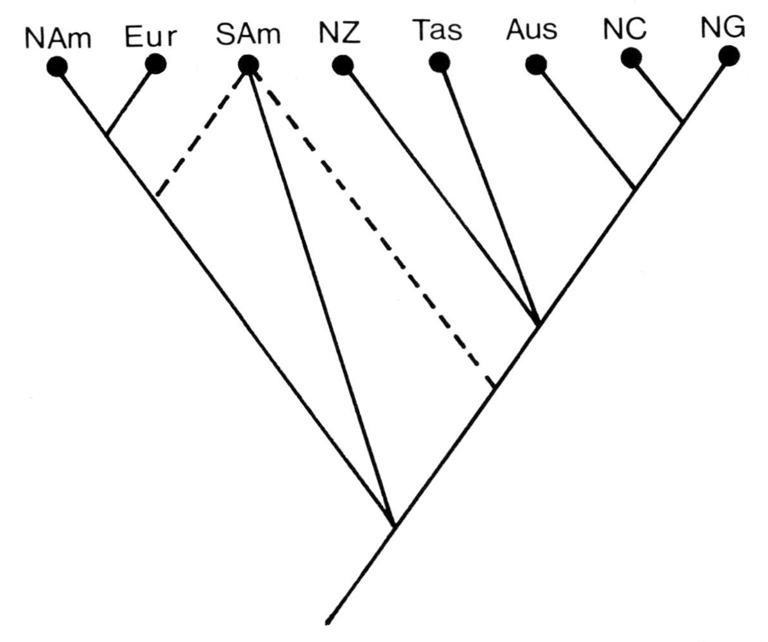

Fig. 19. General patterns for area relationship for the cladograms of Figure 17 considering only the resolved patterns.

Conclusion

To conclude, it appears that Patterson's remark at the beginning is justified. Previous claims that the distributional history of the southern beeches is in any way special such as the key to life in the far south cannot be scientifically substantiated. What can be said is that the southern beeches do appear to be a wholly southern group whose precise 'centre of origin,' if it has one, is impossible to locate. Poor resolution in the reduced area cladogram suggests that there were several species groups already extant at the breakup of Gondwanaland and it does appear that continental drift created barriers in at least two of the modern vicariant species groups. I have tried to demonstrate, by using *Nothofagus* as an example, that narrative theories are dependent on many assumptions all of which can be legitimately questioned and which do indeed change with time. Earlier narrative theories assumed a world of stable geography, or one in which present land masses were linked by land-bridges. These assumptions are not now fashionable. Later narrative theories involved continental drift, but, like their predecessors, are coupled with the assumption that animals and plants dispersed over large parts of the earth's surface. This leads to further questions of deciding from where, by what means and by what routes the particular groups of organisms travelled. Narrative theories are often based, as in the case of *Nothofagus,* on imprecise ideas of the interrelationships of the organisms and indeed it is not uncommon for the assumed biogeographic history to determine the scheme of relationships.

On the other hand, the analytical approach attempts to determine, as completely as possible, the phylogenetic interrelationships of the organisms concerned. The distributional data are then added and compared with similar data from other groups. If, as is axiomatic to the analytical approach, there is a common history, caused by geological events, then there should be congruence between many different groups of organisms. Routes of dispersal, and centres of origin are never assumptions in the analytical methods and fossils are never given special importance.

The analytical approach is in its infancy. Phylogenetic relationships are known in so few groups that, so far, analytical historical biogeography lacks precision. *Nothofagus* may not itself be the key to an understanding of the history of southern continents, but any general, historical theory of the flora and fauna of the southern hemisphere must explain the present-day distribution of *Nothofagus*.

Acknowledgements

I extend sincerest thanks to Prof. T. C. Chambers for providing facilities at the Botany School, University of Melbourne where much of this work was carried out whilst in tenure of a Melbourne University Fellowship. Thanks are extended also to Dave Brodin, Peter Forey, Norman Robson, Donn Rosen and Arthur Chater who all gave useful comments on the manuscript, and to Bob Press for drawing some of the figures. Finally, I thank Barbara Joyce, Loveday Hosking and Kathy Kavanagh for typing various drafts of this paper.

Quantitative Phylogenetic Biogeography

M. F. Mickevich

Table of Contents

Introduction

An explicitly cladistic viewpoint was first introduced into biogeography by Hennig's (1966) major treatise and Brundin's (1966) vigorous application of cladistic principles to biogeographic problems. Prior to this beginning, however, it was customary in a systematic study to include a map showing the distributions of taxa. The association of taxa and areas was considered central to theories of differentiation of taxa, and to theories of the establishment of biotic communities. Biogeographic theories—theories of the origin of the distribution patterns themselves—were obtained by intuitively combining a classification or a phylogeny with whatever evolutionary theory the systematist deemed reasonable according to the information on hand. A lucid example of this approach is found in Burns' (1964) study on skipper butterflies. Cladistics, of course, can be recognized from a historical viewpoint as having originated with the acceptance of scientific principles into systematics. Yet, a means for obtaining a biogeographic map consistent with recent advances in cladistic principles and methods has still not been developed.

Biogeography, always of interest to systematists, is now a rich science because it offers deductive theories which make predictive statements concerning the nature of biogeographic maps, that is, of the origins of distributions. Vicariance biogeography predicts that due to similar factors which control all the cladogenetic events in a set of areas one should observe repeated patterns relating the phylogenies and the localities in different groups of organisms (Croizat et al., 1974). Other biogeographic theories deny, or at least do not imply or require, the

Table I

Five taxa (A–E) and four characters (1–4), each with two states (0, 1)

Taxa	Characters 1	2	3	4
A	1	1	1	1
B	1	1	1	0
C	0	1	1	1
D	0	0	1	1
E	0	0	0	0

possibility of repeated patterns, offering instead predictions of their own (Darlington, 1957; Brundin, 1966). Biogeographic maps, if constructed according to cladistic methodology, may be viewed as data by which competing theories can be evaluated. A means for evaluating these theories using biogeographic maps, however, does not presently exist.

In what follows, I will develop a procedure for obtaining biogeographic maps and a means for testing biogeographic hypotheses.

Biogeographic Maps

A biogeographic map describes the association between the interrelationships of a group of taxa and the localities at which the taxa are known to occur. Maps can be purely descriptive, but in modern studies a map is a historical hypothesis describing transformations between areas in the context of a biogeographic theory. For example a migration route, such as postulated by Darlington (1957), which includes passage across the Bering strait, or a geotectonic sequence used by vicariance biogeographers to explain one of the "observed general patterns" predicts a sequence of area states. Almost any historical theory can be used to generate sequences in accordance with a given cladogram. All these possible sequences can be referred to as a "biogeographic map." The question of course is how to choose between these historical theories. Since amongst this bewildering array we must begin somewhere, it seems that the only recourse is to refer to the evidence at hand. For biogeographic studies, the evidence at hand consists of the cladogram and the areas occupied by the taxa. Therefore, the method for obtaining a biogeographic map must result in the best fit map according to these data. The best fit statement for any observation on a cladogram is called a transformation series and is obtained by a procedure called Transformation Series Analysis.

Transformation Series Analysis (Mickevich, In prep.) is a new phylogenetic method. It is not opposed to older ideas of phylogenetic analysis, but is an extension and generalization of them, the latest in a series of advances towards solving the problem of obtaining a cladistic character. Hennig (1966) precisely delineated the best fit or most parsimonious sequence of changes for a non-homoplasious character for which the polarity is assumed. Table I describes a set of data for four characters (1–4) for five taxa (A–E). The corresponding cladogram is shown in Figure 1. Here the character states of the stem species are determined by Hennig's precept that the more plesiomorphic of the states shown by two sister taxa is the state of their common stem. In an example of a multi-state character (Fig. 1b), the ancestor of taxa X and Y has state B if the transformation series is defined to be A-B-C (Fig. 1c). Hennig discussed convergent characters,

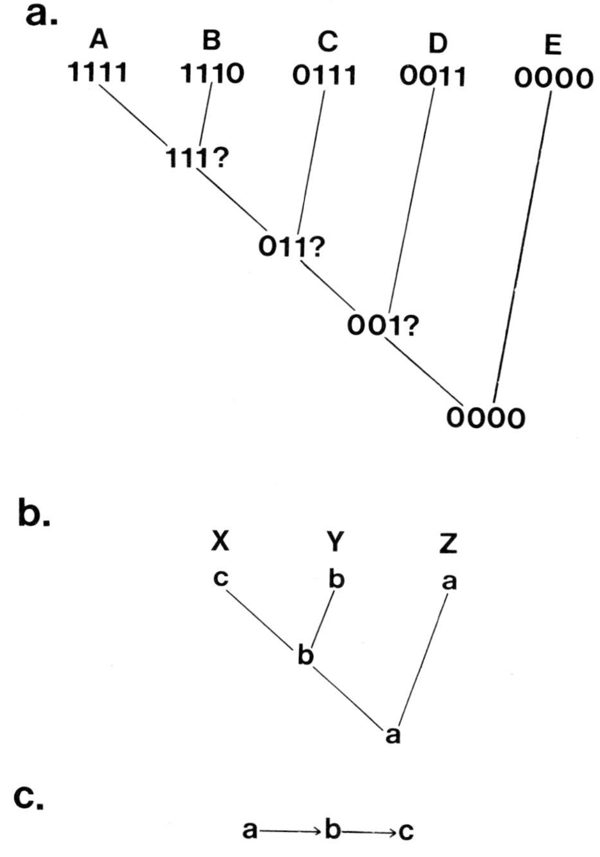

Fig. 1. An example of Hennig's rule for internode character state resolution. **a.** Cladogram for taxa and data in Table I. The "?" represents unresolvable internode states. **b.** Three taxon cladogram with resolved internode states. This resolution is defined by the transformation series in **c.**

pointing out that recoding the character is desirable. However, he did not generalize his procedure to deal with homoplasious characters. Note that character 4 (Table I) has no ancestral state solutions by his criteria in Figure 1a. Nor did Hennig recognise that this procedure resulted in best fit or most parsimonious statements of character change.

Farris (1970) was the first to describe explicitly a general procedure for obtaining the minimum number of changes for a character on a given cladogram. He demonstrated that the median state of the three taxa closest on a cladogram (separated by the least cladistic distance; Farris, 1967a) is the minimum path between the taxa, and that the most parsimonious interpretation of character change on a cladogram can be obtained accordingly. He also demonstrated that polarity of change for a character is not a prerequisite for determining the best fitting character pattern. His procedure is called here *Farris optimization*.

Farris optimization (Fig. 2) results in solutions equivalent to Hennig's if homoplasy is absent. The ancestral states for characters 1–3 (Table I) are identical for both methods as can be seen by comparing Figure 1a and Figure 2a. Farris optimization resolves the fit for character 4 on the cladogram in Figure 2a. This

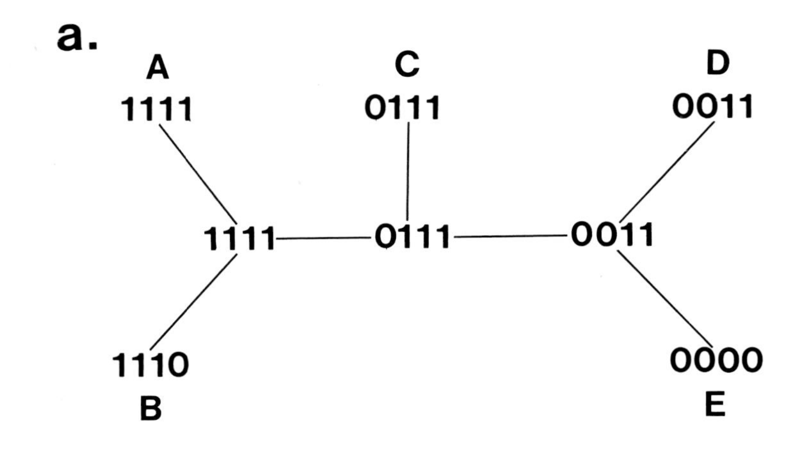

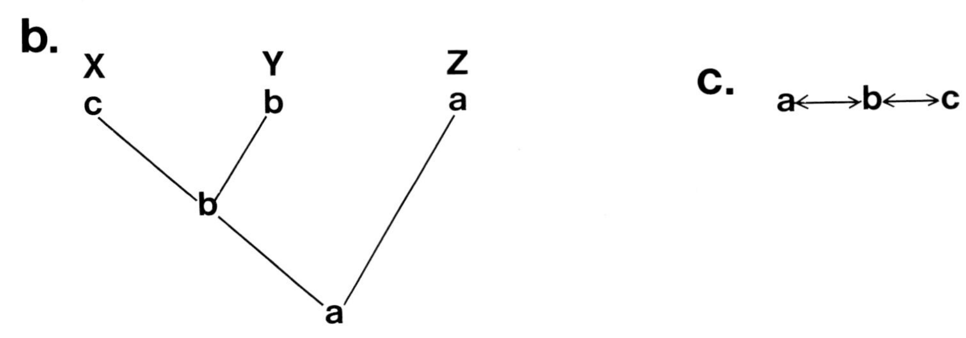

Fig. 2. An example of Farris optimization. **a.** Cladogram with fully resolved internodes. The data are from Table I. This cladogram is undirected in order to represent the lack of dependency on a priori plesiomorphic conditions for Farris optimization. **b.** A three taxon cladogram with resolved internode states under the transition series of c.

procedure frees cladistics from two restraints: absence of homoplasy and predefined plesiomorphic groups and character states. Therefore, if the median condition is defined for a group of taxa the best fit statement or cladogram character can be found. The median is well defined under two conditions: (*1*) if two of the three cladistically closest taxa have the same character state, (*2*) if a character state tree is predefined for the different states (e.g. Fig. 2b). However, this still leaves the problem of defining a cladogram character without the aid of a predefined character state tree which Farris' method assumes.

The transformation series for a character is defined by its relation to the sequence of cladogenetic events of the monophyletic group in which that character evolves. It therefore follows, that when delineating assumptions of polarity and specifications of character state change, we may accept a common character transformation if and only if character states share a cladogenetic event. On a cladogram a *monophyletic group* is defined to be a group which shares an ancestral stem common only to members of that group. Such groups are recognized by internodes sometimes referred to as hypothetical intermediates. Taxa which

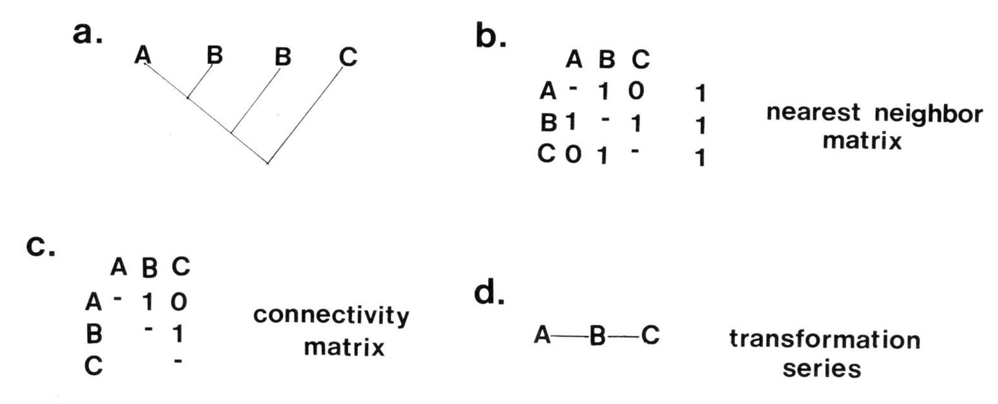

Fig. 3. An example of Transformation Series Analysis. **a.** The cladogram. **b.** The nearest neighbor matrix. **c.** The connectivity matrix. **d.** The transformation series.

have the fewest hypothetical intermediates between them are those that share the most cladogenetic events. These taxa will be referred to as *nearest neighbors*. Character states which are found in nearest neighbors also share the most cladogenetic events. Transformation Series Analysis takes advantage of this idea to infer transformation series. This method is illustrated in Figure 3. The cladogram in Figure 3a has 4 terminal taxa with character states A, B, and C. The nearest neighbor of state A is state B, for it is separated from B by one internode. The nearest neighbors for state B are both A and C because the taxa with state B are equally close to taxa with states A and C (a distance of one internode). The nearest neighbor for state C is B. This information is recorded in a nearest neighbor matrix. This matrix is nonsymmetric. Figure 3b contains all nearest neighbor relationships for Figure 3a. A state is involved in a transformation with (connected to) the state with which it is most often nearest neighbors. Other interpretations are conceivable but would offer less correspondence to the evidence. The matrix which shows states connected to the state or states with which it shares the most cladogenetic events is called a *connectivity matrix*. This is illustrated for this example (Fig. 3a) in Figure 3c. The connectivity matrix defines the transformation series of the character on the cladogram (Fig. 3d). After a transformation series is derived for a character, Farris optimization on the derived sets of changes produces the best fitting cladistic character. This cladistic character can be summarized by a reduced character cladogram (Mickevich, 1978a).

The transformation series is the most parsimonious interpretation of the data and reflects only well substantiated cladogenetic events. It is a most parsimonious or best fitting statement because the states are connected if and only if they are nearest neighbors and if and only if these are most frequently associated. For Figure 3a the transformation series is C-B-A. If non-nearest neighbors were joined in, say, the character state tree A-C-B, Farris optimization would result in the reduced character cladogram A-B-C, but the transformation between A and B would have two assumed steps under the character state tree A-C-B: A transforms through state C to change into state B. This is one more step than is required by the transformation series A-B-C.

Another method which produces best fitting cladistic characters is the Fitch (1971) optimization procedure, which relies on the supposition that any state of a character can transform into any other. Fitch developed his method from the

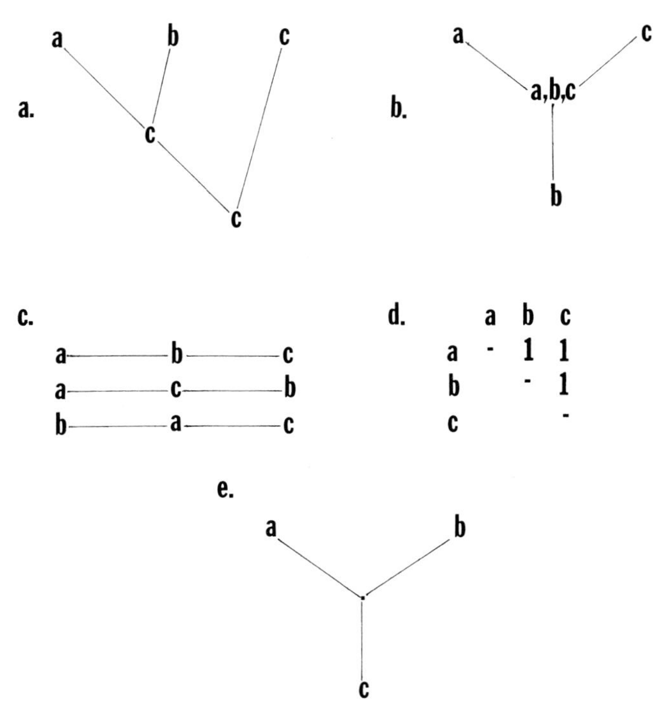

Fig. 4. An example of Fitch optimization and Transformation Series Analysis. **a.** The cladogram and resolved internode states for a directed Fitch optimization. **b.** Undirected Fitch optimization for the same data. **c.** The set of possible transformations which can result from Fitch optimization. **d.** The connectivity matrix. **e.** The transformation series resulting from Transformation Series Analysis.

result that the intersect state of an internode (hypothetical intermediate) yields the minimum change character pattern. The intersect is the mathematical equivalent of the median for unordered characters, and thus Fitch's idea follows naturally from Farris' work. For the example in Figure 3 both the Farris and Fitch procedures produce the same reduced character cladogram A-B-C. Thus Transformation Series Analysis extends the utility of both methods by allowing them to work without prior specifications of transformation series. Very often all three procedures give the same answer. However, for defining an unknown series of character state changes (which Farris optimization often cannot do), Fitch's reconstruction fails for our purposes. It may distinguish character patterns for which there is no definite evidence of synapomorphy (shared cladistic events) and it may yield answers which are highly dependent on predefined plesiomorphic character states.

Two of the drawbacks of Fitch's method are illustrated on Figure 4. Figure 4a shows the results of a directed Fitch procedure. The reduced character cladogram is A-C-B. This series requires two assumptions, that the taxon with state C is plesiomorphic and further that C is the state from which both A and B are derived, rather than an autapomorphic state. Neither assumption can be verified by the evidence on hand. The undirected Fitch procedure (Fig. 4b) accepts any of the three possible character state trees for the 3 characters (Fig. 4c). The connectivity matrix (Fig. 4d) has all three states equally related. This results in a transformation series whose state relationships are graphically represented by Figure 4e.

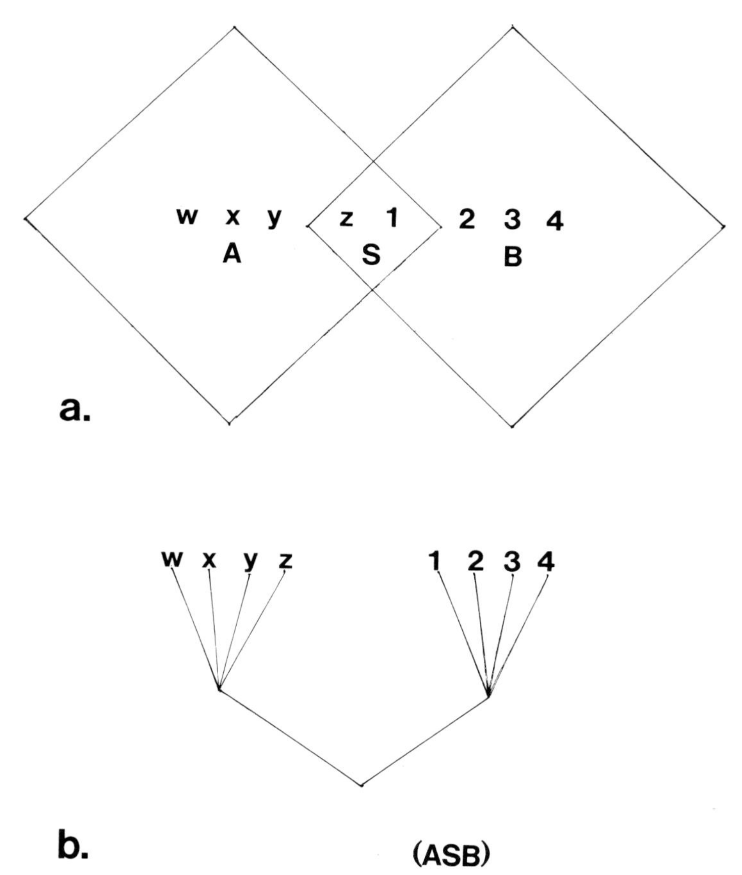

Fig. 5. Overlapping ranges for two taxa. **a.** The ranges and the localities sampled. Localities 1, 2, 3, and 4 are samples for one taxon and localities x, w, y, and z are samples for the other taxon. **b.** The general cladogram implied by ignoring the area of overlap S.

Such a character type expresses the fact that there is no evidence for a definite cladogenetic sequence of character states defined by the cladogram. This type of character, in which every state is equally distant from every other state, is called a non-additive character, by analogy to the type of binary coding (Farris et al., 1970) which can be used to describe these relationships.

Area States

A *biogeographic map* is a transformation series for areas of taxa (ranges) on a cladogram obtained by Transformation Series Analysis. An *area state* is considered to be a set of taxon ranges disjoint from other states. If each terminal taxon has a range which is geographically disjoint from all other taxa under study, the area states would be the actual taxon ranges. If there are overlapping ranges, area states are obtained by uniting overlapping ranges into states separate from every other suite of overlapping ranges.

This definition of area states as separate sets of ranges is chosen to stabilize biogeographic analysis by reducing the number of assumptions that must be made

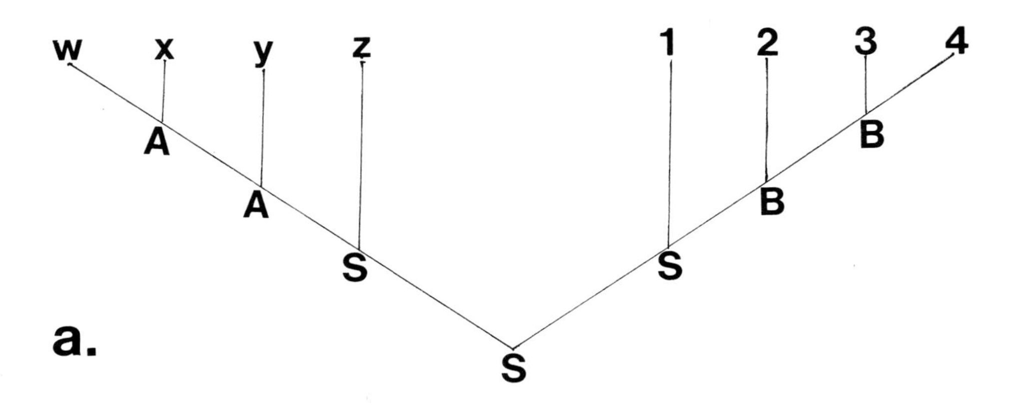

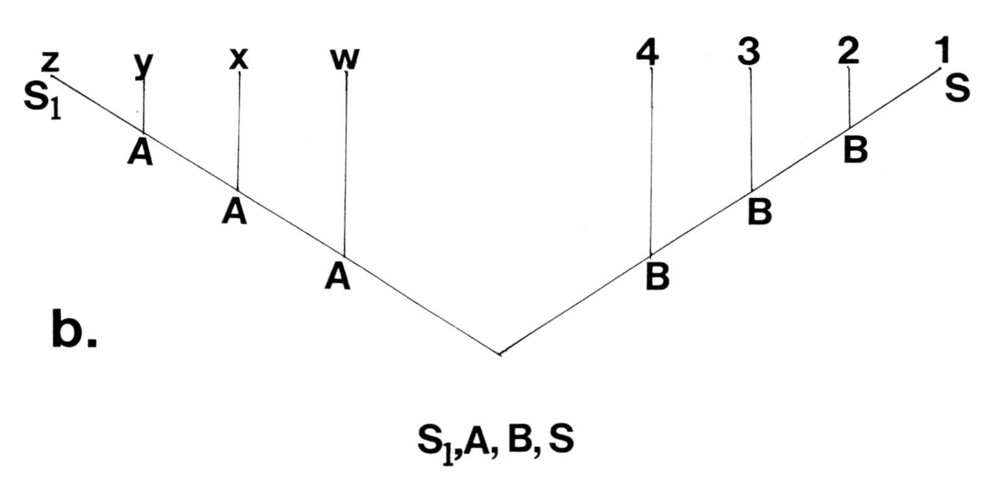

Fig. 6. **a.** Cladogram in which S is a plesiomorphic state. **b.** Cladogram for which S is synapomorphic for both taxa.

concerning the relationships between geography and taxonomy in cases of range overlap. Assume that there are two taxa found in somewhat overlapping areas (Fig. 5a). If the area of overlap is ignored the acknowledged states are A and B. Both states include a small component, S. These state assignments seem superficially to accommodate either of the possible cladograms in Figure 6 because it appears that as long as taxa in A and B form monophyletic groups these area states are adequate representations of reality. However, a closer look at area S will serve to emphasize the possible consequences of such a procedure. If upon subsequent taxonomic study the plesiomorphic populations for both taxa are found in area S, as in Figure 6a, this biogeographic map is A-S-B. Or, perhaps one might find that the plesiomorphic populations are found in the disjoint parts of A and B. The biogeographic map would then be S-A-B-S. Since we are dealing with transformations, the latter map, which shows S to be convergent, supports the recognition of separate areas A and B, for some parts of each have S as

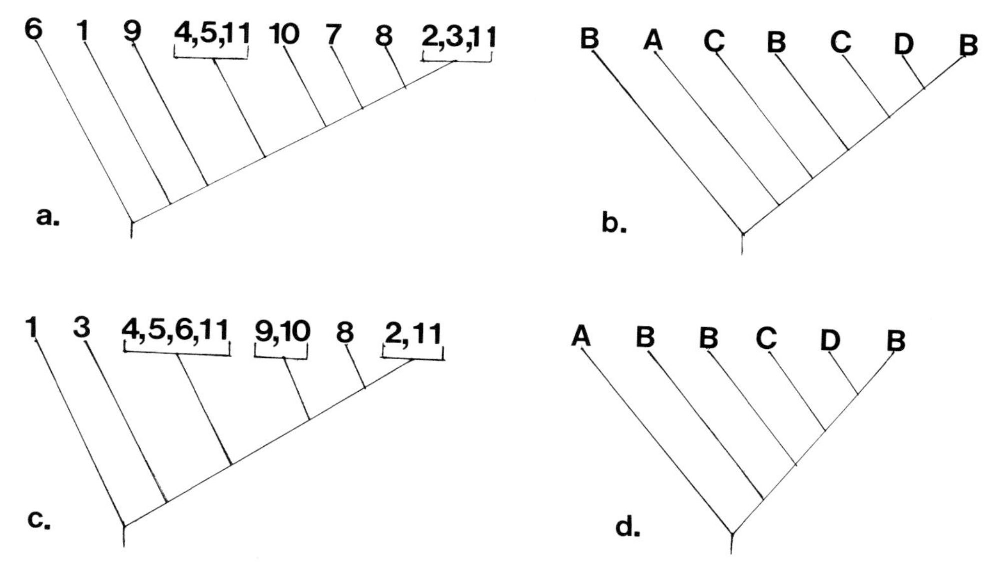

Fig. 7. **a.** Area cladogram for *Heterandria* (Rosen, 1978, Fig. 21). **b.** Area state cladogram for *Heterandria*. **c.** Area cladogram for *Xiphophorus*. **d.** Area state cladogram for *Xiphophorus*.

components. However the former set of area transformations does not support the original areas A and B, but rather disjoint areas A-S, B-S and S. Ignoring areas of overlap forces us to assume that upon further study the plesiomorphic populations (taxa) will always be found in the disjoint areas. For two taxa, as in this example, the possible mistakes are of small consequence. However as the number of taxa with slightly overlapping ranges increases, the number of taxonomic assumptions increases as well. This has the result of forcing conjecture on the pattern of the biogeographic map. In an extreme case, such a course would force the results of a speciation study so that only parapatric speciation could be concluded.

In summary, when there are overlapping ranges in a biogeographic study, the areas of overlap limit the choice of area states. If the areas of overlap are ignored in designating area states, then a pattern of taxonomic relationships is implied which may or may not be true. Such a practice does nothing to improve biogeography for it may produce spurious results.

An Example

A short outline for obtaining a transformation series for area states follows:

1. Place the area states of each terminal taxon on the cladogram.
2. Make all possible Farris optimizations.
3. Specify all nearest neighbor relations for different states.
4. Count the nearest neighbors for different states.
5. Produce a connectivity matrix. Define the transformation series by connecting states which are most often nearest neighbors.

The procedure for obtaining a biogeographic map will be illustrated for Rosen's (1978) cladograms and locality data of *Heterandria* and *Xiphophorus*. First, the locality data must be transformed into a set of area states

Table II

Area state nearest neighbor matrices for *Heterandria* and *Xiphophorus*

	Heterandria				*Xiphophorus*			
	A	B	C	D	A	B	C	D
A		1				1		
B	1		2	1	1		1	1
C	1	4		1		2		1
D		1				1		

Localities	Area states
1	A
2–6, 11	B
9–10	C
8	D

Placing area states of each terminal taxon on the cladogram transforms Rosen's *Heterandria* locality cladogram (Fig. 7a) into the area state cladogram of Figure 7b. Rosen's *Xiphophorus* locality cladogram (Fig. 7c) becomes the area state cladogram of Figure 7d.

The nearest neighbor matrices for the area states are listed in Table II. The connectivity matrix for *Xiphophorus* results in the following biogeographic map

```
    C
    |
A—B—D
```

A and D are joined to B because B is the only nearest neighbor of A or D; C is joined to B because B is nearest neighbor for C more often than it is to A or D. *Heterandria* has the same most frequent nearest neighbors, and so the same biogeographic map. The two sets of taxa seem to agree completely on area relationships. They agree, that is, in their general patterns. The sets of distributions are not identical; the analysis shows that the differences between them are idiosyncratic rather than general.

Testing Theories in Biogeography

All theories in biogeography can deduce area state sequences. A biogeographic map is a best-fitting hypothesis for an individual cladogram, and therefore is only one observation. General theories such as vicariance (Croizat et al., 1974) or dispersal (Brundin, 1966) attempt to make summary statements about all organisms. We can select between general theories by resorting again to considerations of best fitting hypotheses.

The reason for such a procedure can best be illustrated by considering Rosen's data (Fig. 7). Much has been made of the possibility that Rosen's cladograms do not make the same biogeographic statement. According to the criteria presented in this paper, both cladograms produce identical biogeographic maps. However, upon subsequent study a slightly different map might be produced. It has been stated by some systematists that one exception is sufficient to justify dismissing a theory. Such a position is rather naive, for it presumes godlike qualities of the data collector, namely the ability to produce error-free data. Data are, after all, themselves theory-dependent. The only way disparate observations can be rec-

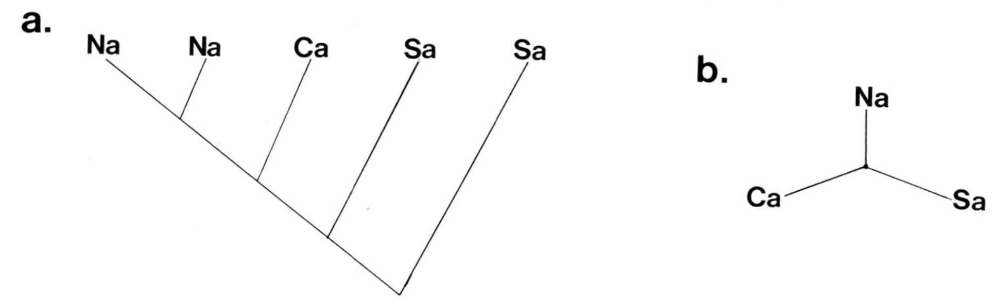

Fig. 8. An example of the properties of a non-bold character. **a.** Cladogram from Cracraft (1974, Fig. 3). **b.** Biogeographic map from a.

ognized as such is under some theory which states that these observations are not predicted, just as a theory is used to confer concordance to a data set. This, of course, is insufficient, for any theory produces concordant and disparate observations for an applicable data set. If we only concern ourselves with the most favored theory, our only conclusion could be that the data fulfill our theory with some exceptions. However, we can choose among theories by estimating the proportion of contradictions for each theory and choosing the one with the fewest contradictions. This is the *best fit theory*.

In biogeography this is accomplished by comparing reduced character cladograms of biogeographic maps with the area sequences predicted by the theories. This is achieved with the use of two phylogenetic concepts: character boldness and character consistency. Boldness is found in characters which define a series of synapomorphies. It is recognized as a feature for some types of characters but not for other types. There are four types of characters (Mickevich, In prep.): additive, nonadditive, disjoint, and convergent.

Additive characters, named for the type of binary coding with which they can be expressed (Farris et al., 1970), place all states in a simply connected sequence. This condition defines a series of synapomorphies. These characters are unambiguous—*bold*—because they specify a precise set of relationships. A-B-C is an example of an additive character. *Convergent characters* have all the characteristics of additive characters, but are distinguished from them by the presence of a predefined state mistakenly considered to be a single quantity. A-B-C-A is a convergent character.

Non-additive characters are not bold. A non-additive character (Fig. 4) has at least three states which are equally closely connected to each other (Fig. 4). A *disjoint character* has subsets within which there are well-defined relationships, but between which no sequence can be determined. Examples are given by Mickevich and Mitter (1981).

Non-bold characters support all or most theories about a set of area states. They are unacceptable for testing biogeographic theories. For example, in Figure 8a the area cladogram (Cracraft, 1973) has been used to support dispersal between North (Na) and South America (SA) through Central America (CA). However, the biogeographic map is non-additive (Fig. 8b). Therefore, any state transformation is possible: SA-NA-CA or NA-SA-CA, which would not support the dispersal sequence and NA-CA-SA, which would.

Character consistency (Kluge and Farris, 1969) measures the fit of a predefined character state sequence to a cladogram. For our purposes it is defined to be the number of transformations predicted by a biogeographic map divided by the

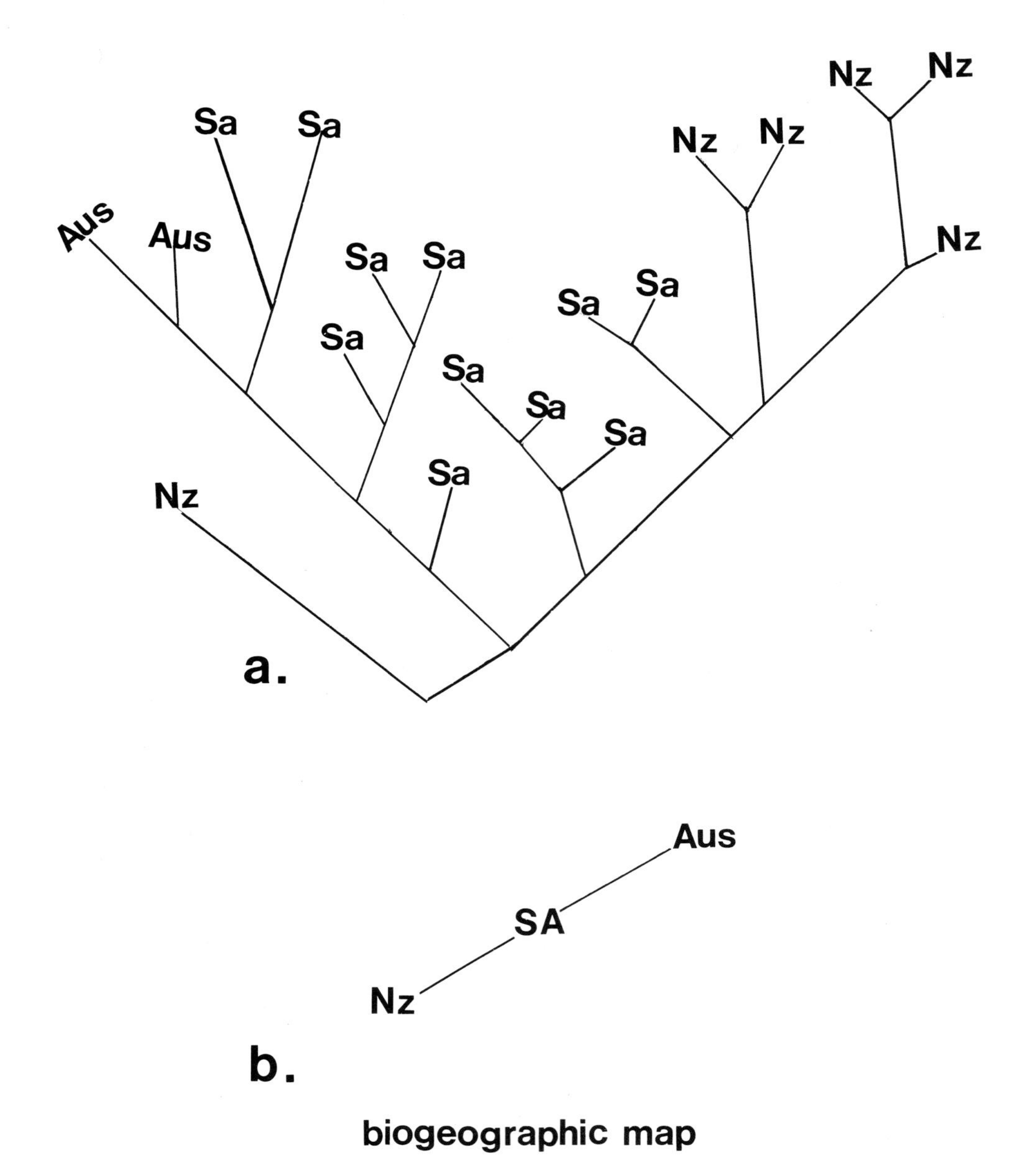

Fig. 9. **a.** Area cladogram from Brundin (1981: Fig. 7). **b.** Biogeographic map for a.

amount of transformations predicted by the theory's area sequence on the reduced character cladogram. This measurement ranges from 1 to 0. The greater the value, the greater the degree of fit of the theory to the data.

This procedure is illustrated by an example from Brundin (1981). Figure 9a is a cladogram with localities taken from Brundin's Figure 7. Figure 9b is the biogeographic map. Brundin uses dispersal theory to predict the area state sequence illustrated in Figure 10b. Consideration of a series of tectonic events results in the area state sequence illustrated in Figure 10c. The sequence of tectonic events

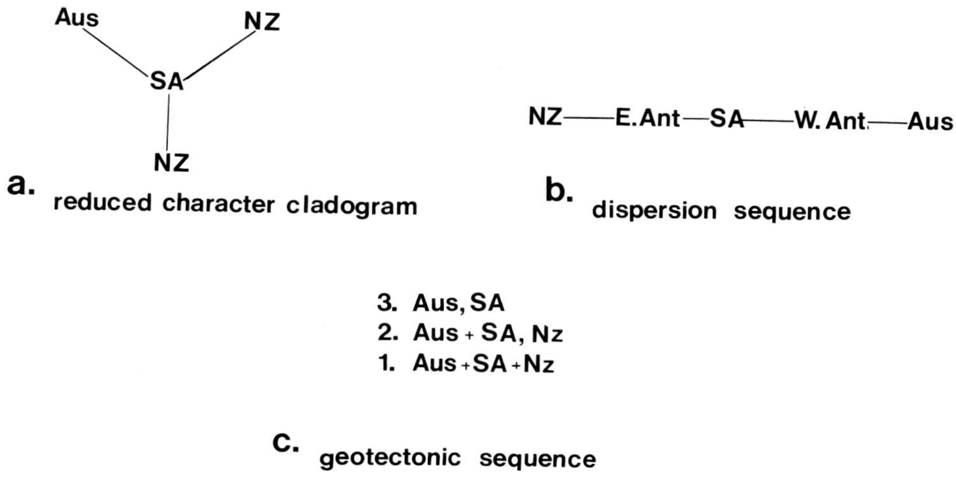

Aus NZ
SA
NZ

a. reduced character cladogram

NZ——E.Ant—SA——W.Ant.—Aus

b. dispersion sequence

3. Aus, SA
2. Aus + SA, Nz
1. Aus + SA + Nz

c. geotectonic sequence

Fig. 10. **a.** Reduced character cladogram for area cladogram in Figure 9. **b.** Dispersal sequence for cladogram in Figure 9. **c.** Geotectonic sequence for cladogram in Figure 9.

is a subset of those involved in the subdivision of Gondwanaland and is considered a vicariant general pattern. Each theory has a means for determining the area sequences predicted according to the theory. I will not argue the correctness of these procedures; it is unimportant for my purposes. The dispersalists attempt to relate geographic distance to cladogenetic events. Vicariance biogeographers attempt to explain area states in the context of a general pattern of area states which is constant over groups within the larger area.

The biogeographic map is Nz-SA-Aus, which specifies one change between state pairings Nz-SA and SA-Aus. Theories consistent with the map will have the same sequence. The reduced character cladogram is drawn in Figure 10a. Notice that there is one extra step on the cladogram. Eastern Antarctica (E.Ant) and Western Antarctica (W.Ant) are part of the dispersal sequence but are not found in the data. Therefore these are deleted from the area state sequence predicted by dispersal theory to form this sequence Nz-SA-Aus. Under dispersal the state pairs SA-Aus and Nz-SA both have one step if found on the reduced character cladogram. The pairing Aus-Nz on the cladogram would be considered a two step change. There are three one step pairings on the cladogram, but the biogeographic map predicts two changes. The consistency of this theory for the data is then 2/3 or 0.67.

The vicariance sequence could be represented as the undirected series

```
    SA
     |
Au—*—Nz
```

For the three areas there is one transformation associated with each state pair. Three steps are obtained for the geotectonic sequence on the reduced character cladogram. Therefore the consistency index or degree of fit of this theory to the data is 0.67.

At this point it is interesting to consider the possible content of a study devoid of hypothesis testing. Both the dispersalist and the vicariance biogeographer might feel that data give a good fit to the preferred theory. If the combination

Aus-Nz were observed a vicariance biogeographer might consider the dispersal theory falsified. Neither view is justified. It is uninteresting to explain any set of data within the context of a single theory. As seen above, any data can be explained in some way by just about any theory. The only important question is that of which theory better explains available data.

Both predicted sequences have a consistency index of 0.67. If one theory's consistency index were greater than the other, some choice between them could be made. However, such is not the case for these data. Further, it should not be supposed that these are the only theories that can be tested. All others which make predictions about the relationships of taxa and their ranges can be assessed. The reason for preferring one theory over another is solely its continued superior consistency. Theories which have been repeatedly shown to be inferior to other theories should be discarded.

Conclusion

This paper presents a method for summarizing biogeographic data for a set of taxa in the form of a biogeographic map. Maps arrived at in this way provide a means of separating general from particular aspects of distribution data and a means of testing alternative biogeographical theories. The method is based directly on the concept of conformity of theories to evidence, and so is free of the underlying assumptions of particular biogeographical theories, or of different theories in evolution. Fit of data or parsimony is the single most important contribution of phylogenetic systematics, for, as can be seen from this paper, it serves as the basis for further development towards a complete scientific approach in classification and systematics.

Acknowledgements

I wish to thank J. S. Farris and C. Mitter for their help, and A. W. F. Edwards for his book, *Likelihood*. This research was supported by grant number DEB-78–24647 from the National Science Foundation.

Widespread Taxa and Biogeographic Congruence

Norman I. Platnick

Questions of congruence, whether of character distributions among taxa (indicating natural groups) or of taxon distributions among areas (indicating natural regions), are central to comparative biology. Hence it is not surprising that three speakers at this meeting (Humphries, Allen, and Brooks) addressed the empirical question of whether (and how much) congruence exists in the area-relationships indicated by cladograms of different groups of organisms. Such questions would be easy to answer if cladograms generally provided ideal data: if all groups of organisms were each represented by one (and only one) taxon in each of the smallest identifiable local areas of endemism in the world. Data approaching this ideal can sometimes be found, if one searches assiduously, but nature is not usually so cooperative. Hence it is not surprising that much lively discussion at the meeting centered on the problems of analyzing less than ideal data. This paper will address the problems posed by widespread taxa (those which occur in two or more of the areas of endemism being investigated). Data drawn from Rosen's (1978, 1979) landmark study of Middle American poeciliid fish distributions will be used as examples, both because they were extensively discussed at the meeting and because they have been commented on in detail in the literature (Wiley, 1980, and various authors in Nelson and Rosen, 1981).

The poeciliid fish genera *Heterandria* and *Xiphophorus* each include monophyletic subgroups containing several species endemic to Mexico and Central America, and there is much similarity in the details of the distribution of those endemic species (Figs. 1, 2). Accordingly, Rosen asked to what extent these different species might show the same pattern of area interrelationships. He proceeded by identifying 11 disjunct areas occupied by the species (areas 2 and 4–11 of Figs. 1 and 2; areas 1 and 3 are allopatric areas in northeastern Mexico and the Isthmus of Tehuantepec, respectively). Areas 4 and 5 are geographically disjunct but are occupied by the same species in each genus and are therefore considered a single area (45) below. Rosen constructed cladograms for the species of each genus and converted these species-cladograms to area-cladograms by replacing each taxon name with the number of the area(s) in which the taxa occur. Each of the resulting area-cladograms included as a superscript addition a putatively hybrid population occupying area 11; the area-cladograms have been redrawn here (Figs. 3, 4) with the hybrids added as branches at the appropriate

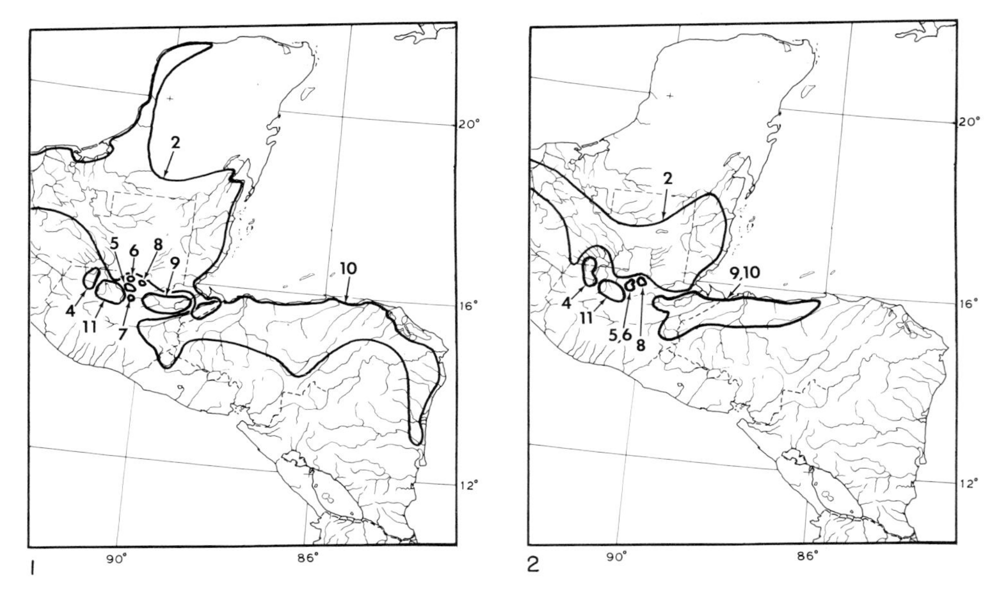

Figs. 1–2. Middle American areas containing endemic species of poeciliid fish. 1. Species of the genus *Heterandria*. 2. Species of the genus *Xiphophorus*. After Rosen (1978).

positions (the node connecting the putatively ancestral populations 45 and 2; Nelson and Platnick, 1980). It should be noted that Rosen's area-cladograms for *Heterandria* are not consistent. In two cases (Rosen, 1978, Fig. 19; Rosen, 1979, Fig. 48), *H. bimaculata* (the taxon in the most apomorphic position) is shown as endemic to area 2; elsewhere (Rosen, 1978, Fig. 21; Rosen, 1979, Fig. 50) that species is shown as widespread in areas 2 and 3. Although the distribution maps (Rosen, 1978, Figs. 1, 16; Rosen, 1979, Figs. 22, 45) support the former pair of cladograms (indicating that no *Heterandria* occurs in area 3), Rosen (pers. comm.) indicates that *H. bimaculata* actually does occur in both areas.

Comparison of the two area-cladograms indicates that *Xiphophorus* is slightly less informative than *Heterandria* because it includes no representative in area 7 and two widespread taxa (rather than one). Areas 9 and 10, for example, each harbor distinct species of *Heterandria* but are occupied by the same taxon of *Xiphophorus*. Such widespread taxa can be analyzed under either of two assumptions (Nelson and Platnick, 1981). Under assumption 1, whatever is true of the widespread taxon in one part of its range (such as area 9) must also be true of the taxon in other parts of its range (such as area 10). Under assumption 2, whatever is true of the widespread taxon in one part of its range need not also be true of the taxon elsewhere. The difference between assumptions 1 and 2 can be visualized easily by supposing that future taxonomic investigation indicated that the populations of the widespread taxon in areas 9 and 10 actually belong to two different taxa. Under assumption 1, it is presupposed that the taxa occurring in areas 9 and 10 (and formerly confused with each other) are actually sister taxa or branch off the cladogram sequentially (i.e., in the order 9, 10, 8, 2 or 10, 9, 8, 2). Under assumption 2, no such presupposition exists, and not only the identity but also the interrelationships of one of the taxa are allowed to have been misconstrued. Theoretical ramifications of these assumptions have been discussed elsewhere (Nelson and Platnick, 1981); assumption 2 seems more realistic not

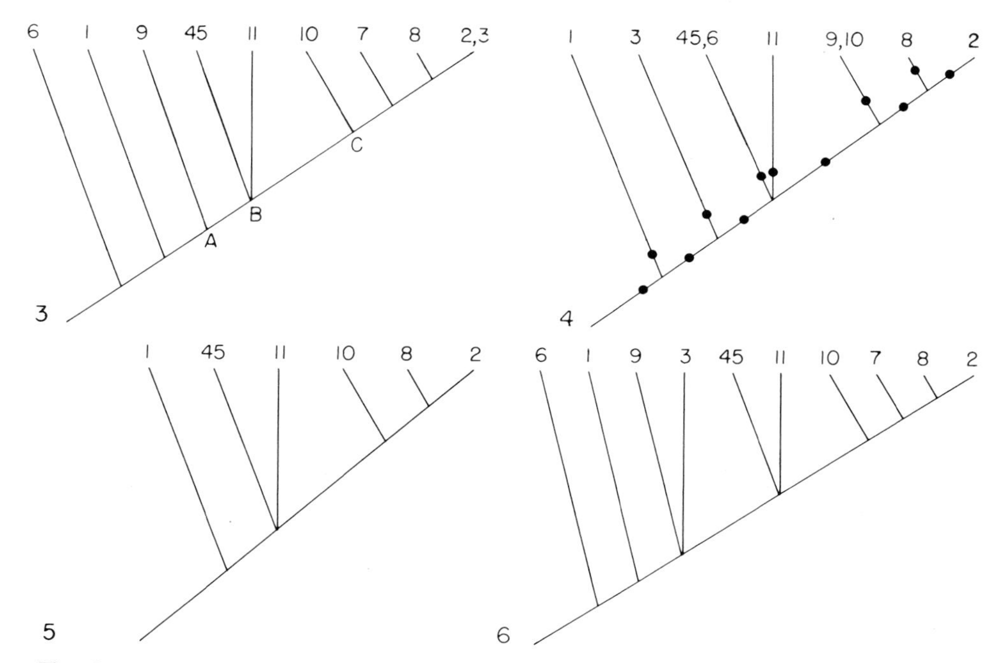

Figs. 3–6. Area-cladograms of the Middle American areas. 3. Area-cladogram indicated by *Heterandria*. 4. Area-cladogram indicated by *Xiphophorus*. 5. Combination of the two area-cladograms under assumption 1. 6. Combination of the two area-cladograms under assumption 2. 3–5 after Rosen (1978).

only in cases of taxonomic error, but also in cases where widespread taxa are due to dispersal or to the failure of a group to differentiate in response to a vicariance event. Discussion here will be limited to Rosen's example, and illustrated by means of the latter historical explanation (failure of a group to differentiate).

Rosen compared the two area-cladograms by a reduction process in which the information on area 7 contributed by the *Heterandria* cladogram and the information on areas 3, 6, and 9 contributed by both cladograms were deleted. The resulting reduced area-cladogram (redrawn in Fig. 5) shows congruence with regard to the interrelationships of six areas. Of the deleted areas, Rosen (1978, p. 179) stated:

> Inspection of the area cladograms for the two groups shows that *Heterandria* has a unique component in area 7 and that the two are incongruent with respect to three areas (3, 6 and 9). In relation to the question of whether each cladogram shows some generality with respect to the other, it is evident therefore that it is not with respect to area 7 (the unique component of *Heterandria*), or areas 3, 6 and 9. . . . One may, of course, ask questions about the significance of the unique or incongruent elements in the original cladograms. Assuming that the original, unreduced cladograms truthfully represent real phylogenies and that the unique or incongruent elements have no generality with respect to other as yet unanalyzed components of the Middle American biota, the incongruent elements would be most simply explained as dispersals. Because of the assumptions required, however, it is evident that such explanations are without significance at the level of general explanation.

Thus, Rosen's analysis of the two widespread *Xiphophorus* taxa was conducted under assumption 1. Given that assumption, the *Xiphophorus* population in area

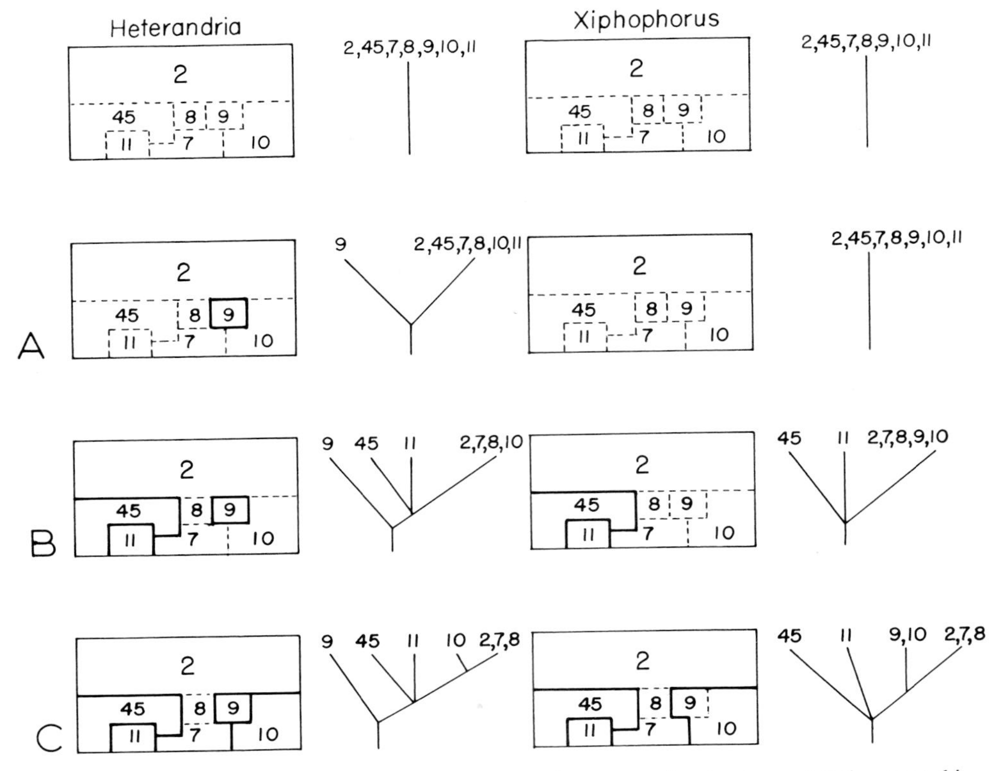

Fig. 7. Schematic maps indicating seven Middle American areas of endemism and their geographic boundaries (dotted lines). The appearance of barriers effective in isolating populations of *Heterandria* (left) and *Xiphophorus* (right) is indicated by the appearance of solid boundaries. Letters in the column at left refer to the events labelled in Fig. 3. Cladograms to the right of each map indicate the relationships at each respective time level.

9 must be most closely related to (and have shared an immediate geographic history with) either the population in area 10, or the population in area 8+2, or both populations together. If so, the information on area 9 supplied by *Xiphophorus* is indeed incongruent with the information supplied by *Heterandria*: that area 9 represents the sister area of six other areas: 45, 11, 10, 7, 8, and at least part of 2+3.

But let us apply assumption 2 instead, and suppose that the information on area 9 supplied by *Heterandria* is actually correct, and that the widespread *Xiphophorus* population has merely failed to respond to one or more vicariance events. For convenience, the *Heterandria* species widespread in areas 2 and 3 will be treated as if it occurred only in area 2; as shown below, this simplification does not affect the results. The *Heterandria* cladogram implies that some event (labelled A in Fig. 3) separated area 9 from the six areas distal to it in Figure 3. If the *Xiphophorus* population in area 9 did not undergo differentiation as a result of that event, a single taxon of *Xiphophorus* would have occupied all seven areas at the time of the subsequent vicariance event shown by *Heterandria* (labelled B in Fig. 3), which may have separated three areas (45, 11, and 2–7–8–10). For the *Xiphophorus* populations in those three areas to have been isolated from each other by such a tripartite division, the population in area 9 (if not itself fragmented by the event) must have been incorporated within one (and only one) of the three areas (Fig. 7; Platnick and Nelson, 1978). Of the three divisions, area 2–7–8–10 is geographically the closest to area 9. If population 9 was included in area 2–7–8–10, then at the time of the subsequent vicariance event (labelled C in Fig. 3), it must similarly have been incorporated within either area 10 or area 2–7–8. The former possibility would result in the cladogram of Figure 4. The sequence of isolations and the resulting cladograms are shown schematically in Figure 7.

In his original study, Rosen (1978, p. 177) noted that the failure of a taxon to differentiate merely limits its informativeness:

> Some populations, whether recognized as species or not, are informative with respect to a history of geographic isolation because they have differentiated. Others are uninformative either because they haven't differentiated, because they are parts of larger populations that span two or more areas (are insensitive to existing barriers), or because taxonomists have thus far failed to detect the ways in which differentiation has occurred (physiologically, developmentally, behaviorally, etc.).

But if it is true that widespread taxa (like the *Xiphophorus* taxon occupying areas 9 and 10) are uninformative, they cannot also be held to be incongruent; the absence of information cannot be incongruent with any information at hand. Under assumption 1, a widespread taxon is always informative to some degree, but it may be (and probably frequently is) *falsely* informative. Under assumption 2, a widespread taxon, by itself, is uninformative, but when different cladograms including widespread taxa are combined, an informative result is possible (indeed, it is possible to derive a fully informative area-cladogram by combining species-cladograms including only widespread taxa). One might ask, of course, whether the result is *truly* informative. Any answer must involve replicability (and its probability due to chance alone), but only a few cladograms are needed to test replicability (Nelson and Platnick, 1981).

With regard to Rosen's data, then, the question is whether there might be a single area-cladogram with which no part of the information provided by either *Heterandria* or *Xiphophorus* disagrees. The *Xiphophorus* cladogram, under assumption 2, allows the population in area 9 (or the population in area 10, but not both) to occur in any of twelve positions (marked with dots in Fig. 4) on a fully resolved area-cladogram. The same leeway is allowed to the population in either area 45 or area 6, and to the *Heterandria* population in either area 2 or area 3 (it was area 3 that was allowed to "float" in the cladogram in the simplification of the *Heterandria* cladogram used in the discussion above). As it turns out, there are three possible area-cladograms for all ten areas which involve no necessary disagreement with either species-cladogram; they can be summarized by a single area-cladogram (Fig. 6) incorporating one trichotomy representing the three dichotomous possibilities at that level (plus one further trichotomy introduced by the hybrids in area 11). The significance of the difference between the results of assumption 1 (Fig. 5) and 2 (Fig. 6) should be obvious: being more informative, Figure 6 allows for a far greater number of comparisons with other groups of organisms and hence many more tests of the generality of the pattern of Middle American area interrelationships displayed by the poeciliids.

Acknowledgements

I am indebted to Drs. D. Brooks, J. Cracraft, J. Farris, V. Funk, C. Humphries, M. Mickevich, C. Mitter, G. Nelson, C. Patterson, D. Rosen, R. Schuh, and E. Wiley for their comments and suggestions.

Summary

D. R. Brooks and V. A. Funk

The first meeting of the Willi Hennig Society revealed diverse and vigorous research programs being pursued by the various participants. One indication of this is the wide variety of interests represented by the subject matter of the papers included in this volume. In addition, this volume marks the first wholesale entry of botanists into contemporary phylogenetics. Application of parsimony methods of cladistic analysis by botanists has not only made possible a common dialogue between botanists and zoologists, it has made zoologists aware of a number of very interesting aspects of phylogenetics unique, or nearly so, to plants.

Because we espouse systematic techniques leading to maximally predictive classifications, we feel justified in making some predictions about the future of cladistics. Although those attending the meetings were generally very happy that phylogenetics is emerging as the consensus approach in systematics, most thought there were still many things to do. Continued careful scrutiny of our methods and refinements where required is the watchword of cladists. In addition, extension of the principles of cladistics to other comparative aspects of biology has a high priority. The most important job phylogeneticists have is the dissemination of information about the various techniques available for cladistic analysis. Concomitant with the continued search for better means of accomplishing our goals, we foresee a period of some consolidation as more and more descriptive papers published in specialty journals routinely use cladistics to summarize data. There is a need for a published bank of data matrices and cladograms for as many taxa as possible using as many different kinds of data as possible.

Cladistics will not go away. It is a better approach than any proposed previously. Whether it represents a paradigm, which will direct research for many decades, or only constitutes an improvement over previous methods, which will lead ultimately to even better methods, will be decided by future events. In either event, as the new consensus it deserves to have its properties scrutinized and applied as broadly as possible. Cladists espouse falsification with a vengeance and only by thorough application of the techniques will any flaws appear and modifications be possible.

Literature Cited

Abramova, A. L. 1969. De mutabilitate *Hydrogrimmia mollis* (Br. et Sch.) Loeske. Novosti Sist. Nizsh. Rast. 6: 257–262.

Allan, H. H. 1961. Flora of New Zealand. Vol. 1. R. T. Owen, Wellington.

Anderson, L. C., R. L. Hartman and T. F. Stuessy. 1979. Morphology, anatomy, and taxonomic relationships of *Otopappus australis* (Asteraceae). Syst. Bot. 4: 44–56.

Anderson, S. and M. K. Evenson. 1979. Randomness in allopatric speciation. Syst. Zool. 27: 421–430.

Anderson, T. W. 1958. Introduction to multivariate statistical analysis. John Wiley and Sons, New York.

Arambourg, C. 1947. Mission scientifique de l'Omo (1932–1933). Bull. Mus. Natn. Hist. Nat., Paris 1: 469–471.

Ashlock, P. D. 1972. Monophyly again. Syst. Zool. 21: 430–438.

Avise, J. C. and F. Ayala. 1976. Genetic differentiation in speciose versus depauperate phylads: evidence from the California minnows. Evolution 30: 46–52.

———**, J. C. Patton and C. F. Aquadro.** 1980. Evolutionary genetics of birds II. Conservative protein evolution in North American sparrows and relatives. Syst. Zool. 29: 323–339.

——— **and R. K. Selander.** 1972. Evolutionary genetics of cave-dwelling fishes of the genus *Astyanax*. Evolution 26: 1–19.

———**, M. H. Smith and R. K. Selander.** 1974. Biochemical polymorphism and systematics in the genus *Peromyscus*. VI. The *boylii* species group. J. of Mammal. 55: 751–763.

———**, D. C. Straney and M. H. Smith.** 1977. Biochemical systematics of sunfish IV. Relationships of centrarchid genera. Copeia 1977: 250–258.

Ayala, F. J. 1975. Genetic differentiation during the speciation process. Evol. Biol. 8: 1–78.

———**, M. Tracey, D. Hedgecock and R. Richmond.** 1974. Genetic differentiation during the speciation process in *Drosophila*. Evolution 28: 576–582.

Bailey, R. M. 1969. Comment on the proposed suppression of *Elipesurus spinicauda* Schomburgk (pisces). Bull. Zool. Nomencl. 25: 133–134.

Ball, I. R. 1976. Nature and formulation of biogeographical hypotheses. Syst. Zool. 24: 407–430.

Ballance, P. F. 1976. Evolution of the upper cenozoic magmatic arc and plate boundary in northern New Zealand. Earth and Planetary Sci. Lett. 28: 356–370.

Bateson, G. 1972. Steps to an ecology of mind. Ballantine Books, New York.

Baverstock, P. R., S. R. Cole, B. J. Richardson and C. H. S. Watts. 1979. Electrophoresis and cladistics. Syst. Zool. 28: 214–220.

Behnke, H.-D. 1969. Die Siebröhren-Plastiden der Monocotyledonen. Vergleichende Untersuchungen über Feinbau und Verbreitung eines characteristischen Plastidentyps. Planta 84: 174–184.

Benveniste, R. E. and G. J. Todaro. 1976. Evolution of type C viral genes: evidence for an Asian origin of man. Nature 261: 101–108.

Blake, S. F. 1915. A revision of *Salmea* and some allied genera. J. Bot. 53: 193–202, 225–235.

Bolick, M. R. and R. K. Jansen. 1981. New combinations in the genus *Salmea* DC. (Compositae: Heliantheae). Brittonia 33: 186.

Brady, R. H. 1979. Natural selection and the criteria by which a theory is judged. Syst. Zool. 28: 600–621.

Bremer, B. 1980a. A taxonomic revision of *Schistidium* (Grimmiaceae, Bryophyta) 1. Lindbergia 6: 1–16.

———. 1980b. A taxonomic revision of *Schistidium* (Grimmiaceae, Bryophyta) 2. Lindbergia 6: 89–117.

———. 1981. A taxonomic revision of *Schistidium* (Grimmiaceae, Bryophyta) 3. Lindbergia 7: In press.

Bremer, K. and H.-E. Wanntorp. 1978. Phylogenetic systematics in botany. Taxon 27: 317–329.

——— **and** ———. 1979. Hierarchy and reticulation in systematics. Syst. Zool. 28: 624–627.

——— **and** ———. 1981. A cladistic classification of green plants. Nord. J. Bot. 1: 1–3.

Brooks, D. R. 1977. Six new species of tetraphyllidean cestodes, including a new genus, from a marine stingray *Himantura schmardae* (Werner, 1904) from Colombia. Proc. Helm. Soc. Wash. 44: 51–59.

———. 1978. Evolutionary history of the cestode order Proteocephalidea. Syst. Zool. 27: 312–323.

———. 1979. Testing the context and extent of host-parasite coevolution. Syst. Zool. 28: 299–307.

———. 1980. Allopatric speciation and non-interactive parasite community structure. Syst. Zool. 29: 192–203.

———. 1981. Classification as the language of empirical comparative biology. *In* Funk, V. A. and D. R. Brooks (eds.). Advances in cladistics: Proceedings of the first annual meeting of the Willi Hennig Society. New York Botanical Garden.

——— **and M. A. Mayes.** 1978. *Acanthobothrium electricolum* sp. n. and *A. lintoni* Goldstein, Henson and Schlicht, 1969 (Cestoda: Tetraphyllidea) from *Narcine brasiliensis* (Olfers) (Chondrichthyes: Torpedinidae) in Colombia. J. Parasitol. 64: 617–619.

——— **and** ———. 1980. Cestodes in four species of euryhaline stingrays from Colombia. Proc. Helm. Soc. Wash. 47: 22–29.

———, ——— **and T. B. Thorson.** 1979. *Paravitellotrema overstreeti* sp. n. (Digenea: Hemiuridae) from the Colombian freshwater stingray *Potamotrygon magdalenae*. Proc. Helm. Soc. Wash. 46: 52–54.

———, ——— **and** ———. 1981a. Cestode parasites in *Myliobatis goodei* Garman (Myliobatiformes: Myliobatidae) from Río de la Plata, Uruguay with a summary of cestodes collected from South American elasmobranchs during 1975–1979. Proc. Biol. Soc. Wash. 93: 1239–1252.

———, ——— **and** ———. 1981b. Systematic review of cestode parasites infecting freshwater stingrays (Chondrichthyes: Potamotrygonidae) in South America, including four new species from Venezuela. Proc. Helm. Soc. Wash. 48: 43–64.

Brotherus, V. F. 1898. *Indusiella,* eine neue Laubmoos-Gattung aus Central-Asien. Bot. Centralbl. 75: 321–322.

———. 1909. Grimmiaceae. *In* A. Engler and K. Prantl (eds.). Die natürlichen Pflanzenfamilien. Ed. 1. Teil 1, Abt. 3. Leipzig, Germany.

———. 1924. Grimmiaceae. *In* A. Engler and K. Prantl (eds.). Die natürlichen Pflanzenfamilien. Ed. 2. Vol. 10: 303–314. Leipzig, Germany.

———. 1925. Ptychomitriaceae. *In* A. Engler and K. Prantl (eds.). Die natürlichen Pflanzenfamilien. Ed. 2. Vol. 11: 6–11. Leipzig, Germany.

Brundin, L. 1966. Transantarctic relationships and their significance as evidenced by chironomid midges. Kungliga Svenska Vetenskapsakadamien Handlingar (f) 11, 1: 1–472.

———. 1981. Croizat's panbiogeography versus phylogenetic biogeography. *In* Nelson, G., and D. E. Rosen (eds.). Vicariance biogeography: A critique. Columbia Univ. Press, New York.

Buck, W. R. 1979. A re-evaluation of the Bruchiaceae with the description of a new genus. Brittonia 31: 469–473.

———. 1980. A generic revision of the Entodontaceae. J. Hattori Bot. Lab. 48: 71–159.

Burbridge, N. T. 1960. The phytogeography of the Australian region. Austral. J. Bot. 8: 75–209.

Burger, W. C. 1975. The species concept in *Quercus*. Taxon 24: 45–50.

Burns, J. M. 1964. Evolution in skipper butterflies of the genus *Erynnis*. Univ. of California Publ. Ent. 37: 1–214.

Camin, J. H. and R. R. Sokal. 1965. A method for deducing branching sequences in phylogeny. Evolution 19: 311–326.

Campbell, R. A. and J. Carvajal G. 1979. Synonymy of the phyllobothriid genera *Rhodobothrium* Linton, 1889, *Inermiphyllidium* Riser, 1955 and *Sphaerobothrium* Euzet, 1959 (Cestoda: Tetraphyllidea). Proc. Helm. Soc. Wash. 46: 88–97.

——— **and** ———. 1980. *Echinobothrium euzeti,* a new cestode from the spiral valve of a Chilean elasmobranch. Proc. Helm. Soc. Wash. 47: 165–167.

Carvajal G., J. 1971. *Grillotia dollfusi* sp. n. (Cestoda: Trypanorhyncha) from the skate, *Raja chilensis,* from Chile, and a note on *G. heptanchi*. J. Parasitol. 57: 1269–1271.

———. 1974. Records of cestodes from Chilean sharks. J. Parasitol. 60: 29–34.

———. 1977. Description of the adult and larva of *Caulobothrium myliobatidis* sp. n. (Cestoda: Tetraphyllidea) from Chile. J. Parasitol. 63: 99–103.

——— **and M. D. Dailey.** 1975. Three new species of *Echeneibothrium* (Cestoda: Tetraphyllidea) from the skate, *Raja chilensis* Guichenot, 1848, with comments on mode of attachment and host specificity. J. Parasitol. 61: 89–94.

——— **and R. J. Goldstein.** 1969. *Acanthobothrium psammobati* sp. n. (Cestoda: Tetraphyllidea: Onchobothriidae) from the skate, *Psammobatis scobina* (Chondrichthyes: Rajidae) from Chile. Zool. Anzeig. 182: 432–435.

——— **and** ———. 1971. *Acanthobothrium annapinkiensis* n. sp. (Cestoda: Tetraphyllidea: Onchobothriidae) from the skate, *Raja chilensis* (Chondrichthyes: Rajidae) from Chile. Zool. Anzeig. 186: 158–162.

Case, S. M. 1976. Evolutionary studies in selected North American frogs of the genus *Rana* (Amphibia, Anura). Ph.D. dissertation, Univ. of California, Berkeley.

———. 1978. Biochemical systematics of members of the genus *Rana* native to western North America. Syst. Zool. 27: 299–311.

———, **P. G. Haneline and M. F. Smith.** 1975. Protein variation in several species of *Hyla*. Syst. Zool. 24: 281–295.

Castello, H. P. 1973. Sobre la correcta posición sistemática de la raya de agua dulce Africana (Chondrichthyes: Dasyatidae) (República Federal del Camerún). Trab. V Congr. Latinoamer. Zool. 1: 67–71.

Castex, M. N. 1968. *Elipesurus* Schomburgk, 1843 (Pisces): Proposed suppression under the plenary powers. N. Z. (S.) 1825. Bull. Zool. Nomencl. 24: 353–355.

———. 1969. Comment on the objections forwarded by R. M. Bailey to the proposed suppression of *Elipesurus spinicauda* Schomburgk (Pisces). N. Z. (S.) 1825. Bull. Zool. Nomencl. 26: 68–69.

Cavalli-Sforza, L. L. and A. W. F. Edwards. 1967. Phylogenetic analysis: models and estimation procedures. Evolution 21: 550–570.

Churchill, S. P. In press. Phylogeny and taxonomy of *Jaffueliobryum* (Grimmiaceae: Musci). Ann. Missouri Bot. Gard. 69.

Cope, E. D. 1879. A stingray from the Green River shales of Wyoming. Amer. Nat. 13: 333.

Cordero, E. H. 1944. Dos nuevas especies de trematodos monogenéticos de los plagióstomos de la costa Uruguaya: *Calicotyle macrocotyle* y *Neoerpocotyle tudes*. Commun. Zool. Mus. Hist. Nat., Montevideo 1: 1–15.

Couper, R. A. 1960. Southern Hemisphere Mesozoic and Tertiary Podocarpaceae and Fagaceae and their palaeogeographic significance. Proc. Roy. Soc. London Botany 152: 491–500.

Coyne, J. A., W. F. Eanes, J. A. M. Ramshaw and R. K. Koehn. 1979. Electrophoretic heterogeneity of alpha-glycerophosphate dehydrogenase among many species of *Drosophila*. Syst. Zool. 28: 164–175.

Cracraft, J. 1972. The relationships of the higher taxa of birds: Problems in phylogenetic reasoning. Condor 74: 379–392.

———. 1973. Continental drift, paleoclimatology, and the evolution and biogeography of birds. Journ. Zool. 169: 455–545.

———. 1975. Historical biogeography and earth history: Perspectives for a future synthesis. Ann. Missouri Bot. Gard. 62: 250–277.

Cranwell, L. M. 1963. *Nothofagus*: Living and fossil. *In* Gressitt, J. K. (ed.). Pacific Basin Biogeography Symposium. Bishop Museum Press, HI.

Craw, R. C. 1978. Two biogeographical frameworks: Implications for the biogeography of New Zealand, a review. Tuatara 23: 81–114.

Crisci, J. V. and T. F. Stuessy. 1980. Determining primitive character states for phylogenetic reconstruction. Syst. Bot. 5: 112–135.

Croizat, L. 1952. Manual of phytogeography: or an account of plant-dispersal throughout the world. W. Junk, Publ., The Hague.

———. 1964. Space, time, form: the biological synthesis. Published by the author, Caracas, Venezuela.

———, **G. Nelson and D. E. Rosen.** 1974. Centers of origin and related concepts. Syst. Zool. 23: 265–287.

Cronquist, A. 1968. The evolution and classification of flowering plants. Thomas Nelson and Sons, London.

———, **A. Takhtajan, and W. Zimmerman.** 1966. On the higher taxa of the Embryobionta. Taxon 15: 129–134.

Crosby, M. R. 1980. The diversity and relationships of mosses. *in* R. J. Taylor and A. E. Leviton, (eds.). The mosses of North America. Amer. Assoc. Adv. Sci., Pacific Div.

Crum, H. 1972a. The dubious origin of *Glyphomitrium canadense* Mitt. J. Bryol. 7: 165–168.

———. 1972b. A taxonomic account of the Erpodiaceae. Nova Hedwigia 23: 201–224.

Cutler, D. 1972. Vicarious species of restionaceae in Africa, Australia and South America. *In* D. H. Valentine (ed.). Taxonomy, phytogeography and evolution. Academic Press, London and New York.

Dahlgren, R. 1980. A revised system of classification of the angiosperms. Bot. J. Linn. Soc. 80: 91–124.

Dailey, M. D. and J. Carvajal G. 1976. Helminth parasites of *Rhinobatos planiceps* Garman, 1880, including two new species of cestodes, with comments on host specificity of the genus *Rhinebothrium* Linton, 1890. J. Parasitol. 62: 939–942.

Darlington, P. J. 1957. Zoogeography: The geographic distribution of animals. John Wiley and Sons Inc., NY.

———. 1965. Biogeography of the southern end of the world. Harvard University Press, Cambridge, MA.

Darwin, C. 1859. On the origin of species. J. Murray, London.

Darwin, S. P. 1976. The genus *Lindenia* (Rubiaceae). J. Arnold Arb. 57: 426–449.

Davis, P. H. and V. H. Heywood. 1973. Principles of angiosperm taxonomy. Krieger Publishing Co., Huntington, NY.

Day, A. 1965. The evolution of a pair of sibling allotetraploid species of cobwebby Gilias (Polemoniaceae). Aliso 6: 25–75.

Deardorff, T. L., D. R. Brooks and T. B. Thorson. 1981. Two species of *Echinocephalus* (Nematoda: Gnathostomidae) from neotropical stingrays. J. Parasitol. 67: 433–439.

Deguchi, H. 1977a. Small male-branches of *Ptychomitrium* (Grimmiaceae) arised from the base of vaginula inside the perichaetial leaf circle. Misc. Bryol. Lichenol. 7: 177–179.

———. 1977b. An undetermined *Racomitrium* species with endogenous gemmae of *Grimmia trichophylla* type. Hikobia 8: 193–196.

———. 1978 (1979). A revision of the genera *Grimmia, Schistidium* and *Coscinodon* (Musci) of Japan. J. Sci. Hiroshima Univ., Ser. B, Div. 2, Bot. 16: 121–256.

DeJong, R. 1980. Some tools for evolutionary and phylogenetic studies. Z. Zool. Syst. Evol. 18: 1–23.

Delevoryas, T. 1979. Polyploidy in Gymnosperms. *In* Lewis, W. H. (ed.). Polyploidy: Biological relevance. Plenum Press, NY.

Delgadillo M., C. 1976. *Indusiella andersonii* sp. nov. in northern Africa. Bryologist 79: 98–100.

deWet, H. 1979. Origins of polyploids. *In* Lewis, W. H. (ed.). Polyploidy: Biological relevance. Plenum Press, NY.

Dietz, R. S. and W. P. Sproll. 1970. Fit between Africa and Antarctica: A continental drift reconstruction. Science 167: 1612–1614.

Ding, Hou. 1960. Thymelaeaceae; Drapetes. *In* C. C. G. J. van Steenis (ed.) Fl. Males. 1, 6: 43–44.

Dixon, N. H. 1931. *Ptychomitriopsis* Dix., gen. nov. Ptychomitriacearum. J. Bot. 69: 284–285.

———. 1932. Classification of mosses. *In* F. Verdoorn (ed.). Manual of bryology. W. Junk, Publ., The Hague.

Drury, D. G. 1974. A broadly based taxonomy of *Lagenifera* section *Lagenifera* and *Solenogyne* (Compositae—Astereae) with an account of the species in New Zealand. New Zealand J. Bot. 12: 365–396.

Duncan, T. and G. Estabrook. 1976. An operational method for evaluating classifications. Syst. Bot. 1: 373–382.

Edwards, S. R. 1979. Taxonomic implications of cell patterns in haplolepidous moss peristomes. *In* G. C. S. Clarke and J. G. Duckett, (eds.). Bryophyte systematics. (Syst. Assoc. Spec. Vol. 14). Academic Press, New York.

Eldredge, N. and J. Cracraft. 1980. Phylogenetic patterns and the evolutionary process. Columbia Univ. Press, NY.

Engler, A. (1879) 1882. Versuch einer entwicklungs-geschiete der Pflanzenwelt. Leipzig.

Estabrook, G. 1971. Some information theoretic optimality criteria for general classification. Math. Geo. 3: 203–207.

———. 1978. Some concepts for the estimation of evolutionary relationships in systematic botany. Syst. Bot. 3: 146–158.

——— **and W. R. Anderson.** 1978. An estimate of phylogenetic relationships within the genus *Crusea* (Rubiaceae) using character compatibility analysis. Syst. Bot. 3: 179–196.

Euzet, L. and J. Carvajal G. 1973. *Rhinebothrium* (Cestoda, Tetraphyllidea) parasites de raies du genre *Psammobatis* au Chili. Bull. Mus. Natn. Hist. Nat., Paris, serie Zool. 101: 779–787.

Farris, J. S. 1967a. The meaning of relationship and taxonomic procedure. Syst. Zool. 16: 44–51.

———. 1967b. Comment on psychologism. Syst. Zool. 16: 643–645.

———. 1969. A successive approximations approach to character weighting. Syst. Zool. 18: 374–385.

———. 1970. Methods for computing Wagner trees. Syst. Zool. 19: 83–92.

———. 1972. Estimating phylogenetic trees from distance matrices. Am. Nat. 106: 645–668.

———. 1973. On comparing the shapes of taxonomic trees. Syst. Zool. 22: 50–54.

———. 1974. Evolution in the *Drosophila obscura* species group. Evolution 28: 158–160.

———. 1976. Phylogenetic classification of fossils with recent species. Syst. Zool. 25: 271–282.

———. 1977. On the phenetic approach to vertebrate classification. *In* Hecht, M. K., P. C. Goody and B. M. Hecht (eds.). Major patterns in vertebrate evolution. Plenum Press, New York.

———. 1979a. On the naturalness of phylogenetic classification. Syst. Zool. 28: 200–214.

———. 1979b. The information content of the phylogenetic system. Syst. Zool. 28: 483–519.

———. 1980. Naturalness, information, invariance, and the consequences of phenetic criteria. Syst. Zool. 29: 360–381.

———. 1981. Distance data in phylogenetic analysis. *In* Funk, V. A. and D. R. Brooks (eds.). Advances in cladistics: Proceedings of the first meeting of the Willi Hennig Society. New York Botanical Garden.

——— **and A. G. Kluge.** 1979. A botanical clique. Syst. Zool. 28: 400–411.

———, ——— **and M. J. Eckardt.** 1970. A numerical approach to phylogenetic systematics. Syst. Zool. 19: 172–189.

———, ——— **and M. F. Mickevich.** 1979. Paraphyly of the *Rana boylii* species group. Syst. Zool. 28: 627–634.

Felsenstein, J. 1973. Maximum-likelihood estimation of phylogenetic trees from continuous characters. Amer. J. Hum. Genet. 25: 471–492.

———. 1978. The number of evolutionary trees. Syst. Zool. 27: 27–33.

Fitch, W. M. 1971. Toward defining the course of evolution: Minimum change for a specific tree topology. Syst. Zool. 20: 406–416.

——— **and E. Margoliash.** 1967. Construction of phylogenetic trees. Science 155: 279–284.

Florin, R. 1940. The Tertiary fossil conifers of south Chile and their phytogeographical significance, with a review of the fossil conifers of southern lands. K. Svenska Vetensk Akad. Handl., 3 ser., 19: 1–107.

Forman, L. L. 1966. On the evolution of cupules in the Fagaceae. Kew Bull. 18: 385–419.

Foulds, L. R., M. D. Hendy and D. Penny. 1979. A graph theoretic approach to the development of phylogenetic trees. J. Mol. Evol. 13: 127–149.

Fox, G. E., E. Stackebrandt, R. B. Hespell, J. Gibson, J. Maniloff, T. A. Dyer, R. S. Wolfe, W. E. Balch, R. S. Tanner, L. J. Magrum, L. B. Zablen, R. Blakemore, R. Gupta, L. Bonen, B. J. Lewis, D. A. Stahl, K. R. Luehrsen, K. N. Chen and C. R. Woese. 1980. The phylogeny of prokaryotes. Science 209: 457–463.

Fryxell, P. A. 1971. Phenetic analysis and the phylogeny of the diploid species of *Gossypium* L. (Malvaceae). Evolution 25: 554–562.

Funk, V. A. 1981. Special concerns in estimating plant phylogenies. *In* Funk, V. A. and D. R. Brooks (eds.). Advances in cladistics: Proceedings of the first meeting of the Willi Hennig Society. New York Botanical Garden.

———. In press. Systematics of *Montanoa* Cerv. in Llave & Lex. (Asteraceae: Heliantheae). Mem. New York Bot. Gard.

Gaffney, E. S. 1977. The side-necked family Chelidae: A theory of relationships using shared derived characters. Amer. Mus. Novit. 2620: 1–28.

———. 1979. An introduction to the logic of phylogeny reconstruction. *In* Cracraft, J. and N. Eldredge (eds.). Phylogenetic analysis and paleontology. Columbia Univ. Press, NY.

Garman, S. W. 1913. The plagiostomia. Mem. Mus. Comp. Zool. Harvard Univ. 36: 1–515.

Gatlin, L. L. 1972. Information theory and the living system. Columbia Univ. Press, NY.

Gerst, J. W. and T. B. Thorson. 1977. Effects of saline acclimation on plasma electrolytes, urea excretion, and hepatic urea biosynthesis in a freshwater stingray, *Potamotrygon* sp. Garman, 1877. Comp. Biochem. Physiol. 56A: 87–93.

Gery, J. 1969. The freshwater fishes of South America. *In* E. J. Fittkau, J. Illies, G. H. Schwabe, and H. Sioli (eds.). Biogeography and ecology in South America. W. Junk, Publ., The Hague.

Goldblatt, P. 1979. Polyploidy in Angiosperms: Monocotyledons. *In* Lewis, W. H. (ed.). Polyploidy: Biological relevance. Plenum Press, NY.

Goldstein, L. and R. P. Forster. 1971. Urea biosynthesis and excretion in freshwater and marine elasmobranchs. Comp. Biochem. Physiol. 39B: 415–421.

Goodman, M., J. Czelusniak, G. W. Moore, A. E. Romero-Herrera and G. Matsuda. 1979. Fitting the gene lineage into its species lineage, a parsimony strategy illustrated by cladograms constructed from globin sequences. Syst. Zool. 28: 132–163.

Goodspeed, T. H. 1954. The genus *Nicotiana*. Chron. Bot. 16: 1–536.

Gorman, G. C. and Y. J. Kim. 1976. The *Anolis* lizards of the eastern Caribbean: A case study in evolution II. Genetic relationships and genetic variation of the *bimaculatus* group. Syst. Zool. 25: 62–77.

———, M. Soule, S. Y. Yang and E. Nevo. 1975. Evolutionary genetics of insular Adriatic lizards. Evolution 29: 52–71.

Gottlieb, L. D. 1975. Allelic diversity in the outcrossing annual plant *Stephanomeria exigua* ssp. *carotifera* (Comp.). Evolution 29: 213–225.

Gower, J. C. 1974. Maximal predictive classification. Biometrics 30: 643–654.

Grant, A. and V. Grant. 1956. Genetic and taxonomic studies in *Gilia* VIII. The cobwebby Gilias. Aliso 3: 203–287.

Grant, V. 1953. The role of hybridization in the evolution of the leafy-stemmed Gilias. Evolution 7: 51–64.

———. 1963. The origin of adaptation. Columbia Univ. Press, NY.

———. 1964. Genetic and taxonomic studies in *Gilia*. XII. Fertility relationships of the polyploid cobwebby Gilias. Aliso 5: 479–507.

———. 1971. Plant speciation. Columbia Univ. Press, NY.

Griffith, R. W., P. K. T. Pang, A. K. Srivastava and G. E. Pickford. 1973. Serum composition of freshwater stingrays (Potamotrygonidae) adapted to fresh and dilute sea water. Biol. Bull. 144: 304–320.

Hagen, I. 1909. Forarbejder til en Norsk Løvmosflora. Kongel. Norske Vidensk. Selsk. Skr. (Trondheim) 5: 3–94.

Hanks, S. L. and D. T. Fairbrothers. 1976. Palynotaxonomic investigation of *Fagus* L. and *Nothofagus* Bl.: Light microscopy, scanning electron microscopy and computer analyses. *In* V. H. Heywood (ed.). Botanical Systematics 1: 1–141.

Harborne, J. B. 1977. Introduction to ecological biochemistry. Academic Press, London.

Harrington, H. J. 1965. Geology and morphology of Antarctica. *In* J. Van Miegham, P. Van Oye and J. Schell (eds.). Biogeography and ecology in Antarctica. W. Junk, Publ., The Hague.

Hartman, R. L. and T. F. Stuessy. 1980. The systematics of *Otopappus* (Compositae). ICSEB Abstracts 2: 232.

Hauser, L. A. and T. J. Crovello. 1980. Phylogenetic and phytogeographic studies of the tribes of the Brassicaceae. ICSEB Abstracts 2: 234.

Heiser, C. B. 1973. Introgression re-examined. Bot. Rev. 39: 347–366.

Hennig, W. 1950. Grundzuge einer theorie der phylogenetischen systematik. Deutscher Zentralverlag, Berlin.

———. 1965. Phylogenetic systematics. Annual Rev. Entom. 10: 97–116.

———. 1966. Phylogenetic systematics. Univ. of Illinois Press, Urbana.

———. 1975. "Cladistic analysis or cladistic classification?": a reply to Ernst Mayr. Syst. Zool. 24: 244–256.

Highton, R. and T. P. Webster. 1976. Geographic protein variation and divergence in populations of the salamander *Plethodon cinereus*. Evolution 30: 33–45.

Hill, C. R. and P. R. Crane. In press. Cladistics and the origin of angiosperms. *In* Joysey, K. & A. Friday (eds.). Problems of phylogenetic reconstruction. Academic Press, New York.

Hillier, F. S. and G. J. Lieberman. 1967. Introduction to operations research. Holden-Day, San Francisco.

Hilpert, F. 1933. Eine neue Laubmoosgattung. Hedwigia 73: 68–70.

Hippa, H. 1978. Classification of *Xylotini* (Diptera, Syrphidae). Acta Zool. Fennica 156: 1–153.

Hirohama, T. 1978. Spore morphology of bryophytes observed by scanning electron microscope, IV. Grimmiaceae. Bull. Natl. Sci. Mus., Ser. B (Bot.) 4: 33–42.

Ho, C. Y.-K., E. M. Prager, A. C. Wilson, D. T. Osuga and R. E. Feeney. 1976. Penguin evolution: Protein comparisons demonstrate phylogenetic relationship to flying aquatic birds. J. Mol. Evol. 8: 271–282.

Hooker, J. D. 1853. Introductory essay. *In* The botany of the Antarctic voyage of H. M. Discovery Erebus and Terror in the years 1853–55. II Flora Nova Zelandiae. Published by the author. London.

Hooker, W. J. 1829. *Scouleria*. Number 63. *In* T. Drummond, Musci Boreali Americani. Exsiccata.

Hull, D. L. 1979. The limits of cladism. Syst. Zool. 28: 416–440.

Humphries, C. J. 1981. Cytogenetic studies in *Anacyclus* (Compositae: Anthemideae). Nord. J. Bot. 1: 83–96.

———. 1981. Biogeographical methods and the southern beeches. *in* Forey, P. L. (ed.). Chance, change and challenge. Vol. 2. The evolving biosphere. British Museum (Natural History) and Cambridge University Press.

Jansen, R. K. 1979. The generic relationships of *Spilanthes* (Compositae: Heliantheae). Bot. Soc. Amer. Misc. Ser. Publ. 157: 60.

——— **and T. F. Stuessy.** 1980. Chromosome counts of Compositae from Latin America. Amer. J. Bot. 67: 585–594.

Jardine, N. and R. Sibson. 1971. Mathematical taxonomy. John Wiley and Sons, London.

Johnson, F. M. 1971. Isozyme polymorphism. *In D. ananassae*: genetic diversity among island populations in the South Pacific. Genetics 68: 77–95.

Johnson, L. A. S. and B. G. Briggs. 1975. On the Proteaceae—the evolution and classification of a southern family. J. Linn. Soc. Bot. 70: 83–182.

Johnson, M. S. 1975. Biochemical systematics of the atherinid genus *Menidia*. Copeia 1975: 662–691.

Jolles, J., F. Schoentgen, P. Jolles, E. M. Prager and A. C. Wilson. 1976. Amino acid sequence and immunological properties of chachalaca egg white lysozyme. J. Mol. Evol. 8: 59–78.

Jones, G. N. 1933. Grimmiaceae. *In* A. J. Grout. Moss flora of North America, North of Mexico. 2: 1–65. Published by the author. Vermont.

Junqueira, L. C. V., G. Hoxter and D. Zago. 1968. Observations on the biochemistry of fresh water rays and dolphin blood serum. Rev. Brasil. Pesquis. Med. Biol. 1: 225–226.

Kawai, I. 1965. Studies on the genus *Grimmia*, with reference to the affinity of gametophyte. Sci. Rep. Kanazawa Univ. 10: 79–132.

———. 1968. Taxonomic studies on the midrib of Musci (1). Significance of the midrib in systematic botany. Sci. Rep. Kanazawa Univ. 13: 127–157.

Kidd, K. K., P. Astolfi and L. L. Cavalli-Sforza. 1974. Error in the reconstruction of evolutionary trees. *In* J. F. Crow and C. Denniston (eds.). Genetic distance. Plenum Press, NY.

——— **and L. A. Sgaramella-Zonta.** 1971. Phylogenetic analysis: concepts and methods. Amer. J. Hum. Genet. 23: 235–252.

Klotz, L. C., N. Komar, R. L. Blanken and R. M. Mitchell. 1979. Calculation of evolutionary trees from sequence data. Proc. Nat. Acad. Sci. USA 76: 4516–4520.

Kluge, A. G. and J. S. Farris. 1969. Quantitative phyletics and the evolution of anurans. Syst. Zool. 18: 1–32.

Koponen, T. 1968. Generic revision of Mniaceae Mitt. (Bryophyta). Ann. Bot. Fennici 5: 117–151.

Kovacs, K. J. and G. D. Schmidt. 1980. Two new species of cestode (Trypanorhyncha, Eutetrarhynchidae) from the yellow-spotted stingray, *Urolophus jamaicensis*. Proc. Helm. Soc. Wash. 47: 10–14.

Kummel, B. 1962. History of the earth: an introduction to historical geology. W. H. Freeman and Co., San Francisco.

Laing, C., G. R. Carmody and S. B. Peck. 1976. Population genetics and evolutionary biology of the cave beetle *Ptomophagus hirtus*. Evolution 30: 484–498.

Larrazet, M. 1886. Des pieces de la peau quelques selaciens fossiles. Bull. Soc. Geol. France (3) 14: 255–277.

Lawton, E. 1967. *Grimmia occidentalis*, a new species from western North America. Bull. Torrey Bot. Club 94: 461–463.

———. 1971. Moss flora of the Pacific Northwest. Hattori Bot. Lab., Nichinan, Japan.

———. 1972. The genus *Rhacomitrium* in America and Japan. J. Hattori Bot. Lab. 35: 253–262.

Leedale, G. F. 1974. How many are the kingdoms of organisms? Taxon 23: 261–270.
Lesquereux, L. and T. P. James. 1884. Manual of the mosses of North America. S. E. Cassino & Co., Boston.
Levin, D. 1963. Natural hybridization between *Phlox maculata* and *Phlox glaberrima* and its evolutionary significance. Amer. J. Bot. 50: 714–720.
———. 1975. Interspecific hybridization, heterozygosity and gene exchange in *Phlox*. Evolution 29: 37–51.
——— **and D. M. Smith.** 1966. Hybridization and evolution in the *Phlox pilosa* complex. Amer. Nat. 100: 289–302.
Lewis, H. and C. Epling. 1959. *Delphinium gypsophila*, a diploid species of hybrid origin. Evolution 13: 511–525.
Lewis, W. H. 1979. Polyploidy in Angiosperms: Dicotyledons. *In* Lewis, W. H. (ed.). Polyploidy: Biological relevance. Plenum Press, NY.
Li, W.-H. 1981. A simple method for constructing phylogenetic trees from distance matrices. Proc. Natl. Acad. Sci. USA. 78: 1085–1089.
Lint, H. and C. Epling. 1945. A revision of *Agastache*. Amer. Midl. Naturalist 33: 207–230.
Loeske, L. 1913. Die Laubmoose Europas I. Grimmiaceae. Berlin.
———. 1930. Monografie der Europäischen Grimmiaceen. Biblioth. Bot. 101: 1–236.
Lozano-C., G., J. I. Hernandez-C. and J. E. Henao-S. 1980. El género *Trigonobalanus* Forman, en el Neotrópico—I. Caldasia 12: 517–537.
Lundberg, J. G. 1972. Wagner networks and ancestors. Syst. Zool. 21: 398–413.
Margulis, L. 1970. Origin of eucaryotic cells. Yale Univ. Press, New Haven, CT.
Marsh, O. C. 1877. Notice of some new vertebrate fossils. Amer. J. Sci. Arts 14: 249–256.
Matthew, W. D. 1915. Climate and evolution. Ann. N.Y. Acad. Sci. 24: 171–318.
Matthews, T. C. 1975. Biochemical polymorphism in populations of the Argentine toad *Bufo arenarum*. Copeia 1975: 454–465.
Maxson, L. R. and A. C. Wilson. 1975. Albumin evolution and organismal evolution in tree frogs (Hylidae). Syst. Zool. 24: 1–15.
Mayes, M. A. and D. R. Brooks. 1981. Cestode parasites of some Venezuelan stingrays. Proc. Biol. Soc. Wash. 93: 1230–1238.
Mayr, E. 1969. Principles of systematic zoology. McGraw-Hill, NY.
McAlpine, J. F. and J. E. H. Martin. 1966. Systematics of Sciadoceridae and relatives with descriptions of two new genera and species from Canadian amber and erection of family Ironomyidae (Diptera: Phoroidea). The Canadian Entom. 98: 527–544.
McQueen, D. R. 1976. The ecology of *Nothofagus* and associated vegetation in South America. Tuatara 22: 38–68.
Melville, R. 1973. Continental drift and plant distribution. *In* Tarling, D. H. and S. K. Runcorn (eds.). Implications of continental drift to the earth sciences. Vol. 1. Academic Press, London and New York.
Mickel, J. T. 1962. A monographic study of the fern genus *Anemia*, subgenus *Coptophyllum*. Iowa State J. Sci. 36: 349–382.
Mickevich, M. F. 1978a. Comments on the recognition of convergence and parallelism on Wagner trees. Syst. Zool. 27: 239–242.
———. 1978b. Taxonomic congruence. Syst. Zool. 27: 143–158.
———. 1981. Quantitative phylogenetic biogeography. *In* V. A. Funk and D. R. Brooks (eds.). Advances in cladistics: Proceedings of the first meeting of the Willi Hennig society. New York Botanical Garden.
——— **and M. S. Johnson.** 1976. Congruence between morphological and allozyme data in evolutionary inference and character evolution. Syst. Zool. 25: 260–270.
——— **and C. Mitter.** 1981. Treating polymorphic characters in systematics: A phylogenetic treatment of electrophoretic data. *In* Funk, V. A. and D. R. Brooks (eds.). Advances in cladistics: Proceedings of the first meeting of the Willi Hennig Society. New York Botanical Garden.
Mitten, W. 1869. Musci Austro-Americani. J. Linn. Soc., Bot. 12: 1–659.
Moore, D. M. 1968. The vascular flora of the Falkland Islands. Brit. Ant. Survey 60: 1–202.
———. 1971. Connections between cool temperate floras with particular reference to southern South America. *In* Valentine, D. H. (ed.). Taxonomy, phytogeography and evolution. Academic Press, New York and London.
Moore, G. W., M. Goodman and J. Barnabas. 1973. An iterative approach from the standpoint of

the additive hypothesis to the dendrogram problem posed by molecular data sets. J. Theor. Biol. 38: 423–457.

Müller, C. 1851. Synopsis muscorum frondosorum. Berlin.

Nei, M. 1972. Genetic distance between populations. Amer. Nat. 106: 283–292.

———. 1978a. The theory of genetic distance and evolution of human races. Jap. J. Hum. Genet. 23: 341–369.

———. 1978b. Estimation of average heterozygosity and genetic distance from a small number of individuals. Genetics 89: 583–590.

Nelson, G. 1973. Classification as an expression of phylogenetic relationships. Syst. Zool. 22: 344–359.

———. 1974. Historical biogeography: an alternative formalization. Syst. Zool. 23: 555–558.

———. 1979. Cladistic analysis and synthesis: Principles and definitions, with a historical note on Adanson's *Familles des Plantes* (1763–1764). Syst. Zool. 28: 1–21.

——— **and N. I. Platnick.** 1980. Multiple branching in cladograms: Two interpretations. Syst. Zool. 28: 86–91.

——— **and** ———. 1981. Systematics and biogeography: Cladistics and vicariance. Columbia Univ. Press, NY.

——— **and D. E. Rosen** (eds.). 1981. Vicariance biogeography: A critique. Columbia Univ. Press, NY.

Nevo, E., Y. J. Kim, C. S. Shaw and C. S. Thaeler. 1974. Genetic variation, selection, and speciation in *Thomomys talpoides* pocket gophers. Evolution 28: 1–23.

Nishida, Y. 1978. Studies on the sporeling types in mosses. J. Hattori Bot. Lab. 44: 372–454.

Noguchi, A. 1952. Musci Japonici. Erpodiaceae. J. Hattori Bot. Lab. 8: 5–17.

———. 1954. Musci Japonici. IV. The genus *Ptychomitrium*. J. Hattori Bot. Lab. 12: 1–26.

———. 1974. Musci Japonici. X. The genus *Racomitrium*. J. Hattori Bot. Lab. 38: 337–369.

Noonan, G. R. 1973. The anisodactylines (Insecta: Coleoptera: Carabidae: Harpalini): Classification, evolution and zoogeography. Quaestionès entomologicae g: 266–480.

Nygren, A. 1967. Apomixis in the angiosperms. Handb. Pflanzenphysiol. 18: 551–596.

O'Brien, J. and R. J. MacIntyre. 1969. An analysis of gene-enzyme variation in natural populations of *Drosophila melanogaster* and *D. simulans*. Amer. Nat. 103: 97–113.

Oliver, W. R. B. 1925. Biogeographical relations of the New Zealand region. J. Linn. Soc. Bot. 47: 99–139.

Ostrowski de Núñez, M. 1971. Estudios preliminaires sobre la fauna parasitaria de algunos elasmobranquios del litoral bonaerense, Mar del Plata, Argentina. I. Cestodes y trematodos de *Psammobatis microps* (Günther) y *Zapteryx brevirostris* (Müller y Henle). Physis 30: 425–446.

———. 1973. Estudios preliminaires sobre la fauna parasitaria de algunos elasmobranquios del litoral bonaerense, Mar del Plata, Argentina. II. Cestodes de *Mustelus schmitti* Springer, 1939. Physis 32: 1–14.

Owen, H. G. 1976. Continental displacement and expansion of the earth during the Mesozoic and Cenozoic. Philos. Trans. Series B 281: 223–291.

Parenti, L. R. 1980. A phylogenetic analysis of the land plants. J. Linn. Soc. Biol. 13: 225–242.

Patterson, C. 1981. Methods in paleobiogeography. *In* G. Nelson and D. Rosen (eds.). Vicariance biogeography: A critique. Columbia Univ. Press, NY.

Pickett-Heaps, J. D. 1975. Green algae. Structure, reproduction and evolution in selected genera. Sinauer Associates, Sunderland, MA.

Pilleri, G. and M. Gihr. 1977. Observations on the Bolivian (*Inia boliviensis* d'Orbigny, 1834) and Amazon Bufo (*Inia geoffrensis* de Blainville, 1817) with description of a new subspecies (*Inia geoffrensis humboldtiana*). *In* G. Pilleri (ed.). Investigations on Cetacea. Vol. VIII. Univ. of Basel, Switzerland.

Platnick, N. I. 1977. Parallelism in phylogeny reconstruction. Syst. Zool. 26: 93–96.

———. 1979. Philosophy and the transformation of cladistics. Syst. Zool. 28: 537–546.

——— **and H. D. Cameron.** 1977. Cladistic methods in textual, linguistic and phylogenetic analysis. Syst. Zool. 26: 380–385.

——— **and G. Nelson.** 1978. A method of analysis for historical biogeography. Syst. Zool. 27: 1–16.

——— **and** ———. 1981. The purposes of biological classification. *In* Asquith, P. D. and I. Hacking (eds.). PSA 1978: Proceedings of the sixth biannual meeting, vol. 2. Philosophy of Science Assoc., East Lansing, MI.

Popper, K. R. 1968. The logic of scientific discovery. Harper and Row, NY.

Prager, E. M., D. P. Fowler and A. C. Wilson. 1976a. Rates of evolution in conifers (Pinaceae). Evolution 30: 637–649.

——— **and A. C. Wilson.** 1976. Congruency of phylogenies derived from different proteins. A molecular analysis of the phylogenetic position of cracid birds. J. Mol. Evol. 9: 45–57.

——— **and** ———. 1978. Construction of phylogenetic trees for proteins and nucleic acids: Empirical evaluation of alternative matrix methods. J. Mol. Evol. 11: 129–142.

———, ———, **D. T. Osuga and R. E. Feeney.** 1976b. Evolution of flightless land birds on southern continents: Transferrin comparison shows monophyletic origin of rattites. J. Mol. Evol. 8: 283–294.

Putzer, H. 1969. Uberblick uber die geologische entwicklung Sudamerikas. *In* E. J. Fittkau, J. Illies, H. Klinge, G. H. Schwabe and H. Sioli (eds.). Biogeography and ecology in South America. W. Junk, Publ., The Hague.

Raven, P. H. and D. I. Axelrod. 1972. Plate tectonics and Australasian paleobiogeography. Science 176: 1379–1386.

——— **and** ——— 1974. Angiosperm biogeography and past continental movements. Ann. Missouri Bot. Gard. 61: 539–673.

Rego, A. A. 1977. Cestoides parásitas de *Carcharhinus longimanus* (Poey, 1861). Rev. Brasil. Biol. 37: 847–852.

———, **J. C. Dos Santos and P. P. Silva.** 1974. Estudos de cestoides de peixes do Brasil. Mem. Inst. Oswaldo Cruz 72: 187–204.

——— **and M. T. Mayer.** 1976. Ocorrencia de duas especies de tetrafilideos em tuburao de costa Brasileira e consideracoes sobre os generos *Cylindrophorus, Platybothrium* y *Phoreibothrium* (Cestoda, Teraphyllidea). Rev. Brasil. Biol. 36: 321–328.

———, **J. J. Vicente and N. I. Herrera.** 1968. Sobre dois novos parasitos de peixe da costa do Peru (Cestoda, Tetraphyllidea). Mem. Inst. Oswaldo Cruz 66: 145–149.

Rich, P. V. 1975. Antarctic dispersal routes, wandering continents and the origin of Australia's non-passerine avifauna. Mem. Nat. Mus. 36: 63–126.

Riedl, R. 1978. Order in living organisms. John Wiley and Sons, NY.

Robinson, H. 1971. A revised classification for the orders and families of mosses. Phytologia 21: 289–293.

———. 1974. Notes on the mosses of Juan Fernández and southern South America. Phytologia 29: 116–120.

———. 1978. Studies in the Heliantheae (Asteraceae). XV. Various new species and new combinations. Phytologia 41: 33–44.

Rogers, J. S. 1972. Measures of genetic similarity and genetic distance. Studies in genetics VII. Univ. Texas Publ. 7213: 145–153.

———. 1973. Protein polymorphism, genic heterozygosity, and divergence in the toads *Bufo cognatus* and *B. speciosus*. Copeia 1973: 322–330.

Roivainen, H. 1955a. Contribution to the Fuegian species of the genus *Rhacomitrium* Bridel. Arch. Soc. Zool. Bot. Fenn. Vanamo 9: 85–98.

———. 1955b. *Bucklandia Bartramii,* a new genus and species of leaf moss from Tierra del Fuego. Arch. Soc. Zool. Bot. Fenn. "Vanamo" 9: 98–100.

———. 1972. *Bucklandiella* Rov., nomen novum (Musci). Ann. Bot. Fenn. 9: 116.

Romer, A. S. 1966. Vertebrate paleontology. Univ. Chicago Press, Chicago, IL.

Rosen, D. E. 1975. A vicariance model of Caribbean biogeography. Syst. Zool. 24: 431–464.

———. 1978. Vicariant patterns and historical explanation in biogeography. Syst. Zool. 27: 159–188.

———. 1979. Fishes from the uplands and intermontane basins of Guatemala: Revisionary studies and comparative geography. Bull. Amer. Mus. Nat. Hist. 162: 267–376.

Ross, H. H. 1956. Evolution and classification of the mountain caddisflies. University of Illinois Press, Urbana, IL.

———. 1974. Biological systematics. Addison-Wesley Press, Reading, MA.

Saito, K. 1973. Notes on the Pottiaceae (3). J. Jap. Bot. 48: 161–167.

Sanders, R. W. 1979. A systematic study of *Agastache* section *Brittonastrum* (Lamiaceae, Nepeteae). Ph.D. Dissertation. University of Texas, Austin.

———. 1980. The evolution of flavonoid profiles in *Agastache* (Lamiaceae). ICSEB Abstracts 2: 334.

———. 1981. New taxa and combinations in *Agastache* (Lamiaceae). Brittonia 33: 194–197.

Sarich, V. M. 1969a. Pinniped origins and the rate of evolution of carnivore albumins. Syst. Zool. 18: 286–295.

———. 1969b. Pinniped phylogeny. Syst. Zool. 18: 416–422.

Scanlan, D., L. Maxson and W. Duellman. 1980. Albumin evolution in marsupial frogs. Evolution 34: 222–229.

Schmidt, G. D. 1978. *Phyllobothrium kingae* sp. n., a tetraphyllidean cestode from a yellow-spotted stingray in Jamaica. Proc. Helm. Soc. Wash. 45: 132–134.

Schröter, C. 1913. Geographie der Pflanzen 2. Genetische Pflanzen-geographie (Epiontologie). Handworterbuch and Naturwiss. 4. Jena.

Schuh, R. T. 1981. Willi Hennig Society: report of first annual meeting. Syst. Zool. 30: 76–81.

Schuster, R. M. 1976. Plate tectonics and its bearing on the geographical origin and dispersal of angiosperms. *In* Beck, C. B. (ed.). Origin and early evolution of Angiosperms. Columbia Univ. Press, NY.

Seeman, B. 1862. Lindenia vitiensis. Bonplandia 10: 33–34.

Selander, R. K., M. H. Smith, S. Y. Yang, W. E. Johnson and J. B. Gentry. 1971. Biochemical polymorphism and systematics in the genus *Peromyscus*. I. Variation in the old-field mouse (*Peromyscus polionotus*). Studies in Genetics VI. Univ. of Texas Publ. 7103.

Sgaramella-Zonta, L. A. and K. K. Kidd. 1973. Phylogenetic trees: Computer methodology. *In* N. E. Morton (ed.). Genetic structure of populations. Univ. Press of Hawaii, Honolulu.

Shields, O. 1979. Evidence for initial opening of the Pacific ocean in the Jurassic. Palaeogeography, Palaeoclimatology and Palaeoecology 26: 181–220.

Sloover, J. L. De. 1976. Note de bryologie africaine. VII. *Pseudephemerum, Bryohumbertia, Eucladium, Streptopogon, Ptychomitrium, Rhachithecium, Antitrichia, Pterogonium, Lindigia, Distichophyllum*. Bull. Jard. Bot. Natl. Belgique 46: 427–447.

———. 1977. Note de bryologie africaine. IX. *Andreaea, Racomitrium, Gymnostomiella, Thuidium*. Bull. Jard. Bot. Natl. Belgique 47: 155–181.

Smith, A. C. 1970. The Pacific as a key to flowering plant history. Univ. of Hawaii, Honolulu.

———**, J. C. Briden and G. E. Drewry.** 1973. Phanerozoic world maps. *In* Hughes, N. F. (ed.). Organisms and continents through time. Palaeontology Assoc., London.

Smith, A. J. E. 1978. The moss flora of Britain and Ireland. Cambridge Univ. Press, London, England.

——— **and M. E. Newton.** 1968. Chromosome studies of some British and Irish mosses III. Trans. Brit. Bryol. Soc. 5: 463–522.

Smith, H. W. 1931. The absorption and excretion of water and salts by the elasmobranch fishes. I. Fresh water elasmobranchs. Amer. J. Physiol. 98: 279–295.

———. 1936. The retention and physiological role of urea in the Elasmobranchii. Biol. Rev. 11: 49–82.

Sneath, P. H. A. and R. R. Sokal. 1973. Numerical taxonomy. W. H. Freeman and Co., San Francisco.

Soepadmo, E. 1972. Fagaceae. Flora Malesiana Ser. 1, 7(2): 277–294.

Sokal, R. R. and C. D. Michener. 1958. A statistical method for evaluating systematic relationships. Univ. Kansas Sci. Bull. 38: 1409–1438.

——— **and P. H. A. Sneath.** 1963. Principles of numerical taxonomy. W. H. Freeman and Co., San Francisco.

Stauch, A. and M. Blanc. 1962. Description d'un selacien rajiforme des eaux douces du Nord-Cameroun: *Potamotrygon garouaensis* n. sp. Bull. Mus. Natn. Hist. Nat., Paris 34: 166–171.

Stebbins, G. L. 1950. Variation and evolution in plants. Columbia Univ. Press, NY.

———. 1974. Flowering plants: Evolution above the species level. Belknap-Harvard Press, Cambridge, MA.

———. 1979. Polyploidy in plants: Unsolved problems and prospects. *In* Lewis, W. H. (ed.). Polyploidy: Biological relevance. Plenum Press, NY.

Steere, W. C. 1974. *Grimmia* (*Coscinodon*) *arctolimnia*, a new species from Great Bear Lake, Northwest Territories, Canada. Bryologist 42: 230–234.

Stevens, P. 1980. Evolutionary polarity of character states. Ann. Rev. Ecol. Syst. 11: 333–358.

Stewart, K. D. and K. R. Mattox. 1975. Comparative cytology, evolution and classification of the green algae with some consideration of the origin of other organisms with chlorophylls a and b. Bot. Rev. 41: 104–135.

——— and ———. 1978. Structural evolution in the flagellated cells of green algae and land plants. Biosystems 10: 145–152.

Straw, R. M. 1956. Floral isolation in *Penstemon*. Amer. Nat. 90: 47–53.

Stuessy, T. F. 1977. Heliantheae—systematic review. *In* V. H. Heywood, J. B. Harborne and B. L. Turner (eds.). The biology and chemistry of the Compositae. Academic Press, London.

Swofford, D. 1981. On the utility of the distance Wagner procedure. *In* Funk, V. A. and D. R. Brooks (eds.). Advances in cladistics: Proceedings of the first meeting of the Willi Hennig Society. New York Botanical Garden.

Szidat, L. 1970. Descripción de una nueva especies de la subfamilia Calicotylinae Monticelli, 1903, *Paracalicotyle asterii* n. gen., n. sp. de un cazón (*Mustelus asterias* Rond-Cloquet) del Atlántico del Sud. Neotropica 16: 53–57.

Taktajhan, A. 1969. Flowering plants: Origin and dispersal. Oliver and Boyd, Edinburgh.

Tateno, Y. 1978. Statistical studies on the evolutionary changes of macromolecules. Ph.D. thesis, Univ. of Texas, Houston.

Taylor, F. J. R. 1974. Implications and extensions of the serial endosymbiosis theory of the origin of eukaryotes. Taxon 23: 229–258.

———. 1978. Problems in the development of an explicit hypothetical phylogeny of the lower eukaryotes. Biosystems 10: 67–89.

Thériot, I. 1928. *Jaffueliobryum* gen. nov. Rev. Bryol. n. ser. 1: 192–195, pl. 8.

Thistleton-Dyer, W. 1909. Geographical distribution of plants. *In* Seward A. C. (ed.). Darwin and modern science. Cambridge Univ. Press, Cambridge, MA.

Thorne, R. F. 1976. A phylogenetic classification of the Angiospermae. Evol. Biol. 9: 35–106.

Thorson, T. B. 1967. Osmoregulation in fresh-water elasmobranchs. *In* P. W. Gilbert, R. F. Mathewson and D. P. Rall (eds.). Sharks, skates and rays. Johns Hopkins Press, Baltimore, MD.

———. 1970. Freshwater stingrays, *Potamotrygon* spp.: failure to concentrate urea when exposed to saline medium. Life Sci. 9: 893–900.

———, **C. M. Cowan and D. E. Watson.** 1967. *Potamotrygon* spp.: Elasmobranchs with low urea content. Science 158: 375–377.

———, ——— **and** ———. 1973. Body fluid solutes of juveniles and adults of the euryhaline bull shark *Carcharhinus leucas*, from freshwater and saline environments. Physiol. Zool. 46: 29–42.

——— **and D. E. Watson.** 1975. Reassignment of the African freshwater stingray, *Potamotrygon garouaensis*, to the genus *Dasyatis*, on physiologic and morphologic grounds. Copeia 1975: 701–712.

———, **R. M. Wotton and T. A. Georgi.** 1978. Rectal gland of freshwater stingrays, *Potamotrygon* spp. (Chondrichthyes: Potamotrygonidae). Biol. Bull. 154: 508–516.

Traub, H. P. 1963. Revision of 'The phyla of organisms.' Plant Life 19: 160.

Troncy, P.-M. 1969. Description de deux nouvelles espèces de nematodes parasites de poissons. Bull. Mus. Natn. Hist. Nat., Paris, 2e serie 41: 598–605.

Urist, M. R. 1962. Calcium and other ions in blood and skeleton of Nicaraguan fresh-water shark. Science 137: 984–986.

Van Balgooy, M. M. J. 1975. Pacific plant areas 3. Rijksherbarium, Leiden.

van Steenis, C. G. G. J. 1953. Papuan *Nothofagus*. J. Arnold Arb. 34: 301–373.

———. 1962. The land-bridge theory in botany with particular reference to tropical plants. Blumea 11:235–542.

———. 1971. *Nothofagus*, key genus of plant geography in time and space, living and fossil, ecology and phylogeny. Blumea 19: 65–98.

Vitt, D. H. 1980. A new classification of the Musci. ICSEB Abstracts 2: 379.

Wagner, W. H. 1954. Reticulate evolution in the Appalachian Aspleniums. Evolution 8: 103–108.

———. 1961. Problems in the classification of ferns. *In* Recent advances in botany. Univ. Toronto Press, Canada.

———. 1969. The role and taxonomic treatment of hybrids. BioScience 19: 785–789.

———. 1970. Biosystematics and evolutionary noise. Taxon 19: 146–151.

———. 1980. Origin and philosophy of the groundplan divergence method of cladistics. Syst. Bot. 5: 173–193.

——— **and R. S. Whitmire.** 1957. Spontaneous production of a morphologically distinct fertile allopolyploid by a sterile diploid of *Asplenium ebenoides*. Bull. Torrey Bot. Club 84: 79–89.

Wallace, A. R. 1876. The geographical distribution of animals, with a study of the relations of living and extinct faunas as elucidating the past changes of the earth's surface. 2 vols. Reprinted ed., Hafner Publishing Co., NY. 1962.

Wallace, D. G., M. King and A. C. Wilson. 1973. Albumin differences among ranid frogs: taxonomic and phylogenetic implications. Syst. Zool. 22: 1–13.

Waterman, M. S., T. F. Smith, M. Singh and W. A. Berger. 1977. Additive evolutionary trees. J. Theor. Biol. 64: 199–213.

Watrous, L. and Q. Wheeler. 1981. The outgroup method of phylogeny reconstruction. Syst. Zool. 30: 1–16.

Webster, T. P., R. K. Selander and S. Y. Yang. 1972. Genetic variability and similarity in the *Anolis* lizards of Bimini. Evolution 26: 523–535.

Whittaker, R. H. 1969. New concepts of kingdoms of organisms. Science 163: 150–160.

——— **and L. Margulis.** 1978. Protist classification and the kingdoms of organisms. Biosystems 10: 3–18.

Wijk, R. van der, W. D. Margadant and P. H. Florschütz. 1959–1969. Index Muscorum. Regnum Veg. Vols. 17, 26, 33, 48 and 65.

Wiley, E. O. 1975. Karl R. Popper, systematics, and classification—a reply to Walter Bock and other evolutionary taxonomists. Syst. Zool. 24: 233–243.

———. 1979. An annotated Linnaean hierarchy, with comments on natural taxa and competing systems. Syst. Zool. 28: 308–337.

———. 1980. Phylogenetic systematics and vicariance biogeography. Syst. Bot. 5: 194–220.

———. 1981. Phylogenetics. The theory and practice of phylogenetic systematics. John Wiley and Sons, NY.

Williams, R. S. 1903. Bolivian mosses. Part 1. Bull. New York Bot. Gard. 3: 104–134.

———. 1927. Mosses of Perú collected by the Captain Marshall Field Peruvian Expedition 1923. Field Mus. Nat. Hist. Bot. Ser. 4: 125–139.

Wright, S. 1978. Evolution and the genetics of populations. Vol. 4, Variability within and among natural populations. Univ. of Chicago Press, Chicago, IL.

Wulff, E. V. 1943. An introduction to historical plant geography. Chronica Botanica Company, Waltham, MA.

Yang, S. Y., M. Soule and G. C. Gorman. 1974. *Anolis* lizards of the eastern Caribbean: a case study in evolution I. Genetic relationships, phylogeny, and colonization sequence of the *roquet* group. Syst. Zool. 23: 387–399.

———**, L. L. Wheeler and I. R. Bock.** 1972. Isozyme variations and phylogenetic relationships in the *D. bipectinata* species complex. Univ. Texas Publ. Genet. 7213: 213–227.

Zam, S. G., D. K. Caldwell and M. C. Caldwell. 1970. Some internal parasites from freshwater cetaceans from the upper Amazon River. *In* G. Pilleri (ed.). Investigations on cetacea. Vol. II. Univ. of Basel, Switzerland.

Zimmerman, E., C. Kilpatrick and B. Hecht. 1978. The genetics of speciation in the rodent genus *Peromyscus*. Evolution 32: 565–576.

Zouros, E. 1974. Genic differentiation associated with the early stages of speciation in the *mulleri* subgroup of *Drosophila*. Evolution 27: 601–621.

Program

Monday 13 October 1980

Symposium: "Phylogenetics and Biochemistry" (Chair: M. F. Mickevich)

Analysis of immunological distance data in phylogenetics by J. S. Farris, State University of New York, Stony Brook.

On the utility of the distance Wagner procedure by D. L. Swofford, University of Illinois, Urbana.

Phylogenetic determination of characters in allozyme data by C. Mitter and M. F. Mickevich, American Museum of Natural History.

Symposium: "Phylogenetic Systematics in Botany" (Chair: V. A. Funk)

Classifications as languages of empirical comparative biology by D. R. Brooks, University of British Columbia.

Special concerns in estimating plant phylogenies by V. A. Funk, New York Botanical Garden.

Technical aspects of determining character state polarity by L. Watrous, Field Museum of Natural History, Chicago.

The cladistic approach to plant classification by K. Bremer and H.-E. Wanntorp, Swedish Museum of Natural History and the University of Stockholm.

Cladistics and biogeography of southern beeches by C. J. Humphries, British Museum (Natural History).

Tuesday 14 October 1980

Contributed Papers (Chair: D. E. Rosen)

The work of H. H. Ross and vicariance biogeography by R. Allen, University of Arkansas.

The cladistics of **Agastache** *(Lamiaceae)* by R. W. Sanders, Ohio State University.

Problems in the determination of character state polarity by P. Stevens, Harvard University.

Racial classification of the American subpopulations by craniometrics by P. M. Lin, Wichita State University.

The cladistics of **Spilanthes** *(Compositae)* by R. K. Jansen, Ohio State University.

A phylogenetic analysis of the genera of the Grimmiaceae (Bryophyta) by S. P. Churchill, University of Kansas.

Cladistics and biogeography of **Salmea** *DC. (Compositae)* by M. Bolick, University of Nebraska.

How some wasps become bees: Hennigian and meta-Hennigian methods applied to evolutionary ethology by R. Jander, University of Kansas.

Symposium: "Paleontological Challenges to Cladistics" (Chair: D. Hull)

B. G. Naylor (University of Alberta) and Gareth Nelson (American Museum of Natural History) presented opposing views of the subject of cladistics and paleontology.

Symposium: "Phylogenetic Biogeographic Methods" (Chair: V. Ferris)

Quantitative phylogenetic biogeography by M. F. Mickevich, American Museum of Natural History.

The evolutionary history and biogeography of neotropical freshwater sting-rays (Myliobatiformes) by D. R. Brooks, University of British Columbia.

Phylogenetic relationships and biogeography of the neotropical owl **Glacidium brasilianum** by S. Coats, University of California, Berkeley.

Closing Remarks by N. I. Platnick, American Museum of Natural History, and general discussion.

Index of Selected Topics

Dedicated to Willi Hennig

(1913–1976)

Dr. T. D. Nicholson (American Museum of Natural History) presenting the Gold Medal Award to Dr. Willi Hennig in 1975. (Photo courtesy of The American Museum of Natural History.)

Contributors

M. R. Bolick, W-352 Nebraska Hall, University Museum, Univ. of Nebraska, Lincoln, NE 68508
K. Bremer, Section of Phanerogamic Botany, Swedish Museum of Natural History, Box 50007, S-10405 Stockholm, Sweden
D. R. Brooks, Dept. of Zoology, Univ. of British Columbia, 2075 Wesbrook Mall, Vancouver, B.C. Canada V6T 1W5
S. P. Churchill, Botany Dept., Univ. of Kansas, Lawrence, KS 66045
J. S. Farris, Dept. of Ecology and Evolution, State Univ. of New York, Stony Brook, NY 11794
V. A. Funk, New York Botanical Garden, Bronx, NY 10458
C. P. Humphries, Dept. of Botany, British Museum (Natural History), Cromwell Road, London SW7 5BD, Great Britain
M. A. Mayes, Environmental Sciences Research, Dow Chemical USA, Midland, MI 48640
M. F. Mickevich, Dept. of Ichthyology, American Museum of Natural History, Central Park West at 79th St., New York, NY 10024
C. Mitter, Dept. of Entomology, American Museum of Natural History, Central Park West at 79th St., New York, NY 10024
N. I. Platnick, Dept. of Entomology, American Museum of Natural History, Central Park West at 79th St., New York, NY 10024
R. W. Sanders, Dept. of Botany, Ohio State Univ., 1735 Neil Ave., Columbus, OH 43210
D. L. Swofford, Dept. of Genetics and Development, Univ. of Illinois, Urbana, IL 61801
T. B. Thorson, School of Life Sciences, Univ. of Nebraska, Lincoln, NE 68588
H.-E. Wanntorp, Institute of Botany, Univ. of Stockholm, S-10691 Stockholm, Sweden
E. O. Wiley, Museum of Natural History, Univ. of Kansas, Lawrence, KS 66045

Foreword

In 1950, a German entomologist named Willi Hennig published a systematics textbook. A number of continental European systematists adopted Hennig's principles, which represented a formalized method for constructing a general reference classification. Hennig noted that a phylogenetic system had certain logical attributes and priorities which made it the best choice for a general reference system. He clearly stated the attributes of such a system and suggested some general ways to construct such classifications. In addition, he outlined some of the possible uses of the resultant classifications.

Hennig's work was not presented to North American systematists as a whole until 1959, when Sergius Kiriakoff published the first of a series of notes in *Systematic Zoology* concerning Hennig's work. Almost immediately, published papers appeared which made it clear that Hennig's work had been known to some prominent systematists prior to 1959 but that they had failed to comment on his views in print until the issue was raised nearly a decade after publication of his book. The two competing schools of the day, the "evolutionary taxonomists" and the "pheneticists" or "numerical taxonomists," finally found one area of mutual agreement—Hennig's approach was wrong. Each group disagreed with Hennig for different reasons, but at the 1965 meetings of the Society for Systematic Zoology leading proponents of each school, Ernst Mayr and Robert Sokal, did coin the same term of derision for Hennigian approaches—*cladism*. Emotional responses to phylogenetic systematics have continued. Most recently, Hennigians have been accused of fostering a new era of "McCarthyism" (because of their insistence that hypothesis testing be applied to systematics) or ultra-right wing totalitarianism as well as fostering communist subversion, or left-wing totalitarianism. To be sure, phylogeneticists have added fuel to the fire, for Hennigians have always been willing to confront the issues of the day publicly and in print.

In 1966, the year following the coining of the term cladism, an updated version of Hennig's principles was published in English (reprinted in 1980). By 1969, the number of papers dealing with Hennig's methods and with extensions of the principles involved published in *Systematic Zoology* began rising in response to a rapidly growing interest in the work. Numerous refinements have followed, including the formulation of a rigorous quantitative analytical framework for applying the principles of phylogenetic systematics. From 1959 to 1968, nine papers on the subject appeared in *Systematic Zoology*. In the next five years, that number rose by 28 and in the next seven years the number rose by another 84. Nonderisive versions of the labels (cladistics, cladist, and cladogram) have become terms stated and written with pride. A quick perusal of the literature citations in this volume makes one aware of the major contributors to this most remarkable revolution in systematics and comparative biology—Cracraft, Farris, Kluge, Mickevich, Nelson, Platnick, Rosen, Wiley, and many others.

The principles and applications of Hennigian systematics are not restricted to any particular level of biological organization nor are they bounded by any particular kind of data. Cladistics has emerged as a powerful analytical tool in comparative biology because it offers most informative (least ambiguous) summations of any set of biological observations represented in a consistent, testable, reproducible framework. Systematics has thus become a truly empirical science, capable of assuming its rightful place as the one indispensible branch of biology—the framework of comparisons for a comparative science.

For over a decade phylogeneticists had been addressing objections and criticisms of non-Hennigians both at meetings and in published papers. By the late 1970's it became apparent that no new objections were being raised nor were alternative methods being proposed which truly represented a departure from the old battle-lines. Insofar as continued confrontation with proponents of old dogma would not lead to further refinements and advances in systematics, cladists attending the 13th Numerical Taxonomy Conference at Harvard University in 1979 decided that the time had arrived to assemble a cohesive group of systematists identifiable as Hennigians. Accordingly, on October 12–14, 1980, over 70 systematists from Great Britain, Sweden, Canada and the United States gathered at the University of Kansas, at the invitation of Dr. E. O. Wiley, for two days of presentations marking the inauguration of the Willi Hennig Society.

This volume contains most of the papers presented during the first meeting of the Hennig Society (we have included a list of titles for all presentations given; for a review of the meeting see Schuh, 1981). These proceedings do not constitute a primer of cladistic analysis but a cross-section of current primary research in phylogenetic systematics by botanists and zoologists. It is hoped that the problems encountered and the solutions proposed by those present will result in a better combined understanding of the applications of the principles of cladistics.

We wish to thank Dr. E. O. Wiley for hosting the meeting and Dr. James S. Farris for arranging the symposia. We are most grateful to the New York Botanical Garden and in particular to Dr. María L. Lebrón-Luteyn (Curator of Publications) for undertaking the publication of this volume. And lastly, we appreciate the helpful and prompt reviews given by our associates.

Vicki A. Funk
Daniel R. Brooks
New York and Vancouver
May 1981

Preface

The papers collected in this volume are the products of a rather rare event in systematic biology—a meeting between botanists and zoologists to discuss their common goals and problems. Their goals were two-fold, to discover the relationships existing between organisms and to convey these discoveries to the biological community in a way that the information would be available and useful. Of singular interest is the fact that these systematists, who have diverse backgrounds and who study very different sorts of organisms, could actually understand each other. The botanists could criticize or agree with hypotheses forwarded by the zoologists and vice versa.

This ability to communicate is the result of sharing a similar world view, a view effectively articulated by the late Willi Hennig. Hennig suggested that we ask questions about common ancestry, not on the obscure level of such statements as "the reptiles gave rise to the birds," but on a critical level with such statements as "of all organisms, birds and crocodiles share a common ancestral species unique to them." Hennig outlined an empirical method designed to answer such questions and to arbitrate disputes when more than one answer was available. And, he suggested that we use classifications, the results of such analyses, as evolutionary tools, frameworks for comparisons. Hennig was not the first to use derived homologues (apomorphies) to demonstrate common ancestry. He was not the first to advocate strictly natural classifications. He was the first to effectively produce a system to accomplish these goals and to convince a major part of the systematic community to implement such a system. He was able to do so with the help of such advocates as Lars Brundin, Gareth Nelson and Deter Schlee. Without the help of these systematists and others who were convinced to try Hennig's methods, the revolution would have been a long time coming, if at all.

What is this world view? First, evolution has produced a natural system of relationships among organisms through genealogical descent and modification. Second, it is the job of systematists to discover these relationships. Third, the system of relationships that we discover can be communicated to other biologists by using a language, a classification, that is as natural and therefore as informative as the relationships we discover.

Who could object to such a world view? Almost everyone it would seem. The reactions of Hennig's critics have been as sociologically interesting as Hennig's system is biologically interesting. One group, the pheneticists, claimed that phylogenies could not be reconstructed. When it became apparent that parts of phylogenies have been and continue to be discovered, and in an empirically testable manner, the pheneticists claimed either that such systems were not very practical or that phenetic methods were just as good. Unfortunately for phenetics, neither claim has been substantiated. The evolutionary taxonomists originally adopted the attitude that phylogenies were too complicated to be represented by a hierarchy, especially a Linnaean hierarchy. With the demonstrations that even complex phylogenies could be classified using Linnaean hierarchies, the evolutionary taxonomists then claimed that their system of clades and grades was more informative. With the demonstration that the most informative classifications do not contain grades, the evolutionary taxonomists would seem to be reduced to searching for some unknown system of classification to justify their position.

The contents of this volume represent a significant portion of the papers given at the Lawrence meeting. They can be divided into four topics: (1) biochemical data and phylogenetic analysis, (2) phylogenetic classification, (3) applications of phylogenetics to botany, and (4) biogeography.

Of particular concern to those interested in applying biochemical data to systematic problems are the types of data used and the coding necessary to handle such data. Farris discusses problems associated with distance measures and suggests that analyses should be performed on the "raw" data themselves. Swofford describes an improved distance Wagner algorithm. Mickevich and Mitter discuss the problems of coding allozyme data and suggest solutions involving a minimum loss of information.

The reasons for preferring phylogenetic classification over phenetic or evolutionary classifications have been discussed at length in the recent literature. Although Farris has discussed the information content of phylogenetic classifications, Brooks' paper in this volume is the first full-scale attempt to discuss the subject from a strictly information theoretical point of view. He concludes that phylogenetic classifications describe the characteristics of organisms in the fewest possible symbols and thus are the most informative and lowest entropy classification systems. Bremer and Wanntorp's paper discusses the state of botanical classification. They suggest that botanists abandon their beloved grades and get on with the business of making botanical classifications rigorous and testable.

Besides a few pioneers, such as Bremer and Wanntorp, botanists are just beginning to realize the potential for applying the methods of phylogenetics to their problems. Unfamiliarity has even led to methods being called "cladistic" or phylogenetic when, in fact they are not. However, at least some members of the botanical community are quickly readdressing these problems and botanists are beginning to produce phylogenetic analyses and to participate in the theoretical development of the field. It was particularly nice to see so many botanists at the meetings. Indeed, the zoologists were almost outnumbered. The outcome of all of this botanical participation was a number of interesting presentations, four of which are published here. Funk discusses the special problems she perceives in doing analyses of plants. Sanders, Bolick and Churchill present analyses of three different plant genera.

The last major theme of the meetings which is published in this volume is phylogenetics and biogeography. Biogeography has been a preoccupation with phylogeneticists for the past few years, as phylogeneticists strive to integrate the methods of phylogenetic analysis and vicariance biogeography. Various aspects of this integration are apparent in the four papers published here. Brooks et al. consider the biogeography and coevolution of freshwater stingrays and their helminth parasites. Humphries discusses contrasting views of the biogeography of southern beeches. Mickevich applies her Transformation Series Analysis to the search for biogeographic patterns. Finally, Platnick discusses the ontological status of widespread taxa and their relevance to historical biogeography.

I hope that the Willi Hennig Society will continue to provide a common ground for discussing problems of interest to both systematic zoologists and botanists. Dr. Hennig might have objected to the naming of a Society after him, and some of those who knew him personally might feel the same way. But it is the fate of those who have made fundamental contributions to their fields to receive unsought honors and to be remembered by more than their published scientific accomplishments.

It was in this spirit that the Society was named.

E. O. Wiley
Lawrence, Kansas
June 1981

Contents

Biogeography and Cladistics

Founder's Address

J. S. Farris

Willi Hennig was prolific as an entomological systematist, but his activities did not end there. He was the earliest modern worker to recognize that systematics incorporates general principles that govern the classification of all kinds of organisms, and he made great contributions to formulating those principles as a coherent theory of systematics. His work on principles itself had a guiding principle as well: that systematics—discovery of the natural system of organisms—is a scientific enterprise, to be approached and understood like any other part of empirical science.

That combination of theoretical and practical systematic interests was rare in Hennig's time. It is far less rare today, but it is still a noteworthy characteristic. It distinguishes phylogenetic systematists from those who have not yet grasped the implications of adopting a scientific viewpoint. It is what distinguished this society from others as well. Our members are drawn together by a common interest in systematic principles; but their interest is not only theoretical—it has a practical base. This is the Willi Hennig Society in more than name alone.

The great value of Hennig's contributions and the great influence of his work on modern systematics are derived from this distinctive blend of the theoretical and the practical, the general and the specific. I expect that similar causes will underlie the importance of the Willi Hennig Society. Commonly, systematic meetings are devoted to particular groups. Far too often they afford little opportunity for discussion. What discussion there is, will likely concern specific, rather than general, problems, and even when general problems are discussed the participants will probably show a narrow range of group specialties. At the other extreme are meetings dedicated to classification in general—but with a kind of "generality" that requires excluding any scientific consideration or biological content from classification. Understanding of general principles is unlikely to improve much with the first, and systematics of any kind is less likely still to benefit from the second.

There is certainly good reason for holding meetings on the systematics of particular groups. That need is met well enough by older societies. But there is good reason, too, to bring together systematists of diverse interests. Before the Willi Hennig Society was formed, it was seldom indeed that systematists as diverse as botanists, parasitologists, ichthyologists, and entomologists could come together to discuss their common problems and to profit from others' diverse experience. The advantages of this exchange of ideas have already become clear from this first meeting. I expect that future meetings will continue to benefit systematics, both in particular and in general.

Cladistics and Molecular Biology